AF892606

# Geography of the Physical Environment

The *Geography of the Physical Environment* book series provides a platform for scientific contributions in the field of Physical Geography and its subdisciplines. It publishes a broad portfolio of scientific books covering case studies, theoretical and applied approaches as well as novel developments and techniques in the field. The scope is not limited to a certain spatial scale and can cover local and regional to continental and global facets. Books with strong regional focus should be well illustrated including significant maps and meaningful figures to be potentially used as field guides and standard references for the respective area.

The series appeals to scientists and students in the field of geography as well as regional scientists, landscape planners, policy makers, and everyone interested in wide-ranging aspects of modern Physical Geography. Peer-reviewed research monographs, edited volumes, advance and undergraduate level textbooks, and conference proceedings covering the major topics in Physical Geography are included in the series. Submissions to the Book Series are also invited on the theme 'The Physical Geography of…', with a relevant subtitle of the author's/editor's choice. Please contact the Publisher for further information and to receive a Book Proposal Form.

More information about this series at http://www.springer.com/series/15117

Batchuluun Yembuu
Editor

# The Physical Geography of Mongolia

Springer

*Editor*
Batchuluun Yembuu
Department of Geography
Mongolian National University of Education
Ulaanbaatar, Mongolia

ISSN 2366-8865 ISSN 2366-8873 (electronic)
Geography of the Physical Environment
ISBN 978-3-030-61433-1 ISBN 978-3-030-61434-8 (eBook)
https://doi.org/10.1007/978-3-030-61434-8

© Springer Nature Switzerland AG 2021
This work is subject to copyright. All rights are reserved by the Publisher, whether the whole or part of the material is concerned, specifically the rights of translation, reprinting, reuse of illustrations, recitation, broadcasting, reproduction on microfilms or in any other physical way, and transmission or information storage and retrieval, electronic adaptation, computer software, or by similar or dissimilar methodology now known or hereafter developed.
The use of general descriptive names, registered names, trademarks, service marks, etc. in this publication does not imply, even in the absence of a specific statement, that such names are exempt from the relevant protective laws and regulations and therefore free for general use.
The publisher, the authors and the editors are safe to assume that the advice and information in this book are believed to be true and accurate at the date of publication. Neither the publisher nor the authors or the editors give a warranty, expressed or implied, with respect to the material contained herein or for any errors or omissions that may have been made. The publisher remains neutral with regard to jurisdictional claims in published maps and institutional affiliations.

Cover image by Sonja Weber, München

This Springer imprint is published by the registered company Springer Nature Switzerland AG
The registered company address is: Gewerbestrasse 11, 6330 Cham, Switzerland

# Preface

Mongolia is a unique country for geographers, and it has very diverse geographical features, natural landscapes, a long history, as well as all the geomorphological processes. Even many scientific articles and research papers can be found in a variety of sources, such as the internet, related to Mongolia's geography, but no integrated and systematic academically oriented book that we are planning to compile. The chapters of the book will provide readers with general information and features of the physical geography of Mongolia, as well as new findings of the latest research.

Mongolia is a country that occupies vast areas in the center of the continent and has a rich and ancient history with many natural and geographical features. For these reasons, many foreigners came here out of interest to study the natural features. Especially in the second half of the nineteenth century and the beginning of the twentieth century, the research on Mongolian geography progressed considerably. On the basis of these researches, the Institute for Geography of the Geography of Mongolia (1969) published a full-fledged monograph "Physical Geography of Mongolia" (edited by Sh. Tsegmid), which is still the main source of research.

Since then, significant progress has been made in the field of physical geography research, and many new studies were carried out on the subject, but no complete book has been written. In particular, there are no books in English, and readers are very limited to the information on geography of Mongolia, which is only briefly presented at sessions of tourism organizations and in other reports. In this regard, most of the foreign sources of research on Mongolian geography use creations of foreigners, and materials of local authors are not widely available in English.

For example, with regard to foreign studies on Mongolia and the material used, the work of national authors is very rare, and although in some cases projects are implemented with foreign assistance, this is often only one aspect of the issue that is very common. To this end, this book was used as a result of the studies by many national scientists, and each member of the author group has many years of experience in writing the physical geography textbooks.

Therefore, it should be noted that the purpose of this book is to consider a simple form of writing for readers, students, and beginners, rather than for professional researchers. The primary goal of this book is to provide its readers with a more comprehensive picture of Mongolia and its geographical features.

There are numberless materials related to the physical geography of Mongolia, permafrost, and glaciers in foreign languages, internet sources, articles, encyclopedias, and so on. Owing to the published languages (Russian, German, Chinese, etc.), the geographical names of places are written differently in a number of sources. For instance, the historical area of Zuungar Gobi (in Mongolian) is a desert area located between the Mongol Altai mountains in the north and the Tian Shan in the south; it derives its name from the local tribe of Jungar, or Oirat; it is written as "Jungaria", "Jungar", "Dzungaria", and so on. Similar examples are *Huh Nor* or Khukh nuur (nuur means lake in Mongolian), *Hara Hoto* (in Russian), *Khar Khot* (in Mongolian); and *Ejnii gol* (gol is a river) is in Chinese inner Mongolian dialect and it is written as *Eznii Gol* in this book. Thus, we used the standard of written foreign words in the Mongolian language to unify the different types of names.

It is a challenge to publish a comprehensive work on the physical geography of Mongolia. Voluminous excellent works have been written from various points of view. I believe that the book provides an intermediate depth material coverage that is appropriate for the public.

The data in this book, related to the height of the mountain peak, elevation of the weather stations, and river length, are based on official information of the National Statistical Office and the National Authority of Geodesy and Cartography. In terms of geographical coordination, the places were determined from the GPS data of the Google Maps, while the sources of the climate parameters are based on the National Institute of Meteorology and resource materials of the Ministry of Environment and Nature.

This book consists of 11 chapters, including color maps, pictures, and tables. The main structure of the chapters consisted of abstract and key terms, and then begins with a discussion on the overview of the research history in the field, followed by the main issues depending on the field. In addition, the readers would probably note that the book is extensively illustrated with full-color maps, diagrams, and photos.

Chapter 1 briefly presents the general characteristics of Mongolia and some important concepts. The chapter begins with a discussion of detailed descriptions of geographical location, physical features, climate conditions, water resources, population, and culture. Chapter 2 provides a brief history of the territorial transformation and administration of Mongolia. The chapter summarizes an overview of studies on physical geography of Mongolia.

Geomorphology of Mongolia is the subject matter of Chap. 3 encompassing geological formations, landforms, and relief, a scheme of geomorphological zonation. The last part of the chapter describes the morphological characteristics of each geomorphological region, which have been modified and updated based on the results of research by well-known scientists.

Climate and climate change of Mongolia are addressed in Chap. 4. This chapter begins with an overview of climate research in Mongolia conducted by national and foreign scientists. Furthermore, it shows the driving factors of the Mongolia's climate condition, and the spatial features of the main parameters of climate are introduced. Lastly, it presents about the climate change in the country and its impacts. Chapter 5 covers the hydrography of

Mongolia. In the first section of the chapter, an overview of Mongolian hydrological research is introduced. The geographical background of the surface water is introduced in the second part, describing the morphological, hydrological, and ecological characteristics of rivers and lakes, including the description of the genetic lake types. The third section introduces groundwater resources and its geographical distribution.

Glacial and periglacial processes are presented in Chap. 6. The first part of this chapter relates to the Mongolian glacier study review factors such as climate, depressions, surface elevation, sedimentation, geographical distribution, and latitude correlation. The following section illustrates the distribution of glaciers Mongolian Altai, Khangai, and Khuvsgul mountains, as evidenced by changes in the ancient and modern glacial traces and the extent of the area as a result of climate change impacts, causes, and consequences. The effects of glaciers on the surface caused by the surface forms and the changes in the hydrological network due to the global warming process were discussed.

In particular, the chapter analyzes key geographical studies aimed at exploring the physical features of the natural patterns and conditions related to the geomorphology, geology, hydrography, climate and soil characters, glacier, as well as regional variations.

Chapter 7 presents the permafrost distribution in Mongolia, general characteristics of permafrost, cryogenic features and analyzes the degradation of permafrost. The distribution of permafrost is controlled by the climate in the north part of Mongolia and local environmental factors in the middle of the country or the edge of Siberian permafrost, with continuous, discontinuous, sporadic, and isolated types. Also, some examples of permafrost monitoring in Mongolia and the impacts of permafrost caused by climate change are introduced in this chapter.

Soil and processes of degradation and erosion are discussed in Chap. 8. It includes the introduction of the majority type of soils in the country and its distribution. Mongolia has almost all types of soil in the temperate zone of the world, from the mountain tundra to the over desert. But the same type of soil occurs in the mountains and hills, and it is difficult to classify and diagnose. Major classification of a total of 10 types of soil morphology is written and explained its geographic distribution in the chapter.

Chapter 9 describes the biogeographical characteristics of Mongolia in context of physical geography. Thus, it focused on result issues related to the research of vegetation and zoogeographical regions of Mongolia. Landscape diversity and extreme climatic conditions contribute to the diversity of ecosystems of the country. Mongolia's fauna includes 141 mammal species, 502 bird species, and 79 fish species. More than 3127 species of vascular plants, 574 species of mushrooms, 1056 species of lichens, 2003 species of algae, and 580 species of mushrooms have been found in Mongolia.

Chapter 10 briefly presents the development and compilation of a revised map of a physiographic region of Mongolia and gives an overview of physiographic mapping in Mongolia, and zonation on the natural components has been defined. Due to the different landscape forms and natural conditions, especially from south to the north and also vertical layers caused by high mountain ranges, the distributions and patterns of natural zones are more complicated. They form a series of altitudinal belts in the mountainous area and latitudinal zones across the country from north to eastern plains and plateaus and southern desert regions. In the mountain zones of the north and west, the pattern is more complex because elevation rather than latitude is the dominant factor, and there are striking changes over relatively short distances. Within Mongolia, there are six main environmental zones and belts (with subdivisions): High mountain and mountain taiga, mixed and deciduous forest, and forest steppe, steppe, Gobi (desert-steppe), and desert zones.

Land use and nature conservation issues were dealt with in Chap. 11. It presents agricultural and urban land resources, protected areas, deforestation, and reforestation. The chapter discusses the issues of forest cover, protected areas, and anthropogenic changes in Mongolia.

Ulaanbaatar, Mongolia
2020

Prof. Dr. Batchuluun Yembuu

# Acknowledgements

This book was prepared by professors' team at the Geography Department of the Mongolian National University of Education (MNUE). I wish to express my gratitude to all authors and my colleagues for their cooperation in the process of preparation of this book. I would also like to express my special gratitude to professor Narantsetseg Dorjgotov, Deputy Director of MNUE, for her invaluable support. The authors would like to thank their colleagues who made useful suggestions and shared their research works for this book, in particular Dr. Battogtokh Dorjgotov, Dr. Batkhishig Ochirbat and Dr. Oyungerel Jugder from the Institute of Geography and Geoecology, Dr. Gomboludev, Dr. Nazagdorj, Dr. Davaa from the Institute of Meteorology, Hydrology and Environmental Monitoring, Prof. P. D. Gunin from the Geographic Institute of Russia and Dr. Ogonbayar from Khovd University. In addition, special thanks to my students, Munkh-Erdene Tsogtgerel, Bulga-Erdene, Azbayar and Otgonbat Dambajantsan, who made up most of the maps presented in the book.

Finally, I would like to thank the Springer team for their fruitful collaboration. I am also extremely grateful to Dr. Michael Leuchner, former Publishing Editor Earth Sciences, Geography and Environment, for his enormous editorial efforts and encouragement from the start, and the kind cooperation of Ms. Doris Bleier, Ms. Carmen Spelbos and Mr. Ambrose Berkumans from Springer for their assistance. The anonymous reviewers, who provided useful and wise comments on the book, were appreciated.

Prof. Dr. Batchuluun Yembuu

# Contents


1 **General Geographical Characteristics of Mongolia** . . . . . . . . . 1
Batchuluun Yembuu

2 **Historical Geography: Administrative Division and Research in Physical Geography of Mongolia** . . . . . . . . . 9
Batchuluun Yembuu and Dash Doljin

3 **The Relief and Geomorphological Characteristics of Mongolia** . . . . . . . . . 23
Dash Doljin and Batchuluun Yembuu

4 **Climate and Climate Change of Mongolia** . . . . . . . . . 51
Batchuluun Yembuu

5 **Hydrography of Mongolia** . . . . . . . . . 77
Navchaa Tugjamba

6 **Glaciers of Mongolia** . . . . . . . . . 101
Ser-Od Tsedevdorj

7 **Permafrost in Mongolia** . . . . . . . . . 119
Avirmed Dashtseren

8 **Soils of Mongolia** . . . . . . . . . 135
Sandag Khadbaatar

9 **Biogeographical Characteristics of Mongolia** . . . . . . . . . 161
Khurelbaatar Tsogbadral

10 **Division of the Physiographic and Natural Regions in Mongolia** . . . . . . . . . 177
Dash Doljin and Batchuluun Yembuu

11 **Land Use and Nature Conservation in Mongolia** . . . . . . . . . 195
Bat-Erdene Tsedev

**Index** . . . . . . . . . 213

# Editor and Contributors

## About the Editor

**Batchuluun Yembuu** is a professor in the Department of Geography at Mongolian National University of Education. She has published widely in the areas of the Earth Science, General Physical Geography and Geography Education, with more than 200 publications. She is also the author of several books, such as Earth Science, General Physical Geography, The Dictionary of Physical Geography and so on, and several geography textbooks. She is the founder and an executive director of the Mongolian Association for Geographic Education (MAGE) and editor-in-chief of Geoforum-Mongolia journal.

## Contributors

**Avirmed Dashtseren** is a leading researcher and Head of the Permafrost Division at the Institute of Geography and Geoecology, Mongolian Academy of Sciences. He obtained his Ph.D. degree from the Graduate School of Environmental Science, Hokkaido University in Japan. His research is focused on the state of the marginal of permafrost and glaciers in Mongolia in nexus with the interaction of climate change and ecosystem.

**Dash Doljin** is a professor of geography at Mongolian National University of Education. He was formerly employed at the Institute of Geography and Geo-ecology of the Mongolian Academy of Science for more than 30 years. Besides more than 200 publications and maps related to the physical geography, he had developed updated versions of the physiographic, landscape regionalization of Mongolia, which include the National Atlas of Mongolia (1990, 2009). He is the author of the maps related to physical geography of Mongolia in the Desertification Atlas of Mongolia (2014).

**Sandag Khadbaatar** is an associate professor in the Geography Department of the Mongolian National University of Education. He obtained his Ph.D. degree in Moscow, Russia in 2005, at the Soil and Landscape in the Selenge river basins. His research interests are physical geography, soil science, and geomorphology. He authored and coauthored many books. The most recent one is titled, *Transformation of land ecosystems in the southern part of the Baikal basin*, published in Russia (in Russian) in 2018.

**Bat-Erdene Tsedev** Ph.D. in Geography, Associate Professor of Mongolian National University of Education. His research interest includes GIS, remote sensing and geospatial technologies, and urban land use.

**Ser-Od Tsedevdorj** holds a doctoral degree and is a senior lecturer at the Department of Geography. He teaches the courses on Physical Geography of Mongolia. His current research interests include physical geography, landscape studies and glacier environment. He is a member of Joint Russian–Mongolian complex geographical expedition to the Tavan Bogd mountain massif since 2015.

**Khurelbaatar Tsogbadral** is concurrently the head of the Department of Geography, Mongolian National University of Education. His research interests include tourism and recreational geography. He holds a Ph.D. on the geoecological issues of Mongolia. His current research interests include recreational geography and the environment study.

**Navchaa Tugjamba** obtained M.Sc. in Environmental Sciences from UNESCO-IHE, The Netherlands and holds Ph.D. She has been working in the field of geography research and teaching for around 20 years. Her research results were published in the international peer-reviewed journals and presented at the international level conferences. She has written several books on environmental management, water resources management and tourism education. Her research interests are water and ecosystem management, ecotourism and sustainable development.

# List of Figures


| | | |
|---|---|---|
| Fig. 1.1 | Elevation profile from the west (Mongol Altai) to the east (Dornod plain) through 48°00′N | 3 |
| Fig. 1.2 | Elevation profile from the north to the south through 107°00′E | 3 |
| Fig. 2.1 | Itinerary of N.M. Przhevalsky and Pevtsov in Mongolia | 13 |
| Fig. 2.2 | Itinerary of Kozlov and others' expeditions in Mongolia | 17 |
| Fig. 2.3 | Itinerary of Potanin and Obruchev expeditions in Mongolia | 17 |
| Fig. 2.4 | Itinerary of Andrews expedition in Mongolia | 19 |
| Fig. 3.1 | Relief and elevation of Mongolia | 25 |
| Fig. 3.2 | Hypsometric curves of the surface in Mongolia | 25 |
| Fig. 3.3 | Paleogeomorphological scheme of the early and middle Jurassic of Mongolia | 27 |
| Fig. 3.4 | Paleogeomorphological scheme of Mongolia. Neogene | 29 |
| Fig. 3.5 | Volcanic type of relief | 34 |
| Fig. 3.6 | Geomorphological regions of Mongolia | 35 |
| Fig. 3.7 | Hypsographic curve of the Mongol Altai | 36 |
| Fig. 3.8 | Overview of the main morphological structures of the Mongol Altai | 37 |
| Fig. 3.9 | Glacier circus | 38 |
| Fig. 3.10 | Orographic scheme of Khangai mountains range | 39 |
| Fig. 3.11 | Scheme of the ancient glaciation of Khangai mountains | 40 |
| Fig. 3.12 | Basalt canyon in Chuluut river (own photo). Valley basalts of the Chuluut river (I) and Tagin Gorkhi (II) | 41 |
| Fig. 3.13 | Khagiin Khar lake of glacial origin in the central part of Khentii mountains | 42 |
| Fig. 3.14 | Mid-altitude mountains in the Numrug river basin | 43 |
| Fig. 3.15 | Alpine type relief in the desert region | 44 |
| Fig. 3.16 | Buurug Del sand dunes | 45 |
| Fig. 3.17 | Alluvial fans and sediments in the southern slope of Nemegt mountain | 46 |

Fig. 3.18 Granite massif . . . . . . . . 47
Fig. 3.19 Relief types of the Eastern Mongolia. . . . . . . . 48
Fig. 3.20 Geological-geomorphological profile through the Eastern-Mongolian Plain Types of rocks: 1—intrusive, 2—effusive basic, 3—metamorphic, 4—sedimentary . . . . . . . . 48
Fig. 4.1 The spatial distribution of the yearly amount of global irradiation (kWh/m$^2$) . . . . . . . . 56
Fig. 4.2 **a** Variations of the sun altitude at solar noon, **b** Annual mean temperature at two stations: Rinchinlhumbe (51°7′N) and Zamiin-Uud (43°44′N) . . . 56
Fig. 4.3 The variation of the sunshine hours depends on the latitudes . . . . . . . . 57
Fig. 4.4 Geographic distribution of mean air temperature in January . . . . . . . . 58
Fig. 4.5 The regional variations of the average minimum temperature (°C): Ulaangom and Dalanzadgad . . . . . . . . 58
Fig. 4.6 Geographic distribution of mean air temperature in July . . . . . . . . 59
Fig. 4.7 Dates of daily mean temperatures exceeding 10 °C. . . . . 59
Fig. 4.8 Spatial distribution of the mean annual precipitation between 1960 and 1990. . . . . . . . 61
Fig. 4.9 Spatial differences in precipitation between Khangai (Baitag) and Trans-Altai Gobi (Baitag) regions . . . . . . . . 61
Fig. 4.10 Stable snow covered days . . . . . . . . 62
Fig. 4.11 Spatial distribution of air pressure in January . . . . . . . . 63
Fig. 4.12 Spatial distribution of air pressure in July . . . . . . . . 63
Fig. 4.13 Average wind speed (m/s) in foreststeppe and desert regions (2000–2011) . . . . . . . . 64
Fig. 4.14 Wind directions and speed. . . . . . . . 65
Fig. 4.15 Distribution of dust storms/1960–2008/. . . . . . . . 65
Fig. 4.16 Climatic regions in Mongolia . . . . . . . . 67
Fig. 4.17 Regional differences of the climate . . . . . . . . 69
Fig. 4.18 Mean temperature trend in different distinct ecological zones: West (Ulaangom);North (Murun); East (Choibalsan); and South (Dalanzadgad) in Mongolia . . . 70
Fig. 4.19 Multi-year trend of number of days when the daily highest temperature is below 25 °C. . . . . . . . 70
Fig. 4.20 Precipitation change by regions . . . . . . . . 71
Fig. 4.21 Drought frequency in Mongolia. . . . . . . . 73
Fig. 5.1 Hydrological drainage basins of Mongolia Modified for the educational purpose . . . . . . . . 81
Fig. 5.2 Ulz river valley . . . . . . . . 81
Fig. 5.3 Stream order of Ulz river basin . . . . . . . . 82
Fig. 5.4 Annual runoff distribution of the Khentii Mountain Rivers, by the percentage. . . . . . . . 85
Fig. 5.5 The Mongol Altai Rivers' annual runoff distribution . . . 85
Fig. 5.6 Typical annual hydrograph of the Ulz river. . . . . . . . 86

Fig. 5.7 Some rivers' water fed, by percentage . . . . . . . . . . . . . . 86
Fig. 5.8 Lakes of Mongolia Modified for the educational purpose . . . . . . . . . . . . . . . . . . . . . . . . . . . . . . . . . . . 90
Fig. 5.9 Bathymetric map of Khuvsgul Lake . . . . . . . . . . . . . . . . 93
Fig. 5.10 Mineral springs of Mongolia Modified for the educational purpose . . . . . . . . . . . . . . . . . . . . . . . . . . . . 94
Fig. 5.11 Shargaljuut hot spring . . . . . . . . . . . . . . . . . . . . . . . . . 96
Fig. 6.1 Glaciations near the Tavan Bogd mountain–Sapojnikov . . . . . . . . . . . . . . . . . . . . . . . . . . . . . . . . . 102
Fig. 6.2 Glaciers distribution of Mongol Altai, Khangai and Khuvsgul mountains . . . . . . . . . . . . . . . . . . . . . . . . . 105
Fig. 6.3 **a** Distribution of the number of glaciers morphological types. **b** Distribution of glacial area in morphological types . . . . . . . . . . . . . . . . . . . . . . . . . 106
Fig. 6.4 The melting of the Kozlov glacier and air temperature . . . . . . . . . . . . . . . . . . . . . . . . . . . . . . . . . . . 106
Fig. 6.5 Munkhkhairkhan mountain glacier . . . . . . . . . . . . . . . . . . 107
Fig. 6.6 Glacier of Turgen mountains . . . . . . . . . . . . . . . . . . . . . . 108
Fig. 6.7 Munkhsaridag, Khuvsgul mountain range . . . . . . . . . . . . 110
Fig. 6.8 Dynamics of glacier massif areas . . . . . . . . . . . . . . . . . . . 112
Fig. 6.9 Glacier area change in the Kharkhiraa and Turgen mts. with area of glacier in 2000 . . . . . . . . . . . . . . . . . . . 112
Fig. 6.10 Khar nuur- Tsengel soum, Bayan-Ulgii aimag . . . . . . . . . 113
Fig. 7.1 Permafrost distribution and zones in Mongolia . . . . . . . . 121
Fig. 7.2 Sorted circles on the north slope of the Tsengelkhairkhan uul in Altai mountain . . . . . . . . . . . . . 123
Fig. 7.3 Hummock area in the valley of Zuunsalaa, Ulaanbaatar city . . . . . . . . . . . . . . . . . . . . . . . . . . . . . . . 123
Fig. 7.4 Thermokarst lake in the Darkhad depression, with an approximately diameter of 200 m . . . . . . . . . . . . 124
Fig. 7.5 Closed-system pingo at the Darkhad depression, with approximately 12 m height and 150 m diameter . . . 125
Fig. 7.6 Temporal variation in active layer thickness at the selected monitoring sites in Mongolia . . . . . . . . . . . . . . 126
Fig. 7.7 Permafrost temperatures from the monitoring sites in the Altai, Khangai, Khuvsgul and Khentii mountains with different permafrost zones . . . . . . . . . . . . . . . . . . . 127
Fig. 7.8 Mean annual ground temperatures at the monitoring sites in the Altai and Khangai mountains . . . . . . . . . . . . . 129
Fig. 7.9 Mean annual ground temperatures at the monitoring sites in the Khentii and Khuvsgul mountains . . . . . . . . . . 130
Fig. 7.10 The current situation of the old sum center of Khyargas sum in the northern Mongolia . . . . . . . . . . . . . 131
Fig. 7.11 Result of permafrost thawing under the wooden house in Chandmani-Undur sum in north part of Mongolia . . . 132
Fig. 8.1 Soil vertical zones . . . . . . . . . . . . . . . . . . . . . . . . . . . . . 138
Fig. 8.2 Soil belt and zones . . . . . . . . . . . . . . . . . . . . . . . . . . . . 139

Fig. 8.3 Light kastanozem layers . . . . . . . . . . . . . . . . . . . . . . . . . 143
Fig. 8.4 Distribution of Kastanozem in Mongolia . . . . . . . . . . . . . 143
Fig. 8.5 Desert soils and profile (Shinejinst soum, Bayankhongor) . . . . . . . . . . . . . . . . . . . . . . . . . . . . . . . 145
Fig. 8.6 Soil salinization in the oasis Ekhiin-Gol, at a depth of 100 cm (Na content in water extract, mgE/100 g) . . . 146
Fig. 8.7 Soil types . . . . . . . . . . . . . . . . . . . . . . . . . . . . . . . . . 147
Fig. 8.8 Geographic distribution of soil . . . . . . . . . . . . . . . . . . . . 149
Fig. 8.9 New classification of soil types in Mongolia based on the FAO . . . . . . . . . . . . . . . . . . . . . . . . . . . . . 152
Fig. 8.10 Soil-geographic zoning of Mongolia . . . . . . . . . . . . . . . . 156
Fig. 8.11 Water erosion vulnerability map . . . . . . . . . . . . . . . . . . . 158
Fig. 8.12 Soil wind erosion vulnerability map . . . . . . . . . . . . . . . . 159
Fig. 9.1 Phytogeographical regions of Mongolia . . . . . . . . . . . . . . 164
Fig. 9.2 Geographical regions of vegetation in Mongolia . . . . . . . 165
Fig. 9.3 Comparative study in Mongolian vascular plants . . . . . . . 166
Fig. 9.4 Distribution of some endemic species in Mongolia. Endangered and threatened plant species . . . . . . . . . . . . . 168
Fig. 9.5 Some endemic plants of Mongolia . . . . . . . . . . . . . . . . . . 168
Fig. 9.6 Routes of A.G. Bannikov 1942–1945 . . . . . . . . . . . . . . . 169
Fig. 9.7 Zoogeographical regions and sections of Mongolia . . . . . 170
Fig. 9.8 Species composition of Mongolian fauna . . . . . . . . . . . . 170
Fig. 9.9 Mammals diversity of Mongolia Inventory of Mongolia's mammal diversity . . . . . . . . . . . . . . . . . . . . . 173
Fig. 9.10 Distribution of rare mammals . . . . . . . . . . . . . . . . . . . . . 174
Fig. 9.11 Distribution of rare birds: modified from Ecosystems of Mongolia, Atlas . . . . . . . . . . . . . . . . . . . . . . . . . . . . 175
Fig. 10.1 Physiographic Regions of Mongolia . . . . . . . . . . . . . . . . 180
Fig. 10.2 Natural zones and belts in Mongolia . . . . . . . . . . . . . . . . 184
Fig. 10.3 Annual mean temperature and precipitation variations by the natural zones . . . . . . . . . . . . . . . . . . . . . . . . . . 186
Fig. 10.4 Coniferous forest in the Darkhad depression . . . . . . . . . . 187
Fig. 10.5 Khangai mountain . . . . . . . . . . . . . . . . . . . . . . . . . . . . 188
Fig. 10.6 Khongor sand dunes . . . . . . . . . . . . . . . . . . . . . . . . . . . 190
Fig. 10.7 Khermen Tsav, Gobi Desert . . . . . . . . . . . . . . . . . . . . . 191
Fig. 10.8 Altitudinal belts in the Aj Bod mountain . . . . . . . . . . . . 192
Fig. 11.1 Land use in Mongolia . . . . . . . . . . . . . . . . . . . . . . . . . . 196
Fig. 11.2 Land use structure in Mongolia . . . . . . . . . . . . . . . . . . . 197
Fig. 11.3 Arid regions pasture, Uu Durulj, Govi-Altai aimag, Western Mongolia . . . . . . . . . . . . . . . . . . . . . . . . . . . . 198
Fig. 11.4 The built-up area of Ulaanbaatar city . . . . . . . . . . . . . . . 200
Fig. 11.5 2011 Landsat TM image of the selected part of Ulaanbaatar city . . . . . . . . . . . . . . . . . . . . . . . . . . . . . . . 201
Fig. 11.6 Classified image using 2011 data sets . . . . . . . . . . . . . . 201
Fig. 11.7 Digitized topographic map of Ulaanbaatar area of 1969 . . . . . . . . . . . . . . . . . . . . . . . . . . . . . . . . . . . . . 202
Fig. 11.8 Location of protected areas . . . . . . . . . . . . . . . . . . . . . . 206

Fig. 11.9 The protected area of the National Park Gorkhi-Terelj, central Mongolia . . . . . . . . . . . . . . . . . . . . . . . . . . . . . . 206
Fig. 11.10 The protected area of the Natural Monument Eej khairkhan, Western Mongolia . . . . . . . . . . . . . . . . . . . . . . 207
Fig. 11.11 The protected areas covered by the international convention . . . . . . . . . . . . . . . . . . . . . . . . . . . . . . . . . . . . 209

# List of Tables


| | | |
|---|---|---|
| Table 4.1 | Main characteristics of climatic regions | 66 |
| Table 4.2 | The annual air temperature increase rates, between 1961–2011 (ten stations in each with highest and lowest rates) | 70 |
| Table 4.3 | The precipitation changes between 1961 and 2011 (mm) (selected weather stations with highest and lowest rate of changes) | 71 |
| Table 4.4 | Current content of soil organic matters (g/m$^2$) and its future changes (%) | 73 |
| Table 5.1 | Percentage of the area and water resources of the three main hydrological basins in Mongolia | 83 |
| Table 5.2 | Total numbers of rivers, lakes and mineral waters in Mongolia | 83 |
| Table 5.3 | Rivers runoff components of Mongolia | 84 |
| Table 5.4 | The average water temperature of the rivers in Mongolia (°C) | 88 |
| Table 5.5 | Water temperature of Ulz river and Khukh lake | 88 |
| Table 5.6 | Mean values of morphologic elements of lakes | 89 |
| Table 5.7 | Statistics of Mongolian lakes | 89 |
| Table 5.8 | Water balance of the lakes, mm/year | 94 |
| Table 5.9 | Some hot springs of Mongolia | 95 |
| Table 5.10 | Groundwater resources assessment results | 97 |
| Table 6.1 | Morphometric indicators of Mongolian glaciers | 104 |
| Table 6.2 | The classification of glacier massive | 106 |
| Table 6.3 | Change of the Munkhkhairkhan mountain glaciers, 1990–2006 | 108 |
| Table 6.4 | Late Pleistocene ELA reconstruction in the Khentii mountains | 111 |
| Table 6.5 | Total glacier areas in Kharkhiraa and Turgen mts. and Kharkhiraa river basin | 114 |
| Table 6.6 | Definition of some glacial lakes | 115 |
| Table 8.1 | Characteristics of forest soil | 141 |
| Table 8.2 | Genetic horizon thickness and humus content in loamy kastanozem | 142 |

Table 8.3 Genetic horizon thickness and humus content in sandy loam, sandy kastanozem . . . . . . . . . . . . . . . . . . 142
Table 8.4 Checklist of the soil classification . . . . . . . . . . . . . . . . 148
Table 8.5 Soil classification of Mongolia . . . . . . . . . . . . . . . . . . 150
Table 8.6 Soil classification criteria. . . . . . . . . . . . . . . . . . . . . . . 152
Table 8.7 New classification of soil types in Mongolia. . . . . . . . 153
Table 9.1 Main vegetation types by the natural zones and belts. . . . . . . . . . . . . . . . . . . . . . . . . . . . . . . . . . . 162
Table 9.2 Vegetation regions, area percentages and number of species . . . . . . . . . . . . . . . . . . . . . . . . . . . . . . . 165
Table 10.1 The physiographic regionalization of Mongolia. . . . . . 178
Table 10.2 The physiographic regions by areas. . . . . . . . . . . . . . . 181
Table 10.3 The classification of natural zones and belts . . . . . . . . 184
Table 10.4 The natural zones and belts (area and percentage) . . . 184
Table 10.5 Monthly average temperature in each natural zones and belts . . . . . . . . . . . . . . . . . . . . . . . . . . . . . . . . 185
Table 10.6 Dates and annual sum of daily mean temperature exceed 10 °C. . . . . . . . . . . . . . . . . . . . . . . . . . . . . . . 186
Table 10.7 Climatic parameters in selected places in the steppe zone. . . . . . . . . . . . . . . . . . . . . . . . . . . . . . . . . . . 189
Table 10.8 Climatic parameters in selected places in the semi-desert and desert zones . . . . . . . . . . . . . . . . . . . . . 192
Table 11.1 Land use structure in Mongolia. . . . . . . . . . . . . . . . . . 197
Table 11.2 Area changes of an integrated land database . . . . . . . . 198
Table 11.3 Agricultural land use . . . . . . . . . . . . . . . . . . . . . . . . . 199
Table 11.4 Arable land use . . . . . . . . . . . . . . . . . . . . . . . . . . . . . 199
Table 11.5 The total areas for each class in different years, evaluated from multitemporal GIS and RS data sets. . . . . . . . . . . . . . . . . . . . . . . . . . . . . . . . . . 202
Table 11.6 Land with forest resources by region, aimag and the capital . . . . . . . . . . . . . . . . . . . . . . . . . . . . . . . . 203
Table 11.7 Network of protected areas . . . . . . . . . . . . . . . . . . . . . 205
Table 11.8 Protection levels in protected areas . . . . . . . . . . . . . . . 207
Table 11.9 Ramsar sites . . . . . . . . . . . . . . . . . . . . . . . . . . . . . . . 208
Table 11.10 Properties inscribed on the World Heritage . . . . . . . . . 209
Table 11.11 Man and Biosphere site. . . . . . . . . . . . . . . . . . . . . . . . 209
Table 11.12 Ecoregions in Mongolia . . . . . . . . . . . . . . . . . . . . . . . 210

# 1 General Geographical Characteristics of Mongolia

Batchuluun Yembuu

**Abstract**

This chapter provides an overview of the geographical location of the country, as well as its location, climate, topography, water resources, soil and natural resources. Situated in the transition zone between the Siberian taiga and the Asian deserts and steppes, Mongolia has diverse geographical features. In the northwest, mountain ranges are overgrown with forests, lakes and rivers, while arid and semi-arid areas stretch across the eastern part of the country. Mongolia's continental climate is characterized by extreme temperatures and low precipitation strong seasonal variations. End of this chapter briefly describes the population and cultural features of Mongolia.

**Keywords**

Geographical location · Physical features · Climate · Water resources

B. Yembuu (✉)
Department of Geography, Mongolian National University of Education, Ulaanbaatar, Mongolia
e-mail: batchuluun@msue.edu.mn

## 1.1 Geographical Location and Territory

Mongolia is a landlocked territory and located in the heart of the Asian continent, bordering Russian Federation with 3,543 km in the north and China with 4,709 km in the south (NSO 2018). The total length of the Mongolian border is 8,252 km. According to the Agency for Land Management and Land Use, Geodesy and Cartography (2018), Mongolia has a total area of 1,564,116 $km^2$ or about 1% of the earth's land surface. Mongolia is the 19th largest country in the world by area (ALMGC 2018). The area is roughly the same as that of France, Germany, the Netherlands, Belgium, Spain and Portugal combined. The latitude of the northern point—*Mongol Shariin Davaa* (52°09′N)—is as approximate as Warsaw-Poland, The Hague-South Holland, and the southern point—*Orvog-Gashuunii Bor Tolgoi* (41°35′N)—is approximately the latitude of Barcelona-Spain, Boston-USA and Hokkaido-Japan. The western point of the territory is the mountain peak of *Tavan Bogd* (87°47′E), located at the "meeting point" of three countries: Mongolia, Kazakhstan and China. The eastern point is *Mount Soyolz* (119°57′E), located at the same longitude (about) Nanjing-China and Manila-Philippines. Mongolia extends over an area of approximately 32°10′ (2,392 km) from west to east and about 10°31′ (1,259 km) from the north to the south. The capital city of Ulaanbaatar

© Springer Nature Switzerland AG 2021

B. Yembuu (ed.), *The Physical Geography of Mongolia*, Geography of the Physical Environment,
https://doi.org/10.1007/978-3-030-61434-8_1

(1,310 m) is located at 47° 55′ N, 106° 53 and in the valley of the Tuul River, which was founded in 1350.

The highest point is the Khuiten peak (4,374 m a.s.l) of Mount Tavan Bogd in the west of Altai range and the lowest point of Khukh Nuur (532 m a.s.l) in the northeast is Dornod aimag (NSO 2018). Mongolia has an average height of 1,580 m; about 81.2% of its territory is above 1,000 m and 18.8% below 1,000 m (Tsegmid 1969). Mongolia's capital Ulaanbaatar is located at 1350 m. From this point of view, Mongolia is approximately 800 m above the average height of the earth's surface (840 m) and approximately 600 m above the average height in Asia (Shagdar and Batchuluun 2006).

The country's territory is isolated from the seas, located at about 1,000 km from the nearest part of the Pacific Ocean and the high mountains surrounding it: Transbaikal Ridge and Sayan Ridge in the north, Tien Shan and Altai in the west and Great Khyangan in the east. Thus, the geographical characteristics of the country are characterized by great diversity with latitudinal variations, such as climate and vegetation cover. The unique features of Mongolia's ecosystems are widely recognized in comparison with those of other countries located at the same latitude in the northern hemisphere.

The administrative unit known as the "aimag" (province) is responsible for the provision of public services, including education, at the local level. Administratively, Mongolia is divided into 21 aimags and the capital Ulaanbaatar. Aimags are divided into 330 "soums" (sub-provinces), which are then divided into 1,615 bags, the smallest administrative unit under soums. The capital city of Ulaanbaatar is divided into 9 districts, which consist of 152 khoroos (subdistrict) (NSO 2018). In the historical period, a number of aimags have been changed, such as 13 aimags so far in 1940 and 18 aimags in 1990. The average area of aimags ranges from 20,000 to 165,000 sq. km (Southgobi aimag) and has an average population of 80,000–90,000, with the aimag centers of about 15,000–20,000. According to the Law of Mongolia on Legal Status of Towns and Villages (2006), the urban area of Mongolia is defined as a settlement of more than 15,000 people. 21 aimag centers are considered to be urban areas in this category and several large cities (State Great Khural 2006). Ulaanbaatar is the largest city, with two other major cities, Darkhan and Erdenet, located 230 km north and 370 km northwest of the capital, respectively.

Mongolia is a parliamentary democracy established under the Constitution, which was updated in 1992. In 1961, Mongolia became a full member of the United Nations. The country is a member of many international organizations, including UNESCO, WHO and the World Bank. The 2018 Human Development Report ranks Mongolia 92nd (in the high category) in the HDR index of 189 participating countries (UNDP 2018).

## 1.2 Physical Feature

The physical feature of Mongolia is a mountain mixed with steppe and vast plains, with elevation differences of more than 3,800 m. Especially in the western and northern parts of the country the relief is very complex, with high mountains and vast river valleys and in the north the eastern Sayans lie along the border with Russia. The three main mountain ranges of Mongolia are: the Mongol Altai, Khangai and Khentii. In general, the altitude of the surface decreases from northwest to southeast (Figs. 1.1 and 1.2).

The Mongol Altai is the highest mountain system of the Altai Mountains in Central Asia, which is about 2,000 km in the northwest–southeast direction from the West Siberian Plain to the Gobi Desert through Russia, Kazakhstan, Mongolia and China. The peak of Mount Tavan Bogd, the Altai Ridge, is located at the intersection of the three countries: Mongolia, Kazakhstan and China. In terms of structure, relief and other characteristics, the Altai mountains in Mongolia are divided into the Mongol Altai (northwestern part) and the Gobi-Altai mountains (southeastern part) (Tsegmid 1969).

The Mongol Altai extends over long distances of 800 km to the depression of Lake Alag (Sinitsin 1959) or 1,000 km to the mountain

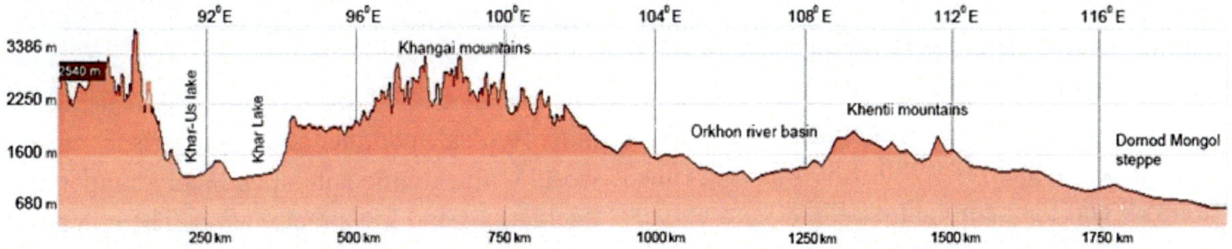


**Fig. 1.1** Elevation profile from the west (Mongol Altai) to the east (Dornod plain) through 48°00′N (developed from the Google Earth)

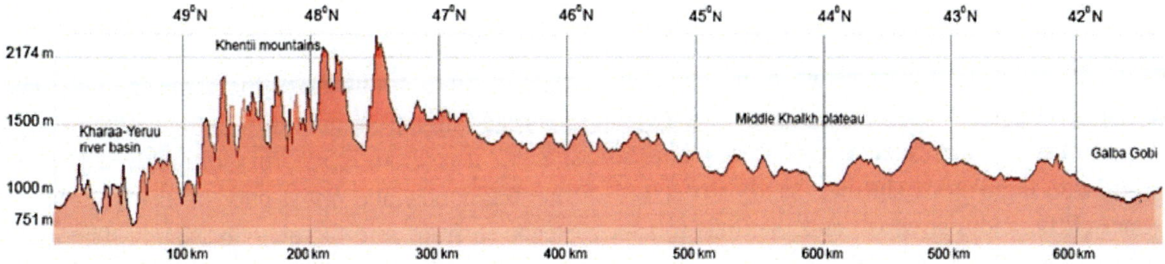


**Fig. 1.2** Elevation profile from the north to the south through 107°00′E (developed from the Google Earth)

Gichgene (Murzaev 1952), which can be considered the end of the Mongol Altai (Tsegmid 1969). The Mongol Altai mountains include at least three parallel ridges separated by intermountain valleys and depressions: Jargalant-Khairkhan, Bumbat-Khairkhan, Bayan, Taishir and Khasagt-Khairkhan, with an average altitude of 2,600–3,000 m, and maximum of 3,500–3,700 m. Mountain Gobi-Altai is also divided into three parallel subgroups: the relatively low Argalant and Nariinkhar mountains (1,900–2,600 m), the high mountains Bayan Tsagaan, Ikh Bogd, Baga Bogd and Arts Bogd (2,453–4,000 m) and Gurvan Saikhan have several branches (2,500 m to maximum altitude), and are no more than 2,400–2,500 m above the sea level (Tsegmid 2013) (see Chap. 3). Most of the glaciers in Mongolia are in the Mongol Altai mountains (see Chap. 6), while the Gobi-Altai mountains have no glaciers dominated by wind erosion.

The second highest mountain range of Khangai occupies most of the central and north-central part of the country, and it takes 750 km in length. The highest peak of Khangai is Otgontenger (3,905 m), which is located only with the glacier of these mountains. The other branches of Khangai have latitudinal directions, such as Khan Khuhii, Bulnai and Tarvagatai. The main characteristics of Khangai are the gentle slopes and edges of the massif, most of them being low mountains, rolling hills, hillocks and mounds with a relative height of about 200 m.

The Khuvsgul mountains are the northernmost mountain ranges in Mongolia and are located on both sides of the Khuvsgul lake depression. Even the appearance of the Khuvsgul ridge is similar to that of Soyan, but geomorphologically it is connected with the Khangai mountains (Tsegmid 1969). The Great Khyangan Ridge (Da Hinggan) rises along the eastern border with China and beyond.

The third mountain range is the Khentii range (400 km) in the northeast of Mongolia, and its direction differs from other mountains, from southwest to northeast, stretching toward Siberia. The highest peak of Asralthairkhan (2,800 m a.s.l) is not high enough with a glacier. Ulaanbaatar is located on the southwestern edge of the ridge. The vast grasslands of the Asian steppe extend over the eastern part of the country with a small gradient of heights, the largest ones are called

Menengiin Tal[1] (pediplain of Menen, 250,000 sq. km) and the rolling and hilly plain of Dornod Mongolia with grasslands. The elevation of the expanding hills and low mountains of the Dornod Mongolian plain is 1,700–1,800 m above the sea level, and the pediplain is 600–800 m above the sea level. These plains extend from the northeast to the southwest toward Khentii (Tsegmid 1969).

Most of southern Mongolia is dominated by the high plateau of the Gobi Desert (more than 400,200 sq. km) with rare vegetation cover, which stretches northwest between the Altai and Khangai ranges. A common feature of this area is also extensive pebble plains, sand dunes, endless plains with large thorny bushes, sandy strips, redstone cliffs and canyons, and Saxual forests. Except for the name of the landscape, the meaning of the word "Gobi" in Mongolia is used as a depression with salt marshes and a dry place with insufficient vegetation cover (see Chap. 10). From this point of view, Mongolia has "Thirty-three Famous Gobies" that distinguish their own geographical characteristics, such as "Shargiin Gobi" and "Nomingin Gobi", and more than 15 sites called "Gobi" that exist in the protected areas of Greater Gobi Desert. Sand researcher T. Baasan (2003) described over 370 places that are called Gobi in Mongolia. Interestingly, despite the arid climate, most of Mongolia's large lakes are located in the Gobi Desert, the Great Lakes depressions and the Valleys of Lakes. The Gobi Desert is the original habitat of two humpback Bactrian camels that adapted animals from the scarce water and cold environment.

Sand dunes are one of the physical features of Mongolia. The largest dune deposit is Mongol Els, which is 600 km long and covers an area of about 3,000 $km^2$, and the fluvio-lacustrine and aeolian sediments are found in this area (Stolz et al. 2012). On the eastern edge of the Mongol Els, the Shurag and Zavkhan rivers converge and form a floodplain on the dune front. Western and northern Mongolia often experience earthquakes, although volcanoes are considered extinct (see Chap. 3).

## 1.3 Climate

The peculiarity of the country's climate is the most typical continental characteristic in the world with extreme annual and daily temperature fluctuations and low precipitation. The climate of the country can be characterized as sharply expressed in four different seasons: long, dry and very cold winters and short, warm summers. Due to Mongolia's internal location and mountainous topography with high altitude, surrounded by high mountains, the climate is generally colder than in other countries of the same latitude. The average annual temperature ranges from 2 to −8 °C; it is about 8.5 °C in Gobi and −7.8 °C in the highlands and most part of the country is negative from mid-October to March. The average temperature in January varies from −20 to −25 °C, and in July from +15 to 22 °C.

The maximum January temperature is −31.1 to −55.3 °C, and the maximum July temperature is +28.5 to +44.0 °C. Winter nights at −40 °C are common in most years (the minimum temperature in the Uvs lake −55 °C basin), while summer extreme temperatures reach +40 °C in the Gobi Desert region. Annual temperature fluctuates at 70–80 °C in most areas. The average annual temperature in the capital city of Ulaanbaatar is −2.9 °C and reaches +33 °C in summer, below −40 °C in winter and around 20 °C diurnal temperature variation. In winter, when it is snowing all over the country, heavy snow falls mainly in mountainous areas. All water surfaces, including deep lakes, freeze on average from November to April.

Despite the harsh climatic conditions, Mongolia is a sunny place and the average sunny days per year is 250–260. Annual average is 3,000 h of sunshine per year, more than in other parts of the world at the same latitude. The short frost-free period is about 100 days on average from the end of May to the end of August.

Precipitation is low, decreasing from the north (350–500 mm) to the south (less than 50 mm) and mainly in summer (85%), with the driest

[1]Tal-plain or.

months from November to March. Annual average precipitation varies between 300 and 350 mm: the wettest are northern forests and large river basins, the most arid are the southeastern plain and the Gobi Desert. In the Mongolian-Altai region, the precipitation is 150 − 300 mm, and 50 − 150 mm in the Gobi Desert, where almost all precipitation is lost due to evaporation.

Most of Mongolia's lowland areas, including Gobi Desert, can be classified as BWk (cold desert) and BSk (cold semi-arid) according to the Koppen-Gieger climate classification system, which are determined by average annual temperatures below 18 °C. In addition, the Dwc climate prevails in the northern part of the country, with dry winters with snow, less than four months above 10 °C and the coldest month below −3 °C. Based on the ratio of precipitation to temperature, Koppen-Gieger can use other types of climate, such as DSk (Batchuluun 2012). In the eastern part of the country, there has been a steady increase in temperature, averaging around 0.6 °C per year, and a slight decrease (12–15%) in summer precipitation. Due to geographical location and topography features, climate change will be much more severe in Mongolia in near future. Since the 1990s, there have been several scientific assessments conducted on climate change (MET 2018). Mongolia has joined the United Nations Framework Convention on Climate Change (UNFCCC) in 1993, the Kyoto Protocol in 1999 and the Paris Agreement on Climate Change in 2016. A detailed analysis of climate and the distribution of climate parameters in Mongolia is provided in Chap. 4.

## 1.4 Water Resources

Mongolia's surface water resources include rivers, lakes, springs and glaciers and are unevenly distributed across the country. Due to extremely high evapotranspiration losses compared to precipitation, water resources scarcity is caused, especially surface water scarcity in the eastern and southern parts of the country. The potential evapotranspiration is about 300–400 mm in the mountainous areas, 600–1,000 mm in the steppe area and 1,000–1,300 mm in the Gobi region. About 90.1% of precipitation evaporates, only 9.9% forms surface runoff and penetrates underground aquifers (Tsegmid 1969; Davaa 2018). Most of the surface runoff (95%) flows out of the country, while a small proportion (5%) flows into lakes and basins in Mongolia (Davaa 2018). The total length of rivers and streams in Mongolia is about 67,000 km, the longest being Orkhon and Kherlen.

Although the average density of the river network in the country is 0.02 km/km$^2$ (Tserensodnom 1969; Davaa 2018), it is much denser in the northern and central parts of the country than in the rest of the territory. About 70% of all surface water resources are formed in mountain ranges, which occupy only 30% of the country's territory. Mongolia's total surface water resources are estimated at 608.3 km$^3$/year, and most of its water is stored in lakes (500 km$^3$/year) and glaciers (62.9 km$^3$/year). Lakes in Mongolia cover only 0.4% of the territory, but retain a significant volume of water. Long-term average annual renewable water resources include 32.7 km$^3$ of surface water and 6.1 km$^3$ of groundwater (Davaa 2018).

Mongolia is located at the intersection of three continental basins: the Arctic Ocean, the Pacific Ocean and the Central Asian Internal Drainage Basin. Approximately half of Mongolia's total river flow is formed in the Arctic Ocean basin. Mongolia's largest river, the Selenge, originates in the northwest of the Khangai mountains and discharges into Lake Baikal, a giant inland sea in Siberia. The main tributaries of the Selenge, Orkhon, Tuul, Eg and Chuluut rivers are part of the Arctic Ocean basin. About 65% of the Mongolian population lives in this basin, and today most of the socio-economic activities in Mongolia are carried out here.

In eastern Mongolia, the two main rivers, the Kherlen and Onon, originate on the eastern slopes of the Khentii mountains. Onon river flows into the Pacific Ocean via the Amur River (between Russia and China). About one-tenth of Mongolia's river flow is formed in this basin. In the western part of Mongolia, the glacier-fed Khovd and Zavkhan rivers eventually flow into the Khar-Us and Khyargas lakes in the Great Lakes depression, respectively. Despite its arid climate, Mongolia contains several thousand freshwater and salt lakes, most of which are located in the country's semi-arid and arid regions. These lakes are considered sensitive indicators of the overall wet state of the basins (Batnasan and Sevastyanov 2002). Mongolia has about 3,500 lakes with an area of more than 1 sq. km (Tserensodnom 1967; Tsegmid 1969). The country's largest lakes are Uvs (3,556.1 $km^2$), Khyargas (1,357 $km^2$) and Khar Us (1,011 $km^2$) (NSO 2018) in the depression of the Great Lakes; Khuvsgul (its surface area is 2,774 $km^2$, length 120 km, average width 30 km) in the northern part of the country, near the Russian border. Approximately 75% of the country's freshwater resources are stored in lakes, and about 70% (380 $km^3$) of this freshwater is stored in lake Khuvsgul (Bogoyavlenskii 1989). It is the largest reservoir of fresh water not only in Mongolia but also in Asia (the second largest lake in Asia). The total volume of Mongolian brackish lakes is 90 $km^3$. The second largest brackish lake is lake Khyargas (after lake Uvs).

## 1.5 Soils, Plant and Animal Life

The latitude, remoteness from any oceans and climatic conditions have determined a special landscape: the southern part of the country is covered with semi-desert and desert (Mongolian Gobi); grassy steppe in the eastern forests and taiga in the north of the country. In the west, mountain ranges and ridges in the north are overgrown with wild forests, large lakes and rivers. In high mountain regions, the content of organic carbon in the soils is 2,180 and 2,660 $g/m^2$ in the steppe region, while the lowest content of organic carbon of 1,330 $g/m^2$ in the soils is observed in desert steppe regions (Batkhishig 2018). Natural vegetation from north to south is gradually moving from cedar and larch forests to steppe vegetation characterized by various feather species and, finally, by very rare desert vegetation. The forest in Mongolia consists of two major types: Boreal conifer forests in the north end and Saxaul forest in the south of the country. Mixed and deciduous forests occupy 7.8% of the country's territory (MEGD 2014; MET 2018) and consist mainly of larch and pine trees in the northern part, which is the southern boundary of the Siberian taiga. The mountainous areas of Khangai and Khentii are covered with pine and larch forests, Khuvsgul with coniferous forests, Mongol Altai and Gobi-Altai with shrub vegetation. Some areas of Gobi are covered by xerophytic (drought-resistant) Saxaul forests, and the oasis found is an underground water source. Chestnut or brown soils dominate most of the eastern plain, while humus soils are slightly saline in arid and semi-arid areas. Especially large areas in the Gobi Desert are covered with pebbles and sometimes different types of sand dunes. The largest sand dune is the Khongor Els, which stretches 400 km in the southern part of the country, with the parallel location of the Gurvan-Saihan Ridge. The distribution and types of flora and fauna differ in the country's natural zones and altitudinal belts. A detailed description is provided in Chap. 10.

## 1.6 Natural Resources

Mongolia is rich in natural resources, including a variety of minerals, wildlife and forests. Major forest areas are in the northern part of the Khuvsgul, Khangai and Khentii ranges. Forests cover 9.6% of the country's territory, about 73% Siberian larch, 11% cedar and 6.5% pine. More than 50 local fish species lived in rivers and lakes in Mongolia, including freshwater taimen in Onon. About 82% of Mongolia's territory is considered a natural pastureland as the main source of livestock grazing, as well as the world's largest grazing ecosystem with a unique natural landscape (MET 2018).

Mongolia is rich in mineral resources, including coal, copper, fluorite, gold, iron ore, lead, oil, tungsten, and so on. The country's mineral deposits in the 1960s and 1980s were mainly explored by the Mongolian–Russian joint geological expedition. Among the fossil fuels rich in anthracite and bituminous coal reserves are Tavan Tolgoi, Khar Tarvagatai, Achit, Baga Nuur and Nalaikh deposits. Coal deposits play an important role in Mongolia by supplying fuel and exporting. Most of the phosphorus deposits extend along the western side of lake Khuvsgul, discovered in the 1980s, and iron ore in the central part of the country.

The large copper and molybdenum deposit area is located on the northern slope of the Khangai Mountains and the Selenge River basin to the lake Khyargas basin (Byamba 2012), the main deposit being Erdenet, which was discovered in 1975. Tsagaan Suvarga and Oyu-Tolgoi in the Gobi region are also considered to be the largest copper and molybdenum deposits. The eastern part of Mongolia also contains large areas of feldspar, including Berkh, Bor Undur, the largest mine. The gold deposits are rich in gold, primarily in the southern and northern regions of Khentii and the Khangai region, as well as in the major river basins.

## 1.7 Population and Culture

Mongolia is one of the lowest population densities in the world. In 2017, the population density was 2 persons per square kilometer at the national level, and in the city of Ulaanbaatar it was 311.3 persons per square kilometer. As of December 2018, the population was 3,233 million (NSO 2018), with an average life expectancy at birth of 69.9 years (75.8 years for women and 66.0 years for men). Population growth rates in 2018 were 1.8%, declining in recent decades (2.5% before 1990), fertility per 1,000 population is 23.8, mortality is 5.0 and net growth rate is 18.8 in 2018 (NSO 2018). Thirty-three percent of the total population is under 14 years of age and 75% is under 35 years of age. As of 2018, about 67.2% of the total population live in urban areas, while the rest live a seminomadic lifestyle (NSO 2018). A third of the total population, or 1,284,500 people, have lived in the capital city of Ulaanbaatar as a result of intensive internal migration since the 1990s.

Mongolia is a homogeneous nation inhabited mainly by Mongols. More than 96.6% of the population are ethnic Mongolians. The only ethnic group, Kazakhs, accounted for 4.8% of the total population (100,120) as of 2018, the majority (over 90%) of which reside in the Bayan-Ulgii (NSO 2018), which is located in the extreme western part of the country. The literary and official language, Mongolian, belongs to the Ural-Altaic language family. The largest group of the population is Khalk (82.4%), the main Mongolian nationality, which mainly resides in the central and northern regions of Mongolia. About 12% of the population speaks different dialects, with more than 20 language groups, such as Durvud, Uuld, Bayad and Torguud, who are descendants of Oirad and Tuva, Uriankhai in western aimags, while Buriad, Barga and Uzemchin, and so on live in eastern aimags of Mongolia.

However, the differences between the different linguistic subgroups were insignificant; they were expressed in the traditions of headdresses and in such insignificant variations in pastoral techniques and Mongolian culture that were relatively uniform across large areas. Halha Mongolian is a standard language, taught in schools and used for all official affairs. Writing is based on the Halha language in Mongolia. The original Mongolian script, also known as Uighurjin, was replaced by Cyrillic in 1944 (Shagdar 1999). Only in Bayan Ulgii aimag, the Kazakh language is used in primary schools and local administrative bodies. The literacy rate as at 2018 is 98.5% (MECSS 2018).

Mongolia is well known for its nomadic traditions. The nomadic lifestyle is still practiced in rural areas of the country. Because of the harsh climatic conditions and short growing season, Mongolia's traditional culture is defined by nomadic lifestyle rather than agriculture. Mongolian nomads breed mainly five main livestock species (goats, sheep, camels, horses and cattle, including yaks), which provide them with food, dairy products, transport and wool. Horses are the most prestigious animals for Mongolians, which is the main mode of transport and children begin to ride at the age of 3. Nomads follow a seasonal pattern, migrating from their herds to new pastures depending on the season. The traditional nomadic residence is the Mongolian Ger, whose mobile dwelling is made of wood and felt. Most Mongols are Buddhists, and Kazakhs are traditionally Muslims.

## References

ALMGC (2018) National integrated report of land database for 2017. Agency for Land Administration and Land Management, Geodesy and Cartography, Ulaanbaatar

Batchuluun Ye (2012) Explainatory dictionary of physical geography, 5th edn. Tsagaan Bambaruush, Ulaanbaatar (in Mongolian)

Batkhishig O et al (2018) Soil and soil cover. In: Nyamdavaa G (eds) Environment of mongolia: the pasture, soil use and conservation, vol 4. Ministry of Environment and Tourism. Munkhiin Useg Publishing, Ulaanbaatar, pp 71–81 (in Mongolian)

Batnasan N, Sevastyanov (2002) Lakes in arid regions of Mongolia and their water resources changes. In: International conference "Geographical study of central Asia and Mongolia. Institute of Geography, Mongolian Academy of Sciences, pp 21–22, Ulnaabaatar

Bogoyavlenskii BA (ed) (1989) Atlas of Khuvsgul lake. Institute of Geography, Moscow. Mongolian and Russian Academy if Sciences. General administration of Geodesy and Cartography, Russia (in Russian)

Byamba J (2012) Plate tectonics of lithosphere. In: Geology and Minerals of Mongolia, vol 4. 494 pp Ulaanbaatar (in Mongolian)

Davaa G (2018) The surface water of Mongolia. Ulaanbaatar (in Mongolian)

MECSS (2018) Education Statistics of 2018. Ministry of Education, Culture, Science and Sports, Ulaanbaatar

MEGD (2014) Mongolia second assessment report on climate change-2014 (Dagvadorj D, Batjargal Z, Natsagdorj L, eds). Ministry of Environment and Green Development, Ulaanbaatar

MET (2018) Mongolia Third National Communication (TNC) under the United Nations framework convention on climate change. (Batjargal Z, ed). Ministry of Environment and Tourism, Ulaanbaatar

Murzaev EM (1952) Mongolian People's Republic. The Description of Physical Geography. State publishing house of geographical literature, Moscow (in Russian)

NSO (2018) National Statistical Book. National Statistical Office, Ulaanbaatar

Shagdar, Batchuluun et al (2006) The physical geography of Mongolia (textbook). Munkhiin Useg Publishing, Ulaanbaatar

Shagdar Sh (1999) The history of Mongolian Education, vol 1. Munkhiin Useg Publishing, Ulaanbaatar (in Mongolian)

Sinitsin BM (1959) The Central Asia. RGS, Moscow (in Russian)

Stolz C, Hülle D, Hilgers A, Grunert J, Lehmkuhl F, Dasch D (2012) Reconstructing fluvial, lacustrine and aeolian process dynamics in Western Mongolia. Zeitschrift für Geomorphologie 56(3):267–300

Tsegmid Sh (1969) Physical geography of Mongolia. Institute of Geography and Cryology, Mongolian Academy of Sciences. State Publishing House, Ulaanbaatar, 405 pp. (in Mongolian)

Tsegmid Sh (2013) The science of geography in Mongolia. Institute of Geograpgy. Mongolian Academy of Sciences, Ulaanbaatar (in Mongolian)

Tserensodnom J (1967) The lakes of Mongolia. State Publishing House, Ulaanbaatar (in Mongolian)

Tserensodnom J (1969) Chapter 3: The water. In: Tsegmid S (ed) The physical geography of Mongolia. The State Publishing House, Ulaanbaatar, pp 133–231 (in Mongolian)

UNDP (2018) Human development indices and indicators: 2018 statistical update. http://hdr.undp.org/sites/default/files/2018_human_development_statistical_update.pdf

# 2 Historical Geography: Administrative Division and Research in Physical Geography of Mongolia

Batchuluun Yembuu and Dash Doljin

**Abstract**

The history of Mongolia's territorial transformation is linked to a nomadic lifestyle that correlates with geographical and climatic conditions. The history of the study of physical geography begins with the ancient explorers Marco Polo; in addition, a great contribution was made by Russian expeditions and geographers (N.M. Przhevalsky, G.N. Potanin, P.K. Kozlov, M.V. Pevtsov, V.A. Obruchev, and others). This chapter presents a brief history of the division of territorial transformations and management in Mongolia and the main geographical studies aimed at studying the physical features of natural structures and conditions associated with the geomorphology, the hydrography, the nature of the climate and soil, the glacier, as well as regional variations. Geographical exploration and research will be presented in three sections: early studies dating back to the thirteenth century; nineteenth–twentieth century studies; and contemporary approaches to the study of physical geography in Mongolia. A more detailed analysis of the physical geographical research in each field of Mongolia's physical geography will be presented in the first section of the following chapters. Despite the large amount of published material on Mongolia abroad (in English and other languages), this chapter does not aim to analyze the history of Mongolian studies, but will focus only on the geographical perspective, focusing more on the main travelers, exploration and expedition results, which has been recognized as a major contribution to the development of physical geography research in Mongolia. As mentioned in the preface, Mongolia's geographical names, written in different styles, will be used here as Mongolian phonetic transcriptions. The period or date given in this chapter is based on Mongolian historical documents.

## 2.1 Territorial and Administrative Transformation

Territorial–administrative transformations of Mongolia were studied by a number of national scientists, including Nazagdorj (1963, 1972, 1991), Sonomdagva (1967), Perlee (1978), Batbuyan (2000) and the first comprehensive map of the "Administration of Mongolia" compiled by Badamjav, D., Budjav, J et al. in 1967 (Batbuyan 2000).

Archeological evidence shows that, since prehistoric times, Mongolia has been home to

B. Yembuu (✉) · D. Doljin
Department of Geography, Mongolian National University of Education, Ulaanbaatar, Mongolia
e-mail: batchuluun@msue.edu.mn

D. Doljin
e-mail: dash@msue.edu.mn

© Springer Nature Switzerland AG 2021

B. Yembuu (ed.), *The Physical Geography of Mongolia*, Geography of the Physical Environment,
https://doi.org/10.1007/978-3-030-61434-8_2

numerous ethnic groups and nomads (Delgerjargal and Boldbaatar 2017). However, due to climatic conditions and other geographical features of the country's territory, migration of nomads from other parts of Asia is limited to Mongolia. In particular, the Altai Mountains in the west and the Gobi Desert in the south have become natural barriers to human migration. According to historical documents, the first powerful Central Asian state of Khunnu (Xiongnu) was founded by nomads and formed the confederation of Modun Shanyu in 209 BC (Delgerjargal et al. 2017). The vast territory of the Khunnu Empire stretched between the Great Khyangan and Manchuria in the east to Tenger-Uul and Zuungar Gobi in the west, beyond Lake Baikal in the north and Lake Alasha and the Ordos of Inner Mongolia in northwest China and its capital in central Mongolia, Orkhon river valley.

In 1206, Chinggis Khan (also known as Genghis Khan) founded the Great Mongol Empire,[1] the world's largest contiguous land empire in history. The Mongol Empire stretched from present-day Eastern Europe and the Persian Gulf in the west to the Korean Peninsula in the east, from Siberia in the north to the Arab Peninsula and Vietnam in the south, and by the end of the thirteenth century occupied about 33 million sq. km[2] (Delgerjargal and Boldbaatar 2017).

There were hierarchical administrative units in the Mongol Empire, such as *Tumen* (tens of thousands of units), *Myangan* (thousands of units), *Zuut* (hundreds of units), *Aravt* (tens of units), and so on. According to the Secret History of Mongolia, there were 95 myangans in the country (Damdinsuren 1957). After Genghis Khan's death in 1227, the empire was divided into four separate sectors of the khanates[3]: The Golden Horde (in Russia), the Chagatai Khanate (in Central Asia), Ilkhanate (in Western Asia and Persia) and the Yuan Dynasty (in China, Mongolia, Korea and Tibet), which was founded in 1271 by Khubilai Khan,[4] who was overthrown by the Ming Dynasty in 1368 (Sukhbaatar 2000).

In the seventeenth to eighteenth centuries, Mongolia was divided into three distinct parts (Southern Mongolia, Khalkh or Central Mongolia and Western or Oirad Mongolia), which were influenced by significant geographical differences. South and Khalkh Mongolia were separated by Gobi, and the boundary between Khalkh and Oirad Mongolia ran along rivers and lakes in the intermountain valleys of the Altai and Khangai ranges (Sukhbaatar 2000). By the time of the Mongol Empire, the territorial division was based on nomadic lifestyle, pastoralism and settlements in Mongolia, but changed in 1725 under the influence of the governance system, during the Manchurian period, from the seventeenth to early twentieth century.

During these territorial changes the heartland of Mongolia has been existing between the Altai in the west and Khyangan in the east and from Siberia to the Tibetan Plateau (Tseden-Ish 2011), and the Mongols' traditional homeland is centered in Mongolia, a vast plateau in Central Asia (Encyclopedia Britannica, Inc 2008).

The name and location of the capital of Mongolia (Ulaanbaatar) was changed several times, which was founded as Urguu (1639) and moved to the current location in the valley of the river Tuul (1778) as Ih-Khuree (Urga to foreigners) became Niislel-Khuree (in 1911). On 26 November 1924, the Constitution of the Great National Khural was adopted, and the Mongolian People's Republic was officially established (Delgerjargal and Boldbaatar 2017). At the same time, Niislel-Khuree was renamed Ulaanbaatar (literally, Red Hero). Today, Ulaanbaatar is the political capital and industrial center of the country, home to most colleges, universities, international airport and the only railway junction linking the country to China and Russia.

Before the 1921 revolution, Mongolia was divided into four aimags (Tusheet Khan, Setsen Khan, Zasagt Khan, Sain Noyon Khan) with

[1]Ikh Mongol Uls.

[2]26,000 sq. km by other sources.

[3]Golden Horde-Altain Ordnii Uls, Ilkhanate-Il Khaadiin Uls, Tsagadai Khanate (also known as Chagatai) Tsagaadain Uls.

[4]Chinggis Khan's grandson, who ascended the throne in 1260.

39 khoshuus.[5] After 1921, the name of the four aimags were changed to mountains, such as the Bogd Khan mountain aimag, Khan Khentii mountain aimag, Khantaishir mountain aimag and Tsetserleg Mandal mountain aimag. In 1925, two aimags were added: the Chandman mountain aimag and Delger Ikh mountain aimag. According to the significant reforms carried out in 1931, there are 13 aimags and 311 soums in the administrative units of Mongolia (Sonomdagva 1967). In 1940, five new aimags were added. These 18 aimags existed until the 1990s as the main administrative unit of Mongolia. Currently, there are 21 aimags and 331 soums in Mongolia.

## 2.2 History of Physical Geography Research

Mongolia's history of geographical research is divided into several stages in different areas: geography (Murzaev 1948, 1952), geology (Marinov 1967, 1970), geomorphology (Jigj 1975), glacier and permafrost studies (Lonjid 1977), soil studies (Dorjgotov 2005) and dune studies (Baasan 2003). According to these stages, the history of geographical research in Mongolia can be divided into six periods (Dash and Mandakh 2011): *(a) the period up to 1921:* 1—the first travelers of the Middle Ages (thirteenth–fourteenth centuries); 2—study tours from the seventeenth century to the 1870s; 3—primarily expeditions from the 1870s to 1921; *(b) from the beginning of the formation of the MNR (1921) to the present day:* 4—from 1921 to 1961; 5—from 1961 to 1991; 6—after 1991.

The following section discusses the early stages of Mongolian research and the main expeditions of the Russian Geographical Society, which have contributed significantly to Mongolian geographical research and the main contemporary approaches to the study of physical geography in Mongolia.

[5]Khoshuu—name of smallest administrative unit.

### 2.2.1 Early Geographical Exploration of Mongolia

The earliest geographical works in Mongolia have a long history of more than 2000 years ago. Numerous ancient maps contain extensive geographic information, most of which is drawn on rocks and in a cave. The first books related to the historical geography were written in Mongolian in the late 1930s, one of which was the "Overview of Mongolian Territory" by Tseveen Jamsran, which included a brief description of physical and economic geography. The second book is entitled "The Whole Territory of Mongolia" by Buyan Chuulgan, a summary of systematized information on Mongolia's territorial division (Sukhbaatar 2000).

The first news about Mongolia were received by Europeans back in the Middle Ages from servicemen (ambassadors, merchants, etc.) and travelers who were attracted by the unknown and mysterious nature of the country. Despite this, however, Mongolia remained for a long time to be a country with little or no access for Europeans.

In 1241, when Mongol troops reached Europe (Hungary and Poland), the Europeans had received the first information about the Mongols. In the twelfth century Europe was really interested in Mongolia (Carruthers 1912). The pioneers of Europeans who traveled to Asia were the first two Franciscan missionaries; John of Plano Carpini (d.1252) left Europe in 1245, while William of Rubruck traveled through Mongolia during 1253–1255 (AfE 2014; Dash and Mandakh 2011; Tsegmid 1969). John of Plano Carpini traveled across Central Asia, described about Mongol customs, beliefs, and practices and provided a brief history of Genghis Khan and the Mongol conquests (Rossabi 2011). His description is very detailed information about the Mongol's customs including wedding ceremony, the clothes of both the men and women and drinking, eating, about the animals, domesticated and wild as well as traditional housing called Ger. However, some descriptions did not correspond to the traditional customs of Mongols and nomadic lifestyle, such as he noted that "their food consists of everything that can be eaten, for they eat dogs, wolves, foxes…”; “they do not use table-cloths or napkins, they have neither bread nor herbs…” (Rossabi 2011). The nomads still do not use the napkins and never eat the dogs and foxes.

Since the thirteenth century stimulated the Europeans to travel to the East, diplomats, merchants, adventurers and wandering monks flocked to Asia, including Mongolia pursuing their practical and missionary goals. In the late thirteenth century, Marco Polo, who had spent seventeen years in China and visited many parts of Asia, described his experiences in his *Travels* (Popova 2008). The accumulation of information about Mongolia and the creation of the necessary prerequisites for research continued until the fifteenth century. This period was marked by a lack of written records based on original sources; much information about Mongols was collected in guides, itineraries and road novels, chronicles, crusade projects, missionaries' and pilgrims' works (Popova 2008). Interest in it has increased significantly since the Mongol-Tatar conquest of Central Asia and Eastern Europe, but later, due to the collapse of the Mongolian feudal empire, this interest noticeably reduced in subsequent centuries, especially in the fifteenth and sixteenth centuries, and travelers visit it rarely. The earliest evidence of Russian presence in Karakorum dates from the thirteenth century (Popova 2008). The first travelers from Russia to Mongolia were Vasily Tyumenets (1615), Ivan Petlin (1618) and Fyodor Baikov (1654) in the seventeenth century. In 1713, E. Trushnikov, a Tobolsk merchant, reached Lake Khukh Nuur.

### 2.2.2 Researches in the Nineteenth Century

Since the second half of the nineteenth century, Mongolia has been the object of geographical research of Russian travelers and naturalists. At the end of this century, the first specialized geological groups took part in the geographical expedition. These Russian geographical expeditions were associated with more geomorphological research in Mongolia, although they were carried out in parallel with major geographical and geological work.

In Russia, a scientific study of Central and Middle Asia was started by N.Ya. Bichurin. His comprehensive description of Mongolia *"Notes on Mongolia"* was published in 1828 and was translated into French. This book included great information about Asian countries including Mongolia, peoples, and places that were previously hard to no access (Popova 2008; Dash and Mandakh 2011). During the period of 1850–1910s, the highest political and military activity of Russia in Central Asia was stated (Martynov and Martynova 2015) and the Russian Geographic Society (RGS)'s expeditions addressed also a geopolitical task as they facilitated the growth of Russia's influence in the region (Yusupova 2008). Also, the main objective of the RGS was gaining more knowledge about Mongolia and scientific descriptions. A particularly important role in the study of Mongolia was played by the expeditions of the RGS, especially those by N.M. Przhevalsky and his followers. Natural science collections that the travelers brought are still of great scientific significance (Przhevalsky 1888). Russian consulate was established in Urga (Ulaanbaatar) in 1858 (Kotkin and Ellaman 1999). A number of outstanding expeditions to Mongolia took place under the leadership of N.M. Przewalski, G.N. Potanin, P. K. Kozlov, V.L. Kotwich, B.Y. Vladimirtsov, V. V., M.V. Pevtsov, V.A. Obruchev and others.

The first expedition was started by N.M. Przhevalsky in 1867. In 1867–1869, Nikolai Mikhailovich Przhevalsky (1839–1888) made his first trip to the Ussurian region. This outstanding traveler undertook four expeditions to Central Asia between 1870 and 1880, covering a total of 30,000 km. In 1870–1873, he went to Mongolia, China and Tibet, then in 1876–1877 to Zuungar and Loh nuur lake; in 1879–1880, he led the First Tibet Expedition and in 1883–1885, the Second Tibet Expedition (Fig. 2.1). He wrote a number of books containing the scientific foundations of his expeditions and detailed and vivid descriptions of local nature, climate, relief, and animal and plant life. Przhevalsky himself modestly

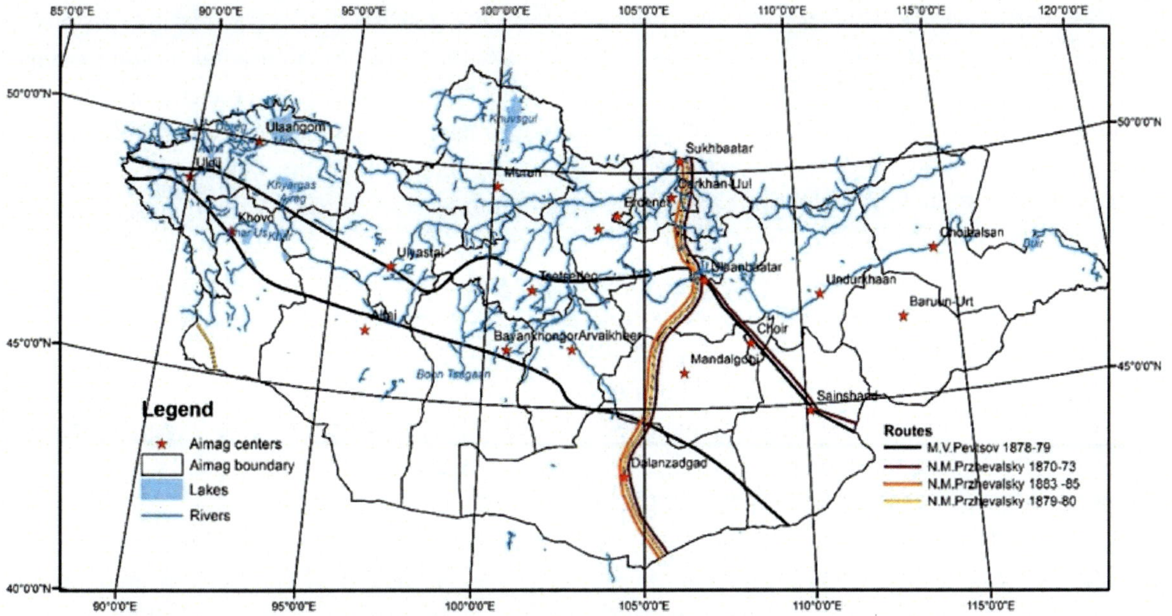


**Fig. 2.1** Itinerary of N.M. Przhevalsky and Pevtsov in Mongolia

described his travel as "scientific exploration"; indeed, the Asian territories he visited had not previously been explored by scholars. It was he who introduced Europeans to Central Asia, stimulated interest in hard-to-reach regions, and thus contributed to the development of extensive and regular expeditionary activities (Popova 2008). One of his most famous works is "*Mongolia and the country of the Tanguts*", which presented the uniqueness of nature and relief of Mongolia, and about the existence on its territory of Siberia (Transbaikal).

N.M. Przhevalsky's pursuit was taken up by his students and followers; the trips were made by Grigory Nikolayevich Potanin (1835–1920), Mikhail Vasilyevich Pevtsov (1843–1902), Vsevolod Ivanovich Roborovsky (1856–1910), Pyotr Kuzmich Kozlov (1863–1935) and Grigory Efimovich Grumm-Grzhimaylo (1860–1936).

G.N. Potanin made a great contribution to the study of Central Asia, particularly Mongolia. He went to northwestern Mongolia and Tuva in 1876–1877 and 1879–1880, to northern China, eastern Tibet and Central Mongolia in 1884–1886 and 1892–1893, and to the Greater Khingan in 1899. In his activities he combined a study of natural history and a study of ethnography. His subsequent publications outlined the wealth of material on culture, folklore and popular arts and crafts of Mongolian and Turkic peoples amassed during his expeditions (Murzaev 1957). If the description of N.M. Przhevalsky's travels is more colorful and fascinating, and the monotonous chalk lands and desert plains did not occupy his attention, G.N. Potanin was interested in everything, and he literally described the details. From the general appearance of the mountains and valleys crossed by him to the traces of glaciation, salt marshes and levels of ancient lakes, everything is strikingly documented in accordance with reality. This is the lasting value of the data of G.N. Potanin, which retained their significance to this day.

A special merit of G.N. Potanin, from the point of view of geomorphologists, is that he was the first to note the large-scale development of the so-called pedestals in the Gobi Mountains and characterized them in a comprehensive and expressive way. G.N. Potanin writes: "*The original view of the local ridges gives their foothills an unusual development, the so-called Pomol pain. It seems that these ridges were submerged in water up to half their height, and under water they were covered with precipitation, smoothed by water currents. Perhaps even the rays surrounding the ridges of the Gobi Altai, as wide panels, were formed and not under water, I used this expression only to make the description more visible".* Beli (бэл-means foothills or piedmont in Mongolian), as noted by G.N. Potanin, is something middle between the plains and low relief hills, which are surrounding by the rocky mountains; the beli has a steep or gentle fall and is divided into two tiers of the lower and upper. The lower clay surface is not covered by anything, and in the upper tier it is covered by gravel and near the rocky bottom of the mountains cut by shallow fan-shaped deposits. Beli are the most barren territories of Mongolia (Murzaev 1948).

Interesting data about the nature and relief of Mongolia were collected during the expedition of M.V. Pevtsov (1878–1879). His expedition crossed the country and returned to the Russian Altai, through the Khangai Mountains and the Great Lakes Basin. M.V. Pevtsov, like G.N. Potanin, made a detailed and interesting description of mountain ranges, rivers and lakes along his route, and on this basis came to important conclusions about a more significant watering of Mongolia in the past. He was the first to point out the young character of the relief of the Gobi Altai mountains and the great role of the newest tectonic movements. Also, of interest to him is the character of the structure of the Khangai surface descending towards southern slopes as numerous lakes are located, which he called a "Valley of the Lakes". He has one of the first (if not the first) physiographic sketches of Mongolia, where much attention is paid to its relief (Matusovsky 1888).

In order to understand Mongolia's terrain, it is also important to explore the possibilities of small expeditions and individual units that, while carrying out specific small regional tasks, carried out more detailed work by capturing the marshes. Such studies have allowed for substantial refinement and completion of the general information previously obtained. The first of these studies was the work of the renowned geologist D.A. Klemet (1892–1896, 1898), who visited many areas of Mongolia over the years and made a number of important observations in the field of geomorphology. Although D.A. Klemet did not leave any generalized works, his instructions on the orography of the Gobi-Altai on the existence of the former intensive volcanic activity and widespread of the ancient moraines on the Khangai plateau, on the modern glaciers in the mountains of Western Mongolia and on the character of the Great Lakes basin dumps have been confirmed in subsequent studies and widely covered in the scientific literature.

### 2.2.3 Researches in the Twentieth Century

At the beginning of the twentieth century, geographical studies in Mongolia were mainly conducted by the Russian Geographical Society. An important place is occupied by the works of V.V. Sapozhnikov (1905–1906, 1908–1909), who studied in detail the western part of the Mongol Altai (Khovd and Khar Erchis basins) and published a detailed work (1911, ed., 1949), which significantly expanded our understanding of the relief of mountains and their glaciers. According to his data, ancient glaciation was widely and diversely developed, large glaciers reached a length of up to 110 km, had a capacity of 500 m (Khovd River basin) and descended to the marks of 1,896 m (Sagsai River). He discovered the hotbeds of modern glaciation in the mountains and many glaciers (the Tavan Bogd Uul massif) were mapped for the first time. The largest glacier (the upper Tsagaan Gol) is named after Potanin (the length Potanin glacier—20.2 km).

While studying the Mongol Altai and the Khangai plateau (1906), I.G. Grane discovered traces of ancient glaciation (Kara moraine, I.G. Grane moraine, etc.). According to his data, the length of glaciers in the Mongol Altai reached even 140–160 km, and the thickness of several 100 m; glaciers descended to the marks of 1300–2400 m, and in Khangai—up to 2100–2600 m. I. G. Grane believed that the mountains of Mongolia survived three or four glaciers, the maximum of which was covered by the upper belt of the mountains of the Mongol Altai; however, this is not confirmed by recent studies. I.G. Grane observed a wide development of plains—almost plains, which he rightly associated with the young rise of these mountains, which are still insufficiently transformed by erosion. According to him, the current relief of the Altai Mountains was formed in the period after the Pliocene urbanization, followed by two deep erosional cuts and glaciation.

New data on the relief of the northern regions of Mongolia during this period were collected by S.P. Peretolchin (1897–1900, 1903), V.L. Komarov (1902), V.S. Yelpatyevsky (1903), L.I. Prasolov (1912), M.A. Usov, I.A. Molchanov and M.K. Korchanov (1913). The first three of them were devoted to the study of Khuvsgul Lake, as well as the adjacent Munkh saridag mountains. They collected interesting information about relief forms, terraces of Lake Khuvsgul, permafrost and ice (V.L. Komarov), as well as about the large ancient glaciation of the Munkh saridag, whose glaciers, as noted by L.A. Yachevskiy (1877), descended to the shore of Lake Khuvsgul (S.P. Peretolchin). According to S.P. Peretolchin, modern glaciation of the Munkh saridag has a close connection with this lake.

The second group of scientists conducted their research in other areas of northern Mongolia. L.I. Prasolov studied soils between Kyakht and Urga (Ulaanbaatar), and the M.A. Usov expedition studied the geological structure of the Khentii plateau. L.I. Prasolov noted on his March the presence of permafrost and related bumps of blowing. The expedition of M.A. Usov, along with great geological results, for the first time made an orographic description of the highlands and established in the mountains clear traces of ancient glaciation (moraines, glacial lakes, etc.). I.A. Molchanov, a participant of this expedition, who specially studied glaciation of the highlands, published an interesting work on this issue (1919). According to his data, the glaciation centers were located in the highest Holtz massifs, with the length of glaciers reaching 18 km and a thickness of 100 m. Glaciers descended to a height of 1,780–1,890 m.

The traveler and geographer P.K. Kozlov (1863–1935) participated in the second Tibetan expedition of Przewalski and fully adopted its widely described method of route exploration, which has always been used in his own expeditions (Kozlov 1963). P.K. Kozlov made an outstanding contribution in the geographical and archeological study in Mongolia through his six big expeditions, during the period 1883–1926 (Murzaev 1948) (Tsegmid 1969). Three of these expeditions were headed by P.K. Kozlov, and the most valuable facts related to Mongolia were collected during the first independent Mongolian-Kama expedition (1899–1901) and the Mongol-Sichuan expedition (1907–1909) which was a landmark for the archeology of the twentieth century in the southern part of Mongolia and Eznii-Gol[6] river basin and the northern part of Inner Mongolia (Tsegmid 1969). Description of relief in the works of P.K. Kozlov, in contrast to G.N. Potanin, attracts more generalized characteristics than details. They covered a large area of the country, and their routes were mostly in new, previously unexplored areas.

During his expedition, Sogoo and Gashuun nuur (lakes) were investigated except for the Eznii-Gol basin. The most significant discovery by Kozlov's team was the ancient capital Khar Khot (Khot is city) of the Tangut State, which had existed for about 250 years (892–1227). The discovered materials of this expedition, including thousands of manuscripts and printed books in Tangut, Chinese and Uighur,[7] were transported to St. Petersburg, which laid the foundations for academic Tangut studies and is a valuable source for historical geography of Mongolia (Dash and Mandakh 2011). According to some scholars (Yusupova 2014), Kozlov's meetings with Dalai Lama were important in terms of Russian–Tibetian relation. He met with Dalai Lama twice, one was in the capital city of Mongolia Urga (now Ulaanbaatar) (Kozlov 2011). Except for the geographical study, the Kozlov's expeditions were important for the policy of the Soviet State (Andreev and Yusupov 2003), and the political background of the Kozlov's expedition was openly reported in the press of that time (Kozlov 1963).

[6]Eznii-Gol (Gol-гол means river in Mongolian).

[7]Traditional Mongolian script (vertical).

During the Mongolian–Tibetan expedition (1923–1926), Kozlov's study was focused on Mongol Altai and Southern Khangai mountains as well as in Khar-Khot.[8] The main results of this expedition were formulated by Kozlov: exploration of the khun tombs in Noyon-Uul, Mongolia; about 3,500 km is mapped with the hypsometric shooting; the mountains, steppes and deserts up to Khar-Khot are studied; steppe lakes with their flora and fauna are explored, the depths are measured; meteorological studies were conducted for three years in Mongolia; zoological and other collections were gathered; ethnographic observations and additional excavations in Khar-Khot were conducted (Murzaev 1949). The discoveries in Noyon-Uul were considered the new era in the archeological study of Turkus in Central Asia which is evidenced by khun tombs (Fig. 2.2). Since 2006, excavations in the Noyon-Uul are performed by Russian–Mongolian archeological expedition (The Academy of Science of Mongolia 2007). The expedition collected countless facts and materials about the geography of Mongolia, mainly Kozlov's diaries, which are still used as a historical source.

S.A. Kondratiev[9] began his study during the Mongolian–Tibetan expedition with Kozlov in 1924. The Mongolian scholar committee (now Mongolian Academy of Sciences) sent S.A. Kondratiev to a "scientific tour" to survey the mounds in the Noyon-Uul within 100 km to the north from capital city-Urga (Kondratiev 2006) (Martynov and Martynova 2015).

Interesting ideas about the Gobi desert's relief and its genesis were expressed by geologist A.A. Chernov, a member of the expedition of P.K. Kozlov, who made a route from Khiakht through the valleys of the rivers Tuul and Ongi Gol to the Gurvan Saikhan mountains and Khukh Lake. According to this researcher, the surface of Gobi is strongly eroded. A.A. Chernov notes the extensive depressions in the Gobi desert are separated by low mountains with ancient rocks raised from the bottom of shallow basins. Thus, their origin is closely connected with the genesis of the latter (1910, p. 68). He also wrote that the young sediments of the Gobi depressions, called grabens, were accumulated in shallow, closed basins and most likely belonged to the Pliocene by age.

Another Russian scientist, Vladimir Afanasyevich Obruchev (1863–1956) made a great contribution to the study of the geography and geology of Mongolia. He described his experiences in a large number of popular science books and fascinating science fiction novels. V. A. Obruchev, traveling through Central Asia, only once managed to cross the territory of Mongolia along the Khaalgan road (Fig. 2.3). However, the materials he collected on Gobi were very valuable. It turned out that the young sediments of the Gobi depressions (in one of them he found the remains of the tertiary rhinoceros' teeth) were continental and not sea sediments of the Khan-Khai Sea, as previously thought. This not only radically changed the existing ideas about the nature of Gobi, but also became the reason for a major American expedition, which found in a number of Gobi depressions numerous remains of vertebrate bones of Cretaceous and tertiary animals and thus confirmed the conclusions of V.A. Obruchev.

Richtgofen's hypothesis about the spread and genesis of the loess was also revised. As V.A. Obruchev has shown, despite this hypothesis, the Gobi depression is the geographical feature of the Gobi depression without trees and consists mainly of tertiary sediments. On the basis of the scientific investigation in Central Asia, Mongolia and China, it has not only essentially reconsidered and has specified representations of Richtgofen, but also has proved the new hypothesis about an origin of loesses. Obruchev wrote the first essay on orography and geology of Central Mongolia (1894), in which he named the mountain foothills of the Gobi-Altai as pedestals, noting that "*often the relative height of the pedestal above the lower part of the neighbouring depression is two or even three times higher than the relative height of the ridge above the upper*

[8]Hara-Hoto-in Russian sources.

[9]Nephew of the composer A.S.Arensky, he began to study the folk music of the Mongols.

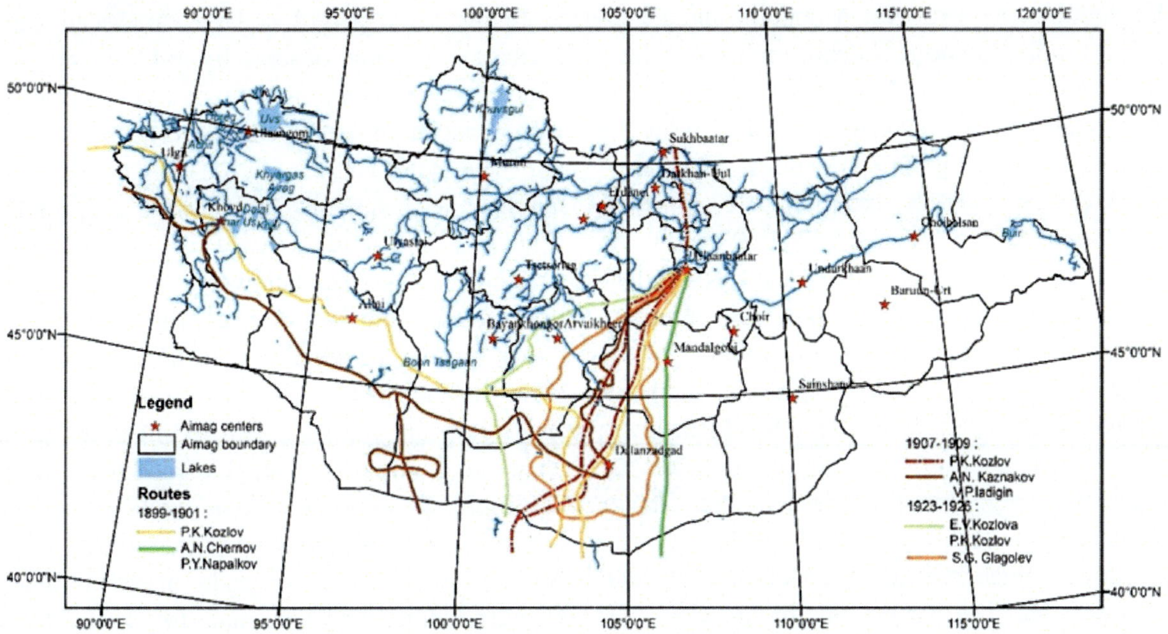


**Fig. 2.2** Itinerary of Kozlov and others' expeditions in Mongolia

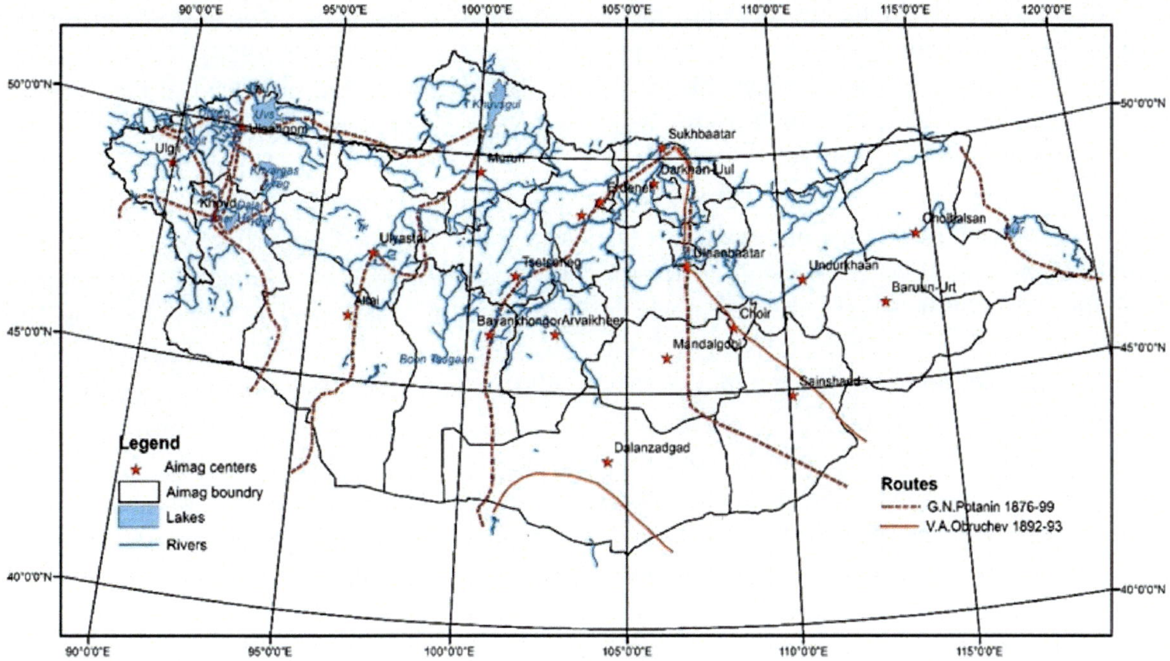


**Fig. 2.3** Itinerary of Potanin and Obruchev expeditions in Mongolia

*boundary of the pedestal*" (1958, p. 422). In his subsequent works he repeatedly returns to the geomorphology of Mongolia, especially to the issue of ancient glaciation.

Except the Russians, not many researchers studied on Mongolia, at the beginning of the twentieth century. The valuable researches of glaciers in Mongol Altai and Khentii range were done by I.G. Grane in 1906, while by Douglas Carruthers (from England) in 1910–1911. Carruthers traveled to the western part of Mongolia, mainly studied Mongol Altai glaciers and he wrote the book "Unknown Mongolia" (Carruthers 1914): A Record of Travel and

Exploration in Northwest Mongolia and Dzungaria 2 vols. (London: Hutchinson 1914) with a foreword by Lord Curzon; H.G.C. Perry-Ayscough and R.B.Otter-Barry, and "With the Russians in Mongolia" (London: John Lane 1914). He compiled the sketch map "*Exploration in Mongolia Zuungar*" (1:7,500,000) and published an article about the west Mongolian-Uriankhai people, their tradition and culture.[10] Both these accounts were preoccupied with the effect of Russian successes on British imperial interests (Kotkin and Ellaman 1999) (Carruthers 1912). In addition, in the study of the lake, the expedition met (Price) the ancient deposits of lakes lying much higher (by 10 m) than its current level, which indicates a reduction in the area of lakes and their drying up. In his work, D. Carruthers highly appreciated the research of Russian travelers on Central Asia and Mongolia.

Understanding of the relief structure of the western regions of Mongolia became more complete after a detailed study by T.E. Grumm Grzhimailo (1903), I.P. Rachkovsky (1903), V. V. Sapozhnikov (1905–1906, 1908–1909), I.G. Grane (1906) and Carruthers (1910–1911). A well-known researcher of Central Asia, G.E. Grumm Grzhimailo, using his observations of a number of areas of Western Mongolia (Mongol Altai, the Valley of the Great Lakes, etc.), which he crossed during his expedition to Tuva, wrote a famous book "Western Mongolia and the Uriankjhai Krai" (1914), summing up our knowledge about nature and geography of the same areas. During the visit to the mountain massifs of Kharhiraa and Turgen, I.P. Rachkovsky discovered a well-defined complex of glacial relief with significant modern glaciers and photographed them on the map. The beginning of complex geological–geographical researches in Mongolia was laid by expeditions of the USSR Academy of Sciences, which carried out long-term works under the general guidance of I.P. Rachkovsky with breaks till 1932–1933.

In the same period, a large American geological expedition headed by R.Ch. Andrews conducted research in Mongolia. Its research, which began in 1922, continued with a break until 1930 (1922–1925, 1928, 1930) and covered mainly the southern Gobi part of Mongolia (Fig. 2.4). The stratigraphy of the Mesozoic and Cenozoic continental deposits of intermountain troughs was studied, and for the first time major stages in the development of the Mongolian relief, its deep penepenilization in the Mesozoic and significant renovation in recent times were identified, and paleogeographic conditions of its development were shown as well as paleogeographical conditions of its development. Andrews and his team made many findings, including dinosaur bones and fossil mammals from the Mongolian Gobi and most notably the first nests full of dinosaur eggs, ever discovered. Andrews' main account of these expeditions are presented in his book "*The New Conquest of Central Asia*" (Andrews 1932). The collections of this expedition are presented at the America ion Museum of Natural History.

Research in the field of physical geography in Mongolia has intensified since 1921 when the Mongolian Science Commission was established. The beginning of complex geological-geographical researches in Mongolia was laid by expeditions of the Academy of Sciences of the USSR which carried out long-term works under the general guidance of I.P. Rachkovsky with breaks till 1932–1933. Along with the specialized geographical researches of these years (biogeographical—A.G. Bannikov, A.A. Yunatov, E.M. Lavrenko, etc., soil—N.D.Bespalov, etc.), it is important to note the general geographical works of E.M. Murzaev, a nature specialist of Mongolia (1947, 1948, 1952). Its routes covered the whole territory of Mongolia with the dense network. The greatest successes and merits of E.M. Murzaev are connected, naturally, by studying the physical geography of Mongolia. However, his works are also of great importance for geomorphology. His detailed research on general geomorphological problems (from 1947 to 1949) is widely known, as well as on a number of individual issues, including the restoration of

[10] Carruthers (1912). Exploration in North-West Mongolia and Dzungaria. *The Geographical Journal, 39*(6), 521–551. https://doi.org/10.2307/1778190.

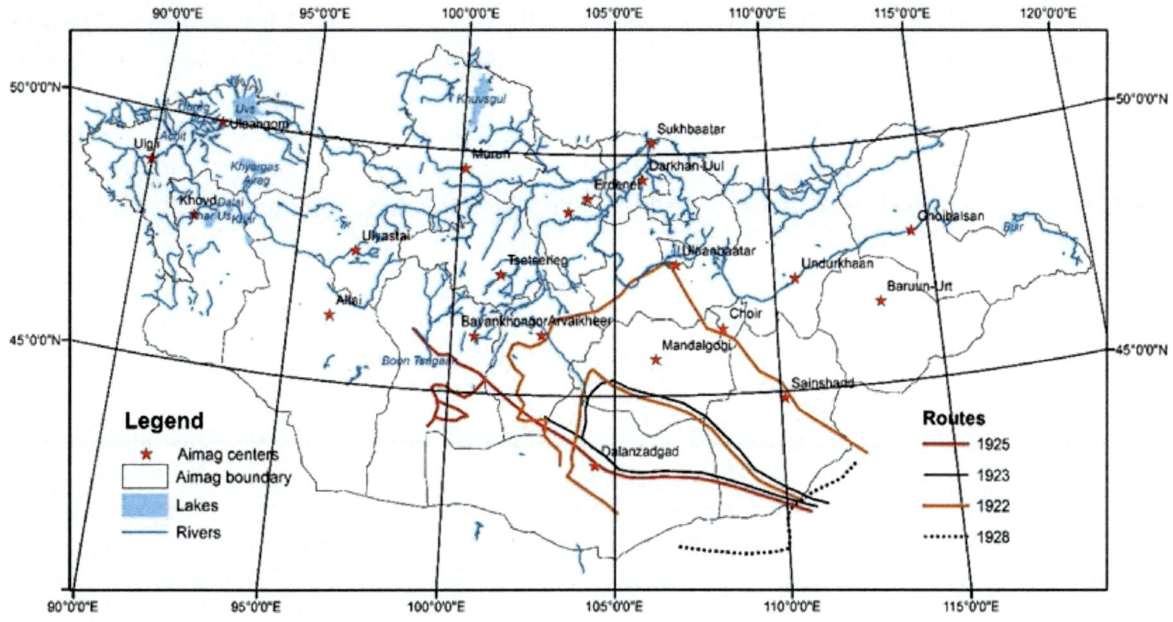


**Fig. 2.4** Itinerary of Andrews expedition in Mongolia

the ancient network of rivers and lakes (1948), the origin of sands and deserts (1947), ancient glaciation of mountains (1949), young volcanism (1948), and paleography of Mongolia (1949).

Hydrological and geomorphological works by N.T. Kuznetsov (1947) on rivers and river flows in Mongolia (1964), restructuring of the river network (1954), last movements (1955), ancient glaciation of the Mongol Altai (1952, 1965) and some issues of its paleography (1954, 1966) were presented. Due to the contribution of V.I. Vlodovets (1950, 1952 and 1955) and V.M. Sinitsyn (1956), as well as V.A. Amantov, who visited Mongolia during these years, it became clear about the geology and relief of the Dariganga volcanic region, and especially about Trans-Altai Gobi, which had previously been a "white spot" on the map of Mongolia.

A special place in the study of Mongolian relief is occupied by the works of I.P. Gerasimov (1952). He and E.M. Lavrenko made a big route along the most natural and typical for Mongolia territories. In fact, this complex study was the first generalized geomorphological work on the relief of Mongolia. Emphasizing the close connection between the relief and the geological structure, structures and the latest tectonic movements, I.P. Gerasimov singles out five major geomorphological regions of Mongolia and, taking into account the history of their development, gives a brief but capacious description of the relief of the terrain and demonstrates the important role of the exogenous factor in the formation of its specific elements (stone shells, etc.).

In the same period, monographs of Z.M. Murzaev (1948, 1952), R.A. Marinova (1957) and V.M. Sinitsyn (1959) were published. The detailed description of the relief of Mongolia in these monographs is equally and inseparably connected with geography and geology. These books are widely used in subsequent sections of this monograph. Next to these monographs is a major work by V.A. Obruchev, published simultaneously on Eastern Mongolia (1954), which contains a general description of its relief and development history.

In the mid-1950s, geomorphologists and physical geographers of Mongolia were trained at the Moscow University in Russia, and later became the main national scientists in Mongolia, whose works formed the basis of this book. In 1960, national studies of physical geography began with the physical zoning and study of individual areas of physical geography; permafrost, soil structure, sand cover, climatology, cave studies, and so on. Intensive Soviet-Mongolian academic relations in the 1950s made a significant contribution to the development of physical geography in Mongolia. The theory of regionalization and the prospects of landscape science contributed to the establishment of integrated physical geography as a new specialization in the evolution of the environment in historical periods. Based on the results of those joint studies, the National Atlas of the Mongolian People's Republic, created by Russian and Mongolian scientists and published in two languages (Mongolian and Russian) in Moscow in 1990, continues to be used as a valuable source of information, and an updated version was published by the Institute of Geography (National Atlas of Mongolia) in 2009.

From 1962, when the Institute of Geography in the Mongolian Academy of Sciences was created, purposeful limnological researches began. Since then, physical geography research has been widely disseminated and national researchers have begun to conduct studies in various fields: geomorphological (Tsegmid, Munkhuu, Jigj, Natsag, etc.), geological (Luvsandanzan, Byamba, etc.), geobotanical (Dashnyam, Ulziikhutag, etc.) and zoogeographical (Dashdorj, Shagdarsuren, etc.) studies and distribution of permafrost (Lonjid, Sharkhuu, etc.), soil cover (Dorjgotov, Avaadorj, etc.), and others. In parallel with the Institute of Geography, researchers from the Institute of Meteorology (established in 1946) and other universities in the field of climate and climate change-related studies (Natsagdorj, Sanjaajamts, etc.), surface water and groundwater (Tserensodnom, Sanjmyatav, etc.) have made valuable contributions, conducting numerous studies.

Joint expeditions in geology, biology and paleontology also made a significant contribution to the development of physical geography of Mongolia. For instance, Joint Mongolian and Russian geocryological expedition (1967–1971) played a considerable role in studying permafrost in Mongolia. As a result of this expedition, Gravis et al. compiled permafrost map of Mongolia (1: 1,500,000). Sharkhuu N et al. carried out subsequent studies on the basis of making ground temperature measurements. Several paleontological expeditions were carried out in the Gobi regions (Bayanzag, Nemegt, Khermen tsav basins etc.) such as the Soviet–Mongolian paleontological expedition (1969–1970) and Polish–Mongolian paleontological expeditions (1970–1971). Nemegt basin fossils were prospected and excavated in Upper Cretaceous deposits (in the Upper and Lower Nemegt Beds). Detailed geological profiles of Upper Cretaceous sediments of Nemegt were drawn by the Polish–Mongolian paleontological expeditions organized in 1970–1971 (Ryszard Gradzinski & Tomasz Jerzyki Ewicz 1972). Main fossiliferous sites, positions, the morphology of the localities and geological profiles and schematic maps were compiled during these expeditions based on detailed research of the certain areas, such as limnology of Uvs lake (Paul 2012). More information and the results of these studies will be presented in other chapters (see Chaps. 3–11).

## 2.3 Contemporary Approaches

The main areas of integrated physical geography research include comprehensive physical and geographical regionalizations, land management, landscape ecology and geographic processes on the surface of the earth. In recent years, there has been an increasing focus on the study of ecosystem services from a geographical perspective, which has expanded the scope of physical geography research. Scientists have used a variety of approaches to understand the patterns and processes associated with complex geographical formations that have evolved from the interaction of different physical and human factors.

Modern physical geography in Mongolia has been influenced by the geographical disciplines of Russia and Eastern Europe. Since the 1960s, integrated studies of physical geography in Mongolia have made remarkable progress in the fields of comprehensive physical geographical regionalization, soil geography, land studies, landscape ecology and land surface geographical processes. The research of contemporary physical geography in Mongolia is deeply affected by the thoughts and theories of Euro-American and Russian geography and joint researches with different countries, like Japan, Korea and German scholars. During the past few decades, under the background of global change and rapid socio-economic transformation, a series of environmental and resources problems have boomed in Mongolia. The research priorities of physical geography are focusing to solve the problems which connect with the responses and adaptation to climate change, water scarcity, desertification, resource management, and environmental pollution and degradation, the management of land use dynamics and agriculture in various biophysical conditions, grazing capacity, ecosystem service research from a geographical perspective; model-based researches, and more integrated studies of physical geography and human geography.

Water-related research is more concerned with the introduction of integrated water resources management (IWRM) in Mongolia. Within this framework, water resources management plans have been developed in most of major river basins, such as Orkhon, Tuul, Ulz, Kharkhiraa and Turgen, etc. (MEGD 2013), with the contribution of hydrological and geographical researches. Spatial and temporal assessment of forest cover changes using remotely sensed data (Tsogtbaatar et al. 2014). Since it is a dry country and has an impact on global warming, one of the main themes of physical geography research is desertification, its problems, impacts and causes. Physical geography studies are used for assessment and mapping of desertification (Mandakh et al. 2015), man-induced impact and combating desertification.

In addition, modern methods are used on research of the physical geography, such as the GIS model, satellite image remote sensing and satellite-based vegetation optical depth (VOD) (Liu YY, Evans JP, McCabe MF, de Jeu RAM, van Dijk AIJM et al. 2013) and normalized difference vegetation index (NDVI), and so on. For instance, trend analysis of MODIS NDVI for detecting land degradation and regeneration (Eckert et al. 2015) and an empirical permafrost model based on GIS calibrated, taking into account the results of ground temperature observations and using a multi-criteria approach to determine the permafrost tolerance zones. Permafrost distribution monitoring and mapping used for tomographic measurements (Jambaljav et al. 2016), as a result of a new map of permafrost in Mongolia (at a scale of 1:1,000,000), were compiled by Jambaljav et al. (2016).

## References


AfE (2014) The Mongols in world history. Asia for Educators, Colombia University. http://afe.easia.columbia.edu/mongols/history/history3.htm. Accessed 20 Jan 2019

Andreev AI, Yusupov TI (2003) P.K. Kozlov and his Mongol-Tibetan expedition of 1923–1926. In: Kozlov PK (ed) The diaries of Mongolian-Tibetan expedition in 1923–1926. Nauka, St. Petersburg, pp 9–19

Andrews RC (1932) The new conquest of Central Asia. American Museum of Natural History, New York

Batbuyan B (2000) The geographical issues of the administration and territorial transformation of Mongolia. Ph.D dissertation. Academy of Science of Mongolia, Institute of Geography, Ulaanbaatar

Bazargur D, Enkhbayar D (2002) Closer definition of geographical names in "Secret History of the Mongols". In: International conference on geographical study of Central Asia and Mongolia. Institute of Geography, Mongolian Academy of Science, Ulaanbaatar, pp 92–97

Carruthers D (1912). Exploration in North-West Mongolia and Dzungaria. Geogr J 39(6):521–551. https://www.jstor.org/stable/1778190?read-now=1&seq=1#page_scan_tab_contents. Accessed 20 Jan 2019

Carruthers D (1914) "Unknown Mongolia": a record of travel and exploration in Northwest Mogolia and Dzungaria 2 vols. Hutchinson, London

Damdinsuren Ts (1957) The introduction of second edition of the secret history of Mongolia. State Publishing House, Ulaanbaatar (in Mongolian)

Dash D, Mandakh N (2011) The history of the development of geographical science. Admon Publishing, Ulaanbaatar (in Mongolian)

Delgerjargal P, Boldbaatar J (2017) The history of Mongolia, 2nd edn. Mongolian National University Press, Ulaanbaatar (in Mongolian)

Delgerjargal P et al (2017) The Khunnu. In: The ancient history of Mongolia. The Institute of Mongolia's Study, National University of Mongolia, Ulaanbaatar (in Mongolian)

Eckert S, Hüsler F, Liniger H, Hodel E (2015) Trend analysis of MODIS NDVI time series for detecting land degradation and regeneration in Mongolia. J Arid Environ (113):16–28

Encyclopedia Britannica, Inc. (2008) https://www.britannica.com/place/Mongolia/Daily-life-and-social-customs#ref394571

Jambaljav Ya, Gansukh Ya, Temuujin Kh, Tsogterdene G et al (2016) Distribution map of Mongolian permafrost, scale 1:1000 000. Institute of Geography, Academy of Sciences, Ulaanbaatar, Mongolia

Jambaljav Ya et al (2016) Long-term research monitoring on permafrost of Mongolia, research report of thematic work. Institute of Geography, Academy of Sciences, Ulaanbaatar

Kondratiev SA (2006) Life and scientific activity of S.A. Kondratiev (1896–1970) in Mongolia and Russia. St. Petersburg Oriental Studies, St. Petersburg

Kotkin S, Ellaman BA (eds) (1999) Mongolia in the twentieth century: landlocked cosmopolitan. M.E. Sharpe.Inc, United States of America

Kozlov PK (1963) Russian traveler in Central Asia: selected works. Publishing House of the USSR Academy of Sciences, Moscow (in Russian)

Kozlov PK (2011) The Mongolia, Amduu and Dead City of Khar Khot (Petrograd, 1923; in mongolian by Baatar Choisuren) (411 pp). Admon Publishing, Ulaanbaatar

Liu YY, Evans JP, McCabe MF, de Jeu RAM, van Dijk AIJM et al (2013) Changing climate and overgrazing are decimating Mongolian steppes. PLoS ONE 8(2). doi:e57599.https://doi.org/10.1371/journal.pone.0057599

Mandakh N et.al (2015) The assessment and mapping of desertification in Mongolia. In: 3rd Scientific conference of UN CCD. Cancun, Mexico. https://doi.org/10.13140/rg.2.1.2994.1525

Martynov ED, Martynova AJ (2015) Science and power in Soviet Russia within the context of "Big Game": Mongol-Tibetan expedition by P.K. Kozlov (1923–1926). Soc Sci:1785–1789

MEGD (2013) Integrated water management plan of Mongolia (Dolgorsuren, et.al). Minitsry of Environment and Green Development, Ulaanbaatar

Murzaev (1948) The Mongolian People's Republic. A physical geographical description (in Russian). State Publication of Geographical Literature, Moscow

Murzaev EM (1957) Middle Asia. An outline of physical geography. Moscow:255 (in Russian)

Murzaev EM (1949) Expedition by P.K. Kozlov to Mongolia (1923–1926), P.K. Kozlov. The travel to Mongolia (1923–1926). State Publishing Geographic Literature, Moscow, pp 3–11

Paul M (2012) Limnological aspects of the Uvs Nuur Basin (octoral dissertation). Dresden

Pevtsov MV (1895) A journey through Eastern Turkestan, Кип Lun, the Northern Edge of the Tibetan Plateau and Jungaria in 1889 and 1890. SPb, pp 10–11

Popova IF (2008) Russian Expeditions to Central Asia at the Turn of the 20th Century. In: Popova IF (ed) Russian expeditions to Central Asia in the Late 19th to early 20th century. Slava, St. Petersburg, pp 11–39

Przhevalsky NM (1888) From Kiakhta to the sources o f the Yellow River. Exploration o f the Northern Borders of Tibet and the road through Lop Nor along the Tarim Basin. SPb, p 15

Rossabi M (2011) The Mongols and global history: a norton documents reader. W.W.Norton & Company. Inc, New York/London

Ryszard G, Tomasz JE (1972) Additional geographical and geological data from The Polish-Mongolian palaeontological expeditions. Palaeontologia Polonica, No, 27. http://www.palaeontologia.pan.pl/Archive/1972-27_17-30_3.pdf

Sonomdagva Ts (1967) The changes of the administration units of Mongolian People's of Republic. State Publishing House, Ulnaabaatar (in Mongolian)

Sukhbaatar, D. (2000). *The Tradition of the Local Adnministration Structure of Mongolia (Changes between 1921–1932).* Ulaanbaatar

The Academy of Science of Mongolia (2007) The annual report of the history and Archaeological Institute of Mongolia. The Academy of Science of Mongolia, Ulaanbaatar

Tseden-Ish B (2011) The history of formation of the boundary of Mongolia. The Institute of International Study, Mongolian Academy of Science, Ulaanbaatar (in Mongolian)

Tsegmid S (1969) Physical geography of Mongolia. Institute of Geography and Cryology. Mongolian Academy of Sciences. State Publishing House, Ulaanbaatar, 405 pp (In Mongolian)

Tsogtbaatar J et al (2014) Spatial and temporal assessment of forest cover changes in Mongolia using remotely sensed data. IUFRO 2014 congress. Salt Lake city, USA. https://doi.org/10.13140/rg.2.1.4567.0168

Yusupova TL (2008) P.K. Kozlov's Mongolia and Sichuan expedition (1907–1909): the discovery of Khara-Khoto

Yusupova TI (2014) Central Asia in the life and discoveries by P.K.Kozlov. The Russian study of Central Asia: historical and modern aspects (to the 150th anniversaiy of P.K. Kozlov) Spb, Politehnica-Service:13–31

# The Relief and Geomorphological Characteristics of Mongolia

# 3

Dash Doljin and Batchuluun Yembuu

**Abstract**

This chapter describes the general characteristics of geomorphology, as well as the evolution in Mongolia. The last part of this chapter discusses geomorphological regionalization in Mongolia and describes the morphological characteristics of the surface, structure and types of relief in each region, which are modified and updated based on the results of research by well-known scientists: E.M. Murzaev, Sh. Tsegmid, E.I. Selivanov, S. Jigj, N.A. Florensov, and S.S. Korjuev. Based on the accepted classification of regions, the regions are determined by their physical geographical factors, and processes also modify their structure, function, and characteristics. Therefore, in the regional descriptions, the factors of formation to the landforms in each macro-region are highlighted. Besides the structural analysis of the regions, some landscape historical data are described.

**Keywords**

Morphotectonic evolution · Landforms Erosional landscapes · Geomorphological region · Mountain typology · Relief roughness

## 3.1 Overview of Geomorphological Research

The history of Mongolia's geomorphology research goes back to the early days of eighteenth century, to the Russian expeditions by Prjevalskii, Pevtsov, Obruchev, and so on. B.A. Obruchev concluded that originally the small-sized depressions' deposits in the Gobi are continental. The first physical geographical regions determined by Sh. Tsegmid are scientifically grounded regions of the landscape divisions of Mongolian by S. Jigj (1975a, b). He distinguished the seven macro-regions in Mongolia as tectonic-erosional, erosional and accumulative, and accumulative relief, the three hierarchical classification of landforms:

The monograph "Neotectonics and geomorphology of Mongolia" by E.I. Selivanov was published in 1972 (Selivanov 1972), which covered the geological evolution of Mongolia and focused on glacial and volcanogenic landforms of the country and quite detailed the descriptions of the geomorphological regions. Aeolian landforms of the eastern part of

D. Doljin (✉) · B. Yembuu
Department of Geography, Mongolian National University of Education, Ulaanbaatar, Mongolia
e-mail: dash@msue.edu.mn

B. Yembuu
e-mail: batchuluun@msue.edu.mn

© Springer Nature Switzerland AG 2021

B. Yembuu (ed.), *The Physical Geography of Mongolia*, Geography of the Physical Environment,
https://doi.org/10.1007/978-3-030-61434-8_3

Mongolia and paleogeographical features of the country were studied by V.I. Chichagov (1999) during the Mongolian–Russian joint geological expeditions and the result of his research "The aeolian landforms of the Eastern Mongolia" was published in Russian. The valuable source of the team work is *"The Geomorphology of Mongolia"* (Florensov and Korjuev 1982) created by Mongolian and Russian geologists and geomorphologists of geomorphological research of Mongolia in 1982. Since the mid-1950's, the geomorphological research was started by the national scholars such as *"The Physical geography and prehistorical glaciers of the Khentii mountain range"* (Tsegmid 1951); "*The structure and geomorphology of the Eastern Khangai"* (Munkhuu 1974); "*Geomorphology of the Selenge river basin* by Natsag (1985); "*Exogenic processes of the Khangai and Khentii mountain ranges"* (Lomborinchin 1989); *"The study on the rift zone morphostructure in Western Mongolia* (Batsaikhan 1999); and so on.

As a result of these studies, a geomorphological map (at a scale of 1:3,000,000) was compiled by Tsegmid et al. (Tsegmid and Vorovev 1990). Although geomorphological surveys were predominantly of a route type, they provided many new and interesting data on terrain and recent sediments. The quaternary sedimentary map (edited by V.E. Murzaeva and E.V. Devyatkin), the geomorphological map of Mongolia (edited by N.A. Florensov and S.S. Korzhuev), and a general series of geological maps (1:1,500,000) were published by the Mongol-Russia joint geological expedition. The results of these studies and the maps are these fundamental sources of geomorphological research in Mongolia to the present day.

One of the geomorphological features of Mongolia is the sand dunes, which are spread in wide areas. S. Baasan published a monograph based on his long research on the origin, morphology, and distribution of sand dunes in Mongolia (Baasan 2006), while fluvial, lacustrine, and aeolian process dynamics were studied by German researchers (Stolz et al. 2012).

Since 1990, the geomorphological research has been increasingly conducted using new research methods, such as GIS, remote sensing, and modeling. For example, the geomorphological mapping of glacial mountains used a combination of remote sensing data (Blomdin et al. 2018), radiometric dating on fluvial sedimentation, and aeolian transport (Klinge and Lehmkuhl 2013). The most recent fundamental physical geographical research is conducted by Ser-Od on mountainous landscapes, as in the case of the Mongol Altai (Ser-Od 2019). The advantages of this research are the issues of landscape use and conservation, which have been studied in detail by use of natural resources and sectors of tourism, agriculture, livestock, and mining.

## 3.2 General Description of Geomorphological Characteristics

Mongolia has a vast territory, occupying 1,564.1 thousand $km^2$. Stretching 1,259 km from north to south and 2,392 km from east to west, its landforms and natural landscapes distinctly change with latitude and longitude. The northern, central, and western parts of the country are primarily mountainous, and the eastern, southern, and southeastern regions are occupied by plains and hollows (Fig. 3.1). The highest point in the country is the peak of Khuiten of Altai Tavan Bogd mountain (4,374 m a.s.l.), and the lowest point is Khukh Lake (560 m a.s.l.) to the east. The average elevation of the country is 1,580 m a.s.l., and approximately 80 % of the territory is occupied by mountains and highlands with elevations above 1,000 m a.s.l. (Dorjgotov 2009) (Fig. 3.2).

The distinctive feature of the country's relief is the arcuate distribution of the central mountain ranges. The mountains in the west run northwest, and in the central and eastern sections, the mountains run along the latitude and to the northeast, respectively. Another peculiarity is that several longitudinal lineaments can be observed in Central Mongolia from the southern end of Khuvsgul Lake Depression. These lineaments cut the arcuate mountain ranges and divide the territory into several different parts. The

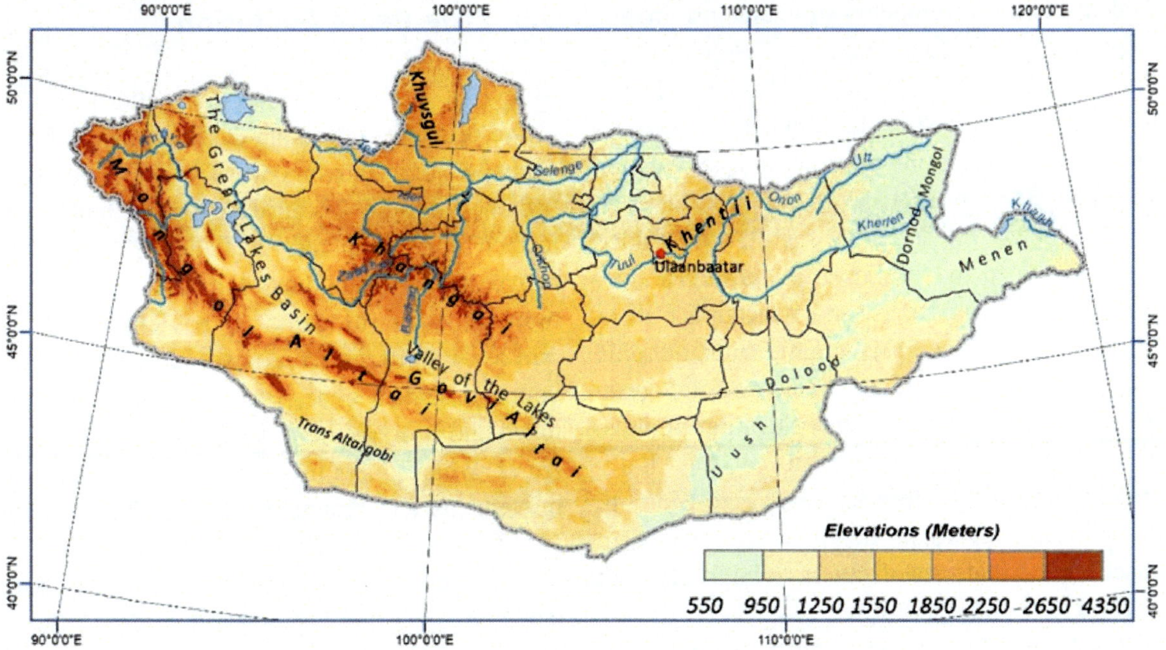


**Fig. 3.1** Relief and elevation of Mongolia

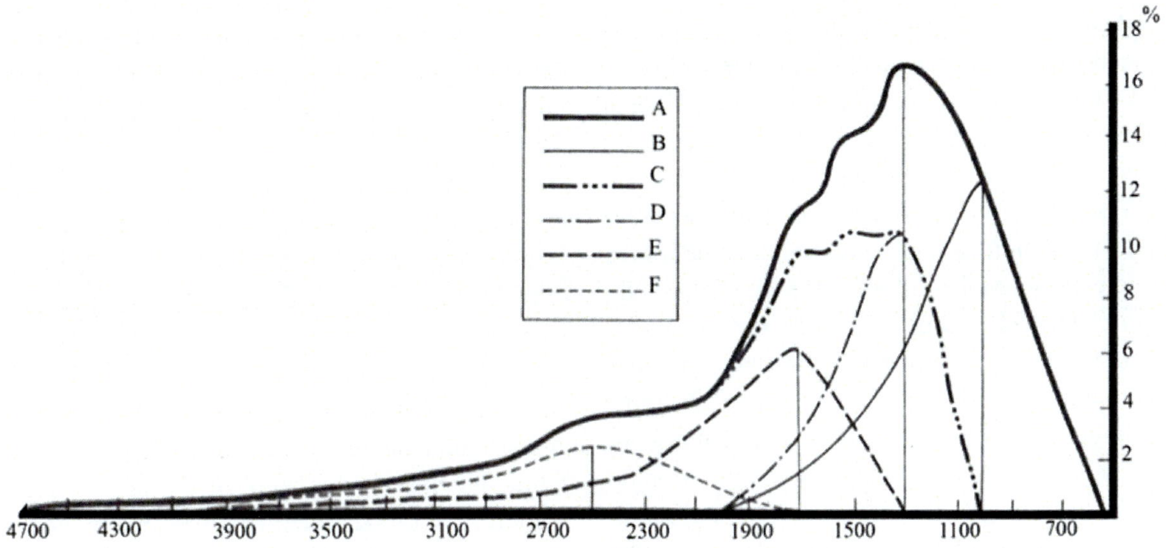


**Fig. 3.2** Hypsometric curves of the surface in Mongolia (Levintov et al. 1978): A—General for the whole territory, B–F for certain areas, B—Plains, C—Mountainous areas, D—Hills and low-mountains, E—Khentii and Khangai plateaus, F—Mongol Altai and Gobi-Altai (Compiled by M.E.Levintov)

formation of large-scale landforms is related to tectonic activity, and the small-scale relief elements are mainly formed through water, wind, and glacial processes (Nazag and Chichagov 1990). Wind activity plays a major role in landform formation, and thus, large sandy lands have formed in different parts of the country. Moreover, over time, large-scale erosion processes occurred throughout Mongolia.

Mongolia can be divided into three folded zones, based on the timing of formation and the surface morphologic and tectonic features: (1) Eastern Sayan and Mongol Altaic; (2) Central Mongolian; and (3) Southern Mongolian. These folded zones are formed by structures originating from five ocean basements (i.e., Proto-Asiatic, Paleo-Asiatic, Neo-Asiatic, Khangai–Khentii, and Paleo-Tethys) after the sequential processes of lateral subduction/collision and subduction accretion (Byamba 2012).

The Eastern Sayan and Mongol Altaic folded zone is formed by the Proto-Asiatic and Paleo-Asiatic ocean basements, and it occupies all of Western Mongolia. The granite metamorphic processes occurred in the Paleo-Asiatic ocean basement and occurred in three distinct periods, including the Late Cambrian to the Early Ordovician, Mid-Ordovician, and Late Silurian. By the Late Devonian, the processes resulted in Caledonian fold stabilization, which formed distinct terrestrial landforms.

The Central Mongolian folded zone consists of multi-period overturned folded structures that formed between the Proterozoic and Jurassic periods. The geological history of this zone is closely related to the Khangai–Khentii oceanic transgression development. Geographically, the center of this zone is located in the Khangai and Khentii mountains and occupies the surrounding structures that are distributed around the mountains.

The southern Mongolian folded zone occupies almost the entire area of the late Paleozoic folded structures, located south of the Mongolian fault, which formed during the Neo-Asiatic ocean development. Geographically, the zone continues approximately 2000 km along the latitude, occupying territories such as from Baruun Khurai hollow to the mountains south of Mongol Altai, from the Baitag mountains to the south-eastern Gobi-Altai mountains, through the southern and eastern Gobi to Nukht Davaa mountain, Gobi Gurvan Saihan, Ikh Shankh, Khaltar Gobi, and the western fringes of the Ikh Khyangan mountains (Byamba 2012).

Mongolia encompasses a unique land with complex geological structures formed during its long geological history, which includes periods of ocean transgression and regression and terrestrial land formation. As a result, the landscape has many different rock types that represent different geological ages.

## 3.3 The Relief Evolution

Mongolia can be divided into after-basement active orogeny and young basement zones based on neotectonic movements. The orogenic zones became isolated because of large faults and formed three separate neotectonic zones: an active lifting zone due to distinct multi-faced movements, a riftogenic zone with active and mid-level uplift, and a relative folding zone. The young basement zone is relatively less differentiated and is divided into slight-uplift and folding zones, which differ in the footprints of the new movements and structures formed (Tumurtogoo and Zaitsev 1990). This surface characteristic is highly reflected in the geomorphic regions of the territory.

Mongolia's geology reflects the development of Central Asia at the end of the Archean period: the Siberian Craton in the north and Tarim, North China and the China-Korean craton in the south. The entire territory of Mongolia was consolidated and was part of the Late Paleozoic Pangaea Supercontinent. Tectonic activity decreased in the middle Jurassic period. The accumulation increased in the ancient lake depressions caused by denudation processes in the middle and late Jurassic period (Fig. 3.3). Further development from Mesozoic to Cenozoic is characterized by orogenesis, magmatism, and sedimentation, especially in southern and eastern Mongolia.

The study of tectonic-geological development in Mongolia has not been completed, although some attempts have been made in the past. Tumurtogoo (2002) identifies three mega-blocks (northern, central, and southern), six super-reliefs and numerous terrains, and describes the geodynamic position and petrology of individual structures. The general development is divided into two parts: from Precambrian to the end of Perm, that is, to the formation of the solid crust and the Mesozoic and Cenozoic (Nikolaeva and Shuvalov 1995). In the Mesozoic and Cenozoic periods, Mongolia's territory has recently been in the geodynamic situation of the platform, where the formation of various overlapping structures was mainly caused by the destruction of the

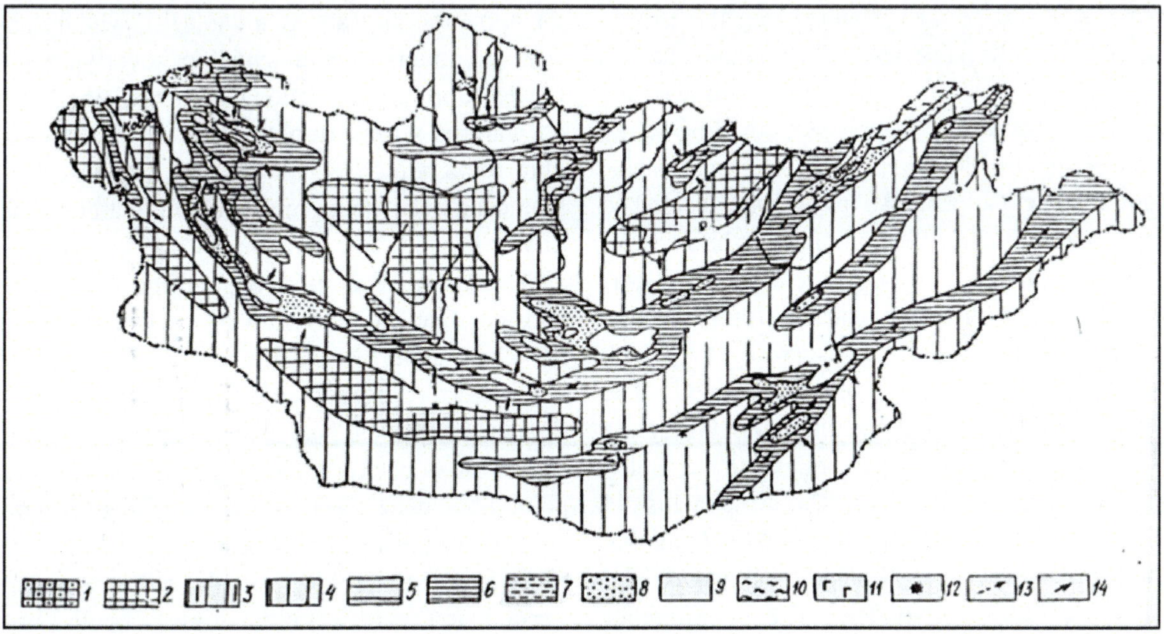

**Fig. 3.3** Paleogeomorphological scheme of the early and middle Jurassic of Mongolia (Nikolaeva, N.V., Shuvalov, V.F, 1995). A. Denudation-tectonic relief: 1: high mountains, 2: mid-altitude mountains, 3: mid and low-altitude mountains, 4: low-altitude mountains, 5: hills and erosional landforms, 6: hilly-plated and plain-fine-bottomed relief, 7: latitudinal and basement plains; B. Accumulated relief and basins of accumulation: 8: alluvial, lake-alluvial, lake-proluvial, alluvial-proluvial plains, 9: lake basins, 10: marine deposits and volcanogenic reliefs: 11: basalt plateaus, 12: volcanic eruption centers, 13: river basins, 14: drift directions

previous continental crust. Large basins were formed mainly in the southern mega-blocks, as well as in some parts of the central ones (Tumurtogoo and Zaitsev 1990).

The modern surfaces of Mongolia are primarily defined by the neotectonic movements that occurred during the Neogene–Quaternary period, which were different across the territory. During this period, active lifting occurred in the Mongolian Altai, Gobi-Altai, Khangai, Khentii, and Khuvsgul mountains, and the eastern parts of the country experienced less intense uplift that preserved the previous basement state stability (Florensov and Korjuev 1982).

During the Miocene and early Pliocene, the Gobi Piedmonts and the Mongol Altai, Khangai, Khentii, and Khuvsgul mountains experienced long-lasting, continuous lifting. Concurrently, the hollows along the Altai mountain fringes actively lowered. This process was responsible for the formation of large hollows, such as Zereg, Ikhes lake, Sharga Gobi, Biger, and Lake Depressions. Evidence of continued lowering far to the southeast is observed in the Miocene-aged sediments found near the Ikh Shankh mountains, east of Dalanzadgad.

In eastern Mongolia, Miocene sediments are found in the western Dariganga Plateau and in Tamsag Depression to the northeast, and are covered by or mixed with basalt sediment, forming different layers. With the uplift of the Altai, Khangai, and Khentii mountains during the early and mid-Pliocene, the Great Lake Depression hollows along the Altai mountains. Lake Depression and depressions in eastern Mongolia continued to experience lowering. The central part of Lake Depression, the current Buun Tsagaan Basin, and Orog and Tsaatsiin Tsagaan lakes were filled with water until the end of the Pliocene (Devyatkin 1981).

During the Miocene and Pliocene, there was volcanic activity in the country that resulted in basalt cover in many areas. For instance, the Miocene eruption epicenters on the southern slopes of Khangai mountains are concentrated along the Bayankhongor fault zone, and the basalt cover reaches Taishir soum in the west and Guchin Us soum in the east. The oldest basalt cover (20 ± 0.5 ma) is found along the eastern shores of Tui River, whereas the youngest basalts (18.9 ± 0.8 and 17.0 ± 2.0 ma) are found on the eastern banks of the Taats River, covering same-age sediments. Moreover, the basalt cover is also found along the Khangai mountains watershed and on a transect of the Chuluut-Northern Tamir and Southern Tamir rivers. There are also sparsely distributed basalt plateaus found in the eastern Khuvsgul Lake Depression, which formed in the same geological period. The Dariganga Volcanic Plateau in eastern Mongolia is also considered a Miocene-age basalt complex (Devyatkin and Smelov 1981). In conclusion, during the Miocene in the present-day Mongolia, three volcanic zones, Khangai, Khuvsgul, and Dariganga, were active. The volcanogenic relief of these zones is similar and primarily consists of a central ancient volcano surrounded by relatively plain base rocks and thick basalt cover.

The second most-recent orogenic period was either in the Late Pliocene or the Eopleistocene. This orogenic period greatly impacted the formation of the current relief in southern Siberia, Baikal, and Mongolia. During this period, the major tendencies of landforms and their structures were defined and obtained their modern appearances. The morphostructures formed in this period of orogenic activity are classified into uplift (mountains and plateaus) and bending (intermountain and inner-mountain depressions) groups. Within this period, five mountainous regions, Mongol Altai, Gobi-Altai, Khangai, Khentii, and northern Mongolian (Khuvsgul mountains, were differentiated. All of these regions formed in the modern Orogenic period, which was very active and resulted in specific complex structures (Devyatkin and Smelov 1974).

The Altai are old folded mountains. The current mountain scenery was formed in the Tertiary (Miocene, Pliocene), when the Altai was lifted again as a whole block (Enkhtaivan 2006). Its morphostructure consists of the uplifted main range of the Mongol Altai mountains; separate mountains divided by Tolbo, Deluun, and many other depressions; and the uplift distributed to the northwest. From a cross-section, this phenomenon appears as a vault-lift, which consists of anticlines and synclines overlying the relict Carboniferous–Paleogene peneplain. There are various sedimentary, abyssal, and metamorphic rocks which developed from Paleozoic to Holocene. The rocks of Devonian, Carbon, and Lower Permian ages are spread only locally.

The Gobi-Altai uplift began later, and some consider that the active lifting only occurred in the Pleistocene. During this period, the Mongol Altai had already passed its independent development stage, as Neogene complex activities started to develop. The high quake (10–12 points) frequency in the Gobi-Altai region indicates that uplift was still occurring. Finally, the Mongol and Gobi-Altai morphostructures experienced uplift in different time periods; the Mongol Altai morphostructure formed in the Neogene, and the Gobi-Altai formed in the Pleistocene. The formation of the Khangai and Khentii mountain plateau is an example of active orogeny. Khentii mountain is part of the Khentii–Daurian consolidated morphostructure. These mountains are separated from each other by piedmonts that formed during the late Pliocene.

The relatively youngest morphostructures in Mongolia are the northern Mongolian and Khuvsgul mountain morphostructures, which consist of longitudinally distributed mountains and intermountain depressions. All of these structures border the southwestern end of the Baikal riftogenic zone. Generally, it is thought that the major formation processes of these morphostructures occurred during the Pliocene and Quaternary.

The low mountains in the Orhon and Selenge Basins as well as the southern Khangai Plateau are related to relatively inactive orogenic processes. These regions are distributed on the outskirts of Khangai and Khentii uplift, and because of slow uplift, they have low differences in height. Conversely, the regions located in the middle of the active uplift and bending zones have a relatively low degree of dissection, preserving their early ages or plate stage look.

Baruun Khuurai Depression, Trans-Altai Gobi, Middle Gobi, South Eastern Gobi, and Dariganga Volcanic Plateau are considered locally developed orogenic regions. During the most-recent tectonic activity, these regions maintained their original structures and plate features. Such stable features are related to plates covered by Cretaceous–Paleogene sediments. In other words, on peneplains with multiple ages, the orogenic processes show random appearances, and the amplitude of tectonic uplift varied from a thousand to several hundred thousand meters.

The regions of most-recent tectonic bending or lowering occurred in Great Lake Depression, Altai Hollows, Lake Hollow, North Eastern Depression, and many other hollows and lowlands. All of these landforms are filled by Cenozoic sediments (from 200–300 to 700–800 m), and more than half of this material consists of Neogene–Quaternary sediments. These newly formed depressions more or less inherited their Mesozoic structures. However, depending on their shape, sediment content, early Cenozoic structural ratio, and plate stage conditions, they can be divided into three different types (Florensov 1968). The first type consists of depressions filled with lacustrine and unconsolidated deposits from the Neogene, Eopleistocene, and Pleistocene (e.g., Uvs and Khyargas Lake Depressions in Great Lake Depression). The second type includes depressions with lineament distributions, such as the hollows along the slopes of Altai mountains and Valley of Lakes. The third type encompasses the youngest depressions, primarily represented in Trans-Altai Gobi.

The northernmost depressions are Uvs and Khyargas Lakes in Great Lake Depression and are filled by 200–300 m of Neogene deposits (Fig. 3.4). Here the thickness of late Pliocene and Quaternary sediments is only several tens of meters. The sedimentation processes in this region are related to surface flow. In Sharga

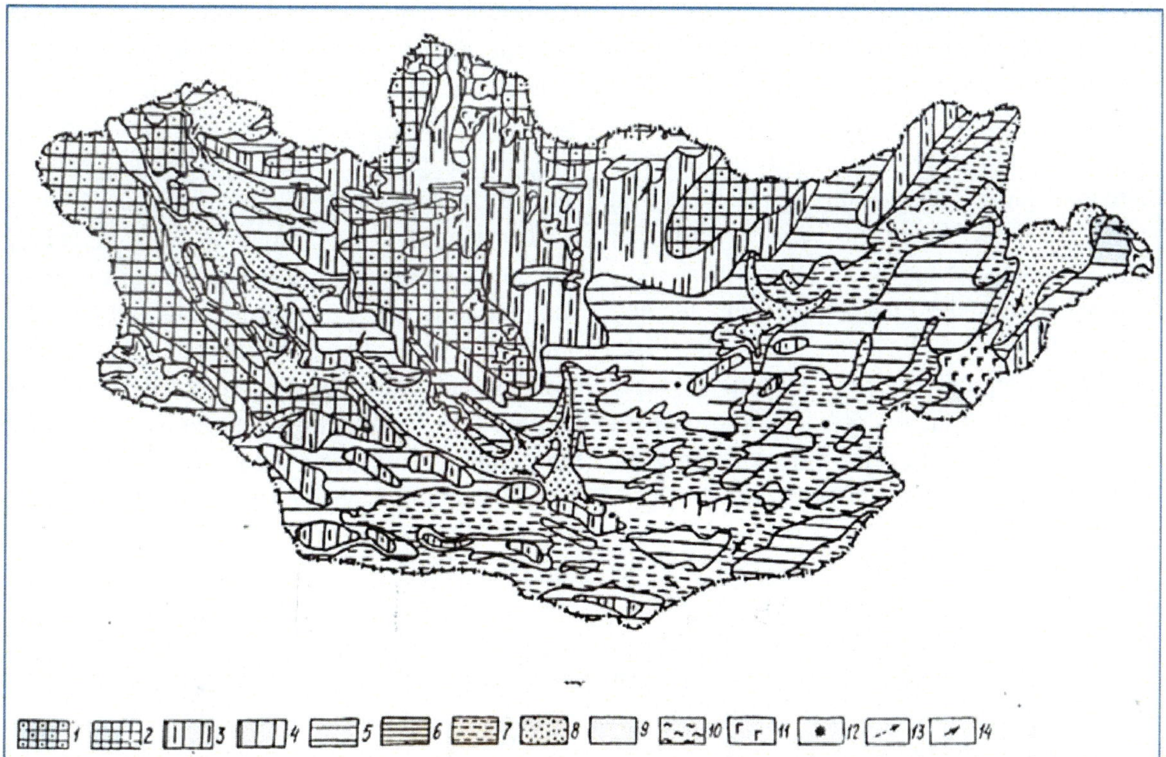


**Fig. 3.4** Paleogeomorphological scheme of Mongolia. Neogene. Denudation-tectonic reliefs: 1—high mountains; 2—medium mountains; 3—mid-altitude and low mountains; 4—low mountains; 5—hills; Denudation reliefs: 6—Hilly-plain reliefs, 7—strata and denudation plains; Accumulated reliefs: 8—alluvial, lake-alluvial, lake-proluvial, alluvial-proluvial plains; 9—lake deposits; 10-sediments of sea floor; Volcanogenic reliefs: 11—basalt plateaus, 12—volcanic eruption centers, 13—rivers, 14—drift directions

Depression, the sediments primarily consist of mid-Miocene lacustrine deposits, and Pliocene sediments are absent from sediment profiles in the central part of the depression. This can be explained by the long-lasting severe drought conditions and the absence of surface flow. Similarly, all depressions south of Altai Mountain developed in the extremely dry conditions since the Paleogene. The basements of these depressions are late Cretaceous sediments with rare Paleogene deposits (Chistyakov et al. 1994). Younger sediments are absent, and in most cases, Cretaceous sediments are exposed at the centers of these depressions because of active wind erosion.

Baikal-type young depressions of the Busiin River, Darkhad, and Khuvsgul are located at the southern end of the Baikal riftogenic zone and are distinct from the southern Mongolian Gobi-type hollows. These hollows formed during the second half of the tectonic period (i.e., Pliocene and Pleistocene). The main features of these hollows are their longitudinal distributions, fault-defined slopes, and relatively thin sediment fill (200–300 m) (Florensov 1975).

Volcanic activity during the Eopleistocene was concentrated in the central part of Khangai, where a basalt complex of this age is located in the Chuluut River Valley and Tariat River Basin. Near the Eopleistocene eruption centers, similar-age basalts cover the Pliocene deposits, thereby raising the terraces. However, the basalt thickness decreases downstream, and the terraces decrease in elevation. According to the detailed study of Corina (1974), three different Pliocene–Eopleistocene age layers can be identified in these locations (Corina 1974).

The Pleistocene and Holocene consisted of relatively short duration landform developments. The main factor impacting landforms changes is climate, which altered the morphostructures and created the modern surfaces. Thus, the Pleistocene was the major period of surface development in Mongolia. The climate changes during the Pleistocene resulted in the formation of new kinds of deposits, e.g., glacial and water-glacial. Moreover, the climate variations (i.e., warming, cooling, moistening, and aridization) changed the bedrock lithology, particularly in rivers and lakes (Devyatkin 1981).

The Pleistocene is considered a major geological period in environmental studies because, in this period, major landforms were created. During the Pleistocene, glacial sediment and glacial landform formation largely took place in the high mountains. The formation of the diluvial and alluvial fan sediments mainly occurred in the piedmonts and in the low- and mid-elevation mountains. Alluvial sediments also accumulated along large river basins and in surrounding lakes and intermountain hollows.

In Mongolia, four centers of early glaciation have been differentiated: the Mongol Altai high mountains, Khangai highland, Khuvsgul mountain, and Central Khentii. The early glaciation forms differ depending on site orography and general climatic settings. The glacial forms in Mongolia are relatively simple compared to those in southern Siberia and mainly include valley glaciation and moraine forms. However, it is important to mention that there was a summit-specific glaciation in the high mountains with peneplains (Dash 2009). The relicts of this glaciation can be seen in Tsast mountain of the Mongol Altai (4,213 m a.s.l.) and in the Munkhkhairhan range (4,231 m a.s.l.).

The largest part of Mongolia is occupied by low and mid-elevation mountains, and the piedmont zone is mainly formed of alluvial fan sediments. This zone occupies all mountainous regions, including high mountains and other convex relief, excluding intermountain hollows, river valleys, and depressions. The main exogenous factor impacting relief in the mountainous zone is surface flow, which has resulted in the formation of large inclined plains at the foot of the mountains. The majority of these alluvial fan sediments were formed in the early and mid-Pleistocene (Timofeev D.A, Chichagov 1974).

In northern Mongolia, Pleistocene climate was humid, and thus, the role of alluvial fan deposits decreased. In this part of the country, it is common to observe diluvium-solifluction deposits in regions adjacent to glacial regions (Kozhevnikov 1968). Such gravel-pebble sediment is mainly distributed in Selenge, Kherlen, and Onon River

Basins, where the bedrock is exposed and loose sandy sediments from the mid-Pleistocene glaciation are concentrated. The diluvium-solifluction sediments of northern Mongolia leveled highly dissected valleys, covered terraces with slope sediments, and led to surface deformation. During the mid- and Late-Pleistocene, this process was accelerated, and small river basins were completely covered by slope sediments.

The lakes in the arid zone of the country provide evidence for large climate variations, which is why many researchers are interested in this topic. In Mongolian arid land paleogeography, it is necessary to distinguish arid and humid phases of development. Most of the research relates the humid phase with the formation of high-latitude and high-elevation mountain regions and the arid phase with the interglacial period. The Great Lake Depression is a suitable area to discern the ratio between the arid and humid phases (i.e., glacial and interglacial periods) (Devyatkin and Murzaeva 1975). This area is a large intermountain depression that includes lakes with both inner and external flows. According to geomorphological investigation of aeolian and colluvial sediments of central Khangai, the Valley of Lakes in the Gobi was slightly wetter, about 1.5 kA, and most of the lakes were large in the early Holocene, about 8.5 kA (Lehmkuhl and Lang 2001; Grunert and Lehmkuhl 2004).

The rivers in this depression, such as the Khovd, Zavhan, Khungui, and Tes, are sourced from the Mongol Altai and Khangai mountains, which were affected by glaciation. Here, isolated complex structures with glacial, surface flow, and lake depressions are formed. The glaciation that caused glacier formation in high mountains decreased the evaporation rate, which resulted in the formation of large lakes during the humid phase. The humid phase is often assumed to have occurred between the end of the interglacial period and the beginning of the glacial period. During the second half of the glacial period, the climate was relatively arid, and thus, the ice caps, glaciers, and surface flow were mainly preserved in the mountains. A relatively humid climate occurred when the mountain glaciers melted. According to Kuznetsov (Kuznetsov 1962), surface water drainage increased by 15–20 % during this period. Thus, in the interglacial period, evaporation intensity increased, resulting in a decline in surface flow and a desiccation of lakes, similar to current conditions.

The water surface of Lake Depression reached its maximum height during the mid-Pleistocene. The water depth was between 1,400 and 1,420 m, and there was one large basin that connected Taats Lake and Buun Tsagaan Lake. However, during the Late-Pleistocene, the lake was divided into two large lakes. Buun Tsagaan Lake had a depth of 1,360 m, and the depth of the second lake (a union of the present-day Orog and Tsaatsiin Tsagaan lakes) was between 1,280 and 1,300 m. By the end of the late-Pleistocene, the basin was divided into three lakes, and the Ulaan lake depression was isolated from other basins (Florensov and Korjuev 1982).

In the mid-Pleistocene, Buir and Dalai lake basins were united. During this period, the northeastern part of the country contained two large lake basins, Buir-Dalai and Tari lake. During the Holocene transgression, the water level in lakes decreased, and many saline lakes developed.

The Selenge, Onon, and Kherlen river basins, located in the north and northeastern Mongolia, have a long history, at least up until the Miocene. This period occupied all of the major rivers and their tributaries (mainly first level). Most of them developed along Mesozoic faults and deeply dissected the bedrock. For instance, the Selenge, Ider, and Orkhon Rivers flow along the northern Mongolian faults in the same direction as present-day surface flow. In the Khangai mountains, the rivers flowing from the southern slopes, e.g., Zavkhan, Baidrag, Tui, and Taats, have radial distributions, indicating the equilateral slope structure of the southern Khangai Plateau.

The river basins in the Khangai mountains developed during the early Cenozoic era, which was proved by different studies. For instance, in the Khanui River Estuary and in the western tributary of the Selenge river, basalt cover dated at $12.5 \pm 1.0$ ma has been identified (Devyatkin and Smelov 1981). This 350–500 m thick basalt lies at the valley bottom. The basalt from the Miocene period came down via the Khangai watershed and formed a specific trap along the North and South Tamir rivers. With the downflow, the height increased from 200–300 to 60–90 m. This provides evidence that the Khangai mountains hydrologic network was developed during the Miocene.

During the early and mid-Pleistocene, most river systems had a specific deposition regime. As a result of large-slope denudation, a high amount of unconsolidated sediment accumulated in riverbeds, establishing highly elevated terraces filled with sandy sediments. In the late-Pleistocene, river dissection was lower compared to the Pliocene. The valleys in northern Mongolia contain 20–25 m erosional and depositional terraces that are dated to the aforementioned period. The deposits of this age in mountainous regions are mainly unconsolidated and form terraces filled with gravel-boulder rocks. On the top of such terraces, there are widespread alluvial fan sediments or solifluction material.

In the mid- and late-Pleistocene, the rivers flowing from the slopes of the Mongol Altai, Khangai, and Khentii mountains contained large amounts of water, indicated by the high number of boulder stones on depositional terraces. Thus, rivers such as the Chuluut and Ider had the ability to move such large sediments. During the end of the Pleistocene and into the Holocene, the rivers actively desiccated (Florensov and Korjuev 1982).

## 3.4 Morphogenetic Types of Relief

The relief of Mongolia is unique because it combines mountains, plains, as well as intermountain and intramountain depressions formed in different geological periods. Thus, these reliefs have distinct forms, shapes, and elevations. S. Jigj (1975a, b) classified the reliefs of Mongolia into five morphogenetic classes:

(1) High mountains with elevations of 2,500–4,500 m and relative heights greater than 1,000 m;
(2) Mid-elevation mountains with elevations of 1,500–2,500 m and relative heights of 600–1000 m;
(3) Low-elevation mountains with elevations of 1,000–1,500 m and relative heights of 100–400 m;
(4) Hummocks with maximum elevations of 1,000 m and relative heights of 20–50 m; and
(5) Rolling plains with elevations of 600–900 m and relative heights of 10–20 m.

However, the altitudes and heights of this classification may vary by region. For instance, in eastern Mongolia, the elevation variations are low, and thus, it is difficult to follow the aforementioned classification system. For example, the elevation of the Zagal Mount in Gurvanzagal soum, Dornod aimag is 840 m a.s.l., and the relative height is 100 m. In this case, it is better to use a height-only classification. According to genesis, the landforms can be classified into three types: tectonic-erosional, erosion-accumulative, and accumulative (Jigj 1975a).

### 3.4.1 The Tectonic-Erosion Type of Relief

The tectonic-erosion type of relief includes the mountains of the Mongol Altai, Khangai, Khuvsgul, Center of Gobi-Altai, and Gobi Tien-Shan systems. All of these ridges have different shapes, elevations, and locations. This type can be further sub-divided into the following classes:

1. Alpine-type high mountains with ancient and modern glacial forms;
2. High mountains with plain summits and relict glaciation;
3. High mountains in the central parts of the Hentii system with ancient glacial forms;
4. Mid-elevation mountains with ancient peneplain relicts and dense dissection;
5. Densely dissected mid-elevation mountains isolated by intermountain hollows;

6. Mid-elevation mountains with pointed summits and steep slopes; and
7. Low-elevation mountains with relict ancient peneplain surfaces.

The tectonic-erosion type also includes the ravines, canyons, and dry riverbeds that dissect high mountains. For instance, Gegeent valley at Baruun Saikhan mountain, Yol ravine, and Dungenee, Byantsagaan Khavtsgait, Kharangat, and Tuvuntiin valleys in the Zuun Saikhan mountain system crosscut the mountains, forming deep ravines and canyons with steep slopes.

### 3.4.2 The Erosion-Accumulation Type of Relief

The erosion-accumulation type of the relief includes hummocks and rolling, pedestal, and layered plains. Depending on the level of surface dissection and inclination, hummocks can be sub-divided into highly dissected and slightly dissected. Furthermore, these classes can be further sub-divided based on rock structure and elevation. The rolling plains are present along the foot of major mountains and are widespread in the eastern part of the country.

Pedestal plains mainly formed in the transition areas between hummocks and rolling plains. In Mongolia, this relief form occupies vast territory and is present between the southern slopes of Khangai mountains and the plain of the Eastern Mongol. For instance, the largest pedestal plain stretches from the Tui River to the northern parts of the Dariganga Plateau and Erdenetsagaan. The current pedestal plain surfaces are characterized by a mixture of rolling hills and flat plains.

The erosion-accumulation-type depressions and hollows include areas occupied by badlands and are widely pronounced in the Gobi. This relief is a result of active weathering and erosion of ancient lacustrine and oceanic deposits. The surfaces are rarely dissected and are composed of small hills and hummocks, making a pedestal. These sediments are rich in organic relicts, such as fossil trees, fossil bones, and animal prints. The major places with such fossils include Khermen Tsav, Nogoon Tsav, Shar Tsav, Algui Ulaan Tsav, Bugiin Tsav, Nemegt, Bayanzag, and Guril corridors. This sediment is primarily composed of red, green, or yellow silt, and in places, there are sand and sandy loam layers. In some locations, this sediment layer is covered by thick sand.

### 3.4.3 The Accumulative Type of Relief

The accumulative type of landforms includes intermountain hollows, river valleys, and plains. By their genesis, accumulative surfaces can be divided into several types: diluvium-alluvial fan plains; slightly rolling, vale-shaped plains; slightly hilly intermountain hollows with flat bottoms; lacustrine-alluvial plain; and lacustrine-saline plain.

The diluvium-alluvial fan plains are common at the foot of mountains along the largest watersheds and mainly form debris flows or deltas. In the Gobi, large inclined plains formed by diluvium-alluvial fan deposits are widely distributed and referred to as “bel”.

Slightly rolling, vale-shaped plains are the most widespread relief type in Mongolia and are represented in all mountainous regions. On the ground, they are narrow gorges and rarely look like large hollows. Slightly hilly intermountain hollows with flat bottoms are generally present near isolated and inner flow basins. Such hollows are filled with Cretaceous and Neogene deposits and can be covered by Quaternary sediments.

Lacustrine-alluvial plains occupy the largest river valleys and depressions. Their surfaces are smooth and rarely they are slightly inclined. The sediments are mainly composed of 100–250 m thick Quaternary lacustrine sediment and alluvium. Lacustrine-saline plains are widely present in steppe and Gobi regions. These plains are composed of Quaternary and Holocene sediments.

### 3.4.4 Volcanic-Type of Relief

In Mongolia, volcanic-type surfaces include ancient inactive volcanoes, and their basaltic plateaus are relatively numerous, particularly in the eastern part of the territory. The Dariganga volcanic plateau is gradually increasing from north to south. There are about 222 cones of ancient volcanoes, the largest of which are Ikh Uul (1,286 m), Baruun-Yargait (1,434 m) in the north of the plateau and Bayan-Tsagaan (1,534 m), Shiliin Bogd (1,777 m), Bogd mountain (1,749 m) and others in the south. Other volcanic type of reliefs includes Altan ovoo, Khorgo Vent, Uran Vent, Shiree Teeg (1,085.0 m), Tuvshin mountain (1,498.0 m), Dulaan mountain (1,527.2 m), Tom Ovoo (1,575.0 m), and Bulagtai (1,591.0 m) (Fig. 3.5).

### 3.4.5 The Micro-Reliefs

The micro-relief is formed by modern exogenic factors called morphosculptures, which include micro-forms mainly determined by surface flow, wind, frost, and glaciers in Mongolia (Yembuu 2014). Flow to surface landforms can include river valleys and high and low terraces. The glacial landforms are hanging valleys, scars, and end, lateral, and medial moraines. The frost landforms include thermokarst, karst, pingo, solifluction terraces, cryogenic faults, and landslides. The aeolian landforms include sand dunes, barchans, sand cover, and knobby sands.

The impact of wind activity on landforms is important. The accumulative mode of wind erosion forms the largest sandy lands. The largest sandy lands in the Gobi region are distributed along the Zavkhan, Khungui, Tes, and Nariin Rivers, which are part of the Great Lake Depression. In the southern parts of the Gobi, some sand masses are distributed as well. These include Khongor Els, Khaltar Els, Ingen Khuuver, Zulganai, Gurvan Tes, and Duut Sands, as well as the Burd Sandy land. These sands resemble a chain of separate barchans and are not fixed by vegetation. Although the aforementioned sands and sandy lands have different origins, they are all wind erosion products.

The sands distributed in the Great Lake Depression originated in the Holocene when Central Asia was under a dry climate regime. The sand distribution coincides with the locations of lakes and other basins that developed during the Mesozoic and Phanerozoic. During recent geological times, the lakes desiccated, and the sands were widely distributed by an active wind regime. This is also related to the uplift of the depression, covered with loose sediments, during the most-recent tectonic activities (Selivanov 1969). B. P. Vipper (Vipper et al. 1976) used pollen to reconstruct the paleoclimate and suggested that climate became drier during the late-

**Fig. 3.5** Volcanic type of relief (Altan Ovoo mountain, in the Eastern plain)

Pleistocene glaciation. Thus, at the end of the Pleistocene, the Great Lake Depression dried up completely and transitioned to a terrestrial regime.

## 3.5 Geomorphological Regions and Main Characteristics

Mongolia can be divided into 11 regions by geomorphological characteristics (National Atlas of Mongolia) 2009: the Mongol Altai, Khuvsgul, Khangai, Khentii, Khyangan, Orkhon-Selenge river basin, Gobi-Altai, Great Lakes depression-valley of Lake, Trans-Altai Gobi, Dundgovi–Umnugovi, and East Mongolia (Fig. 3.6).

### 3.5.1 The Mongol Altai Region

The Mongol Altai mountain ranges, located in the Mongolian Altai, are the largest mountain system and combine glacial rivers, freshwater glacial lakes, hills, and valleys with different shapes and elevations. These are located in the west and stretch more than 800 km from the northwest to the southeast (Fig. 3.7).

The Mongol Altai Mountain system consists of three major ranges. The first range includes Khuiten orgil Altai Tavan Bogda (4,374 m), Nairamdal (4,082 m), Malchin (4,037 m), Burged (4,068 m), Ulgii (4,050 m), Munkh-khairkhan (4204 m), Tsambagarav (4,165 m), Sair Uul (3,984 m), Tsengelhairhan (3,967 m), Must (3,934 m), Aj Bogd (3,802 m), Alag khairhan (3,738 m), and Khukh Serkh (3,700 m) Ranges. The second group includes Sutai (4,090 m), Baatarkhaihan (3,994 m), Dartsag Khuren (2749 m), and Bayan Bumbun (1,947 m) ranges. The third group encompasses mountains, such as Kharkhiraa (4,037 m), Turgen (3,965 m), Tsagaan shuvuut (3490 m), Altan-khukhii (3,351 m), Jargalant Khairhan (3,796 m), Bumbat Khairhan (3,464 m), Darvi Range (2686 m), Taishir Range (3,070 m), and Khasagt Khairhan (3379 m).

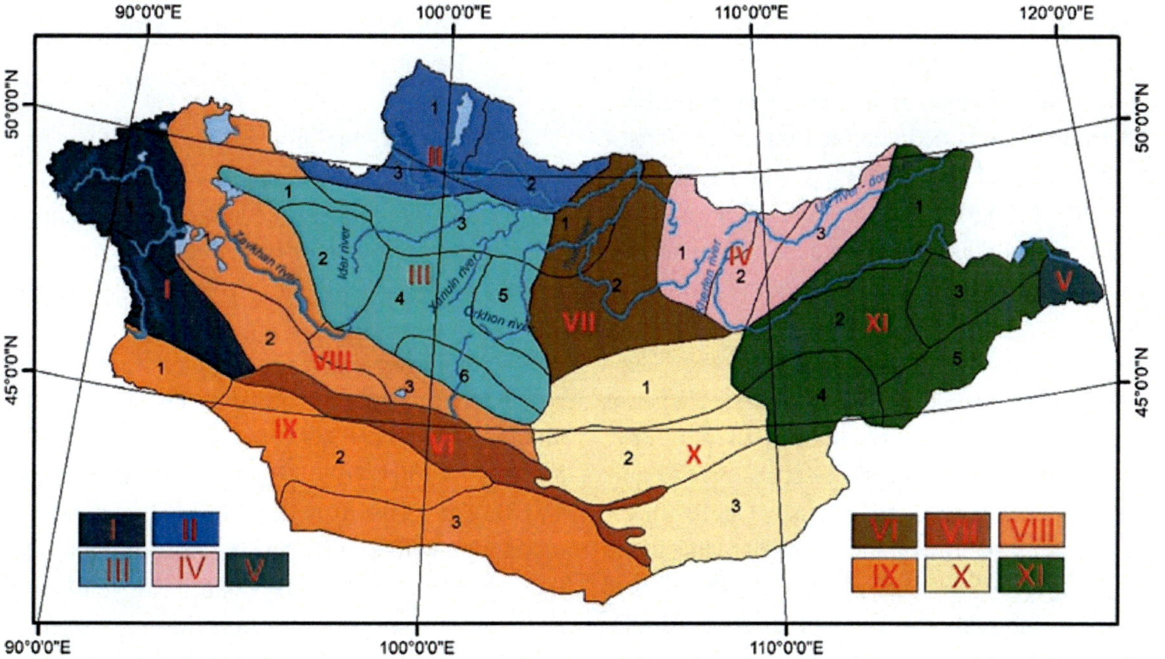


**Fig. 3.6** Geomorphological regions of Mongolia (by Dorjgotov, D. Enkhtaivan, National Atlas of Mongolia; 2009): I—Mongol Altai: 1-North part; 2-South part; II—Khuvsgul: 1-Western; 2-Eastern; 3-Sangelin; III—Khangai: Khan-Khukhii; 2-Uliastai; 3-Bulnai; 4-main ridge of the Khangai; 5-Khanui; 6-southern Khangai; IV—Khentii: Western Khentii; Eastern Khentii; Ereen range: V—Khayngan; VI—Gobi-Altai; VII—Orkhon Selenge low mountains; VIII—Great Lakes depression and Valley of Lake; IX—Trans-Altai Gobi; X—Dundgobi and Umnugobi; XI—Eastern Mongolia

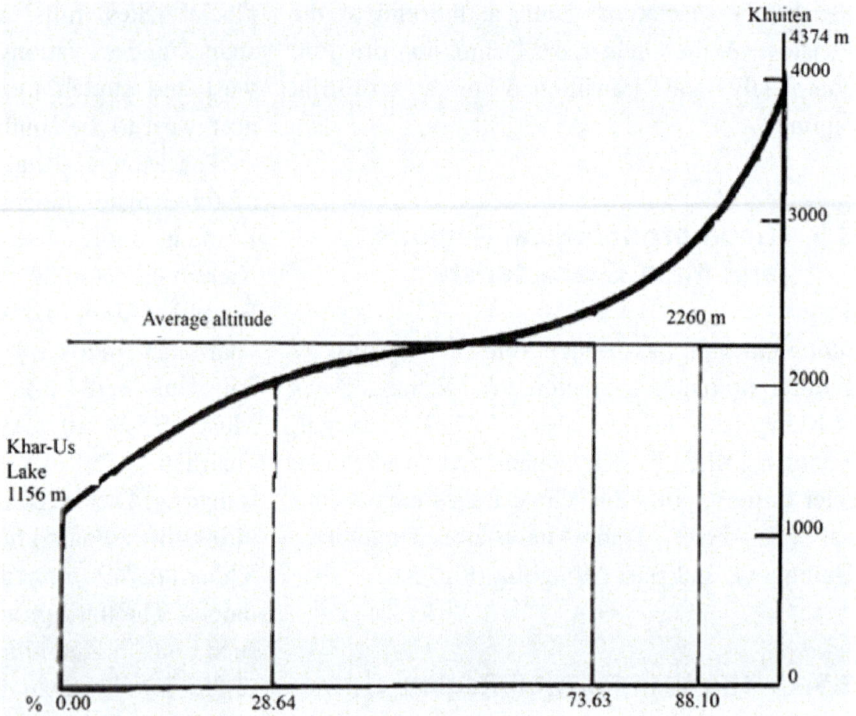


**Fig. 3.7** Hypsographic curve of the Mongol Altai (Hilko and Kurushin 1982)

These mountains often have convex shapes that are aligned with ancient peneplain surfaces, high, sharp peaks, and rocky, steep snowy slopes. The mountain slopes are unequal: the back slopes are short but not sharp, and the front slopes are long with very steep slopes (Dash 2009). The difference in angles between the back and front sides is a distinguishing characteristic of their external forms (Figs. 3.7 and 3.8).).

In the Mongol Altai mountains with elevations above 4,000 m a.s.l., traces of Quaternary glaciation in the foothills are common. Moreover, the largest modern glacier is Potanin, which lies in the Tsagaan River Valley, in the Tavan Bogd Mountains. The length of the glacier is about 20 km, and it has an area of 47.2 $km^2$. Also, among the largest glaciers are Grano, Kozlova, Alexander, and Tsagaan Deglii. In the mountainous region of the Mongol Altai, there are tectonic depressions of various shapes and sizes at different levels, which have similar general directions as the major mountain ranges, such as Lake Tolbo, Tsetseg and Dayan, Achit, Zereg, Ikhes, Darvi, Shargiin Gobi, and Lake Biger. Some of the major river valleys and depressions are often accompanied by local tectonic fractures (Mikhailov and Ostanin 2002) (Fig. 3.9).

### 3.5.2 The Khuvsgul Mountains Region

The Khuvsgul mountains are part of Sayan and Southern Baigal mountains, but have different structures, directions, and variable elevations. The Western Khuvsgul mountains lie along the latitude and are rocky with steep slopes and sharp peaks.

The highest point in the region is the 3,491 m a.s.l Munkh Saridag mountain. There are many mountains above 3,000 m, such as Khoridol Saridag (3,130.0 m), Ulaan Taiga (3,357.0 m), and Bayan (3,093.0 m). Additionally, there are mountains with elevations of 2,500 m, including Ikh Agayagiin nuruu, Byaraanii Range, Upper and Lower Saalig Ranges, and Mount Renchinlkhumbe. These mountains are divided into three large tectonic depressions, Lake Khuvsgul, Darkhad Valley and, and Bus River Basin, which are located along the longitude. This region is characterized by the largest number of relict glaciers,

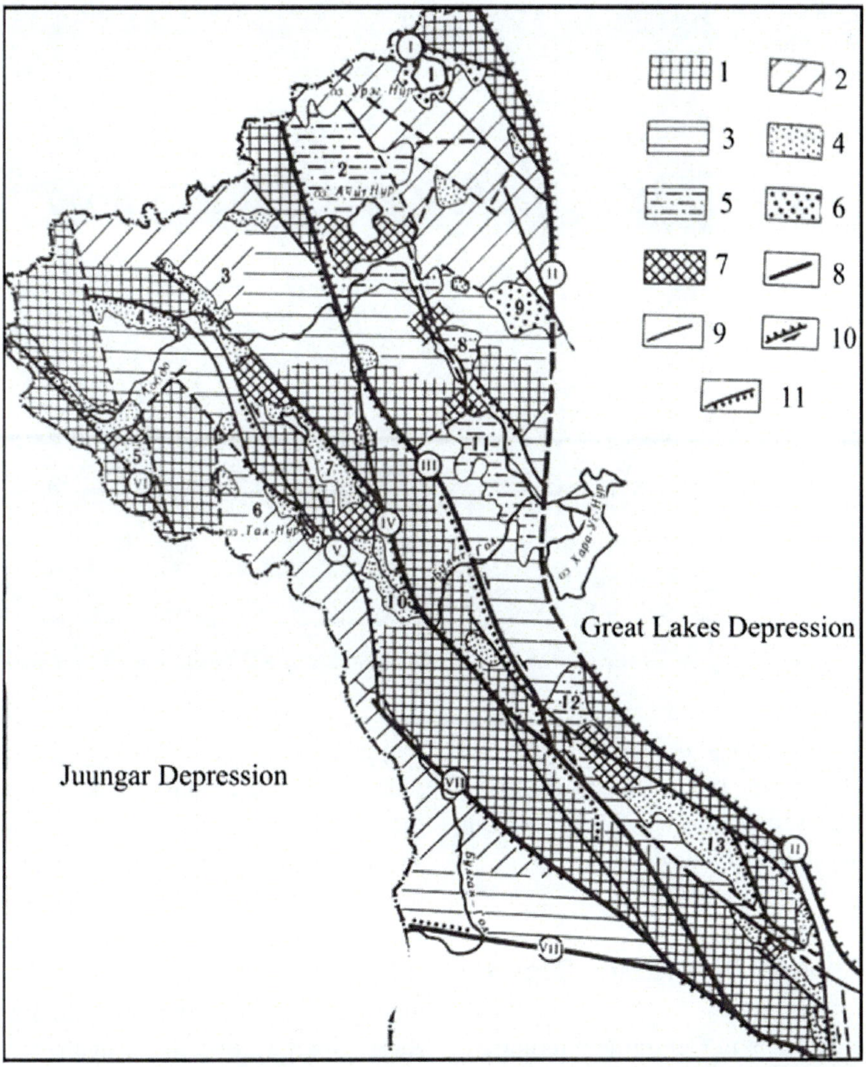


**Fig. 3.8** Overview of the main morphological structures of the Mongol Altai (Hilko and Kurushin 1982): Positive morphological structures: 1—intensively rising high-mountain ridges-blocks formed by deformations of the pre-Cenozoic basement, 2—high and middle altitude mountain ridges and massifs formed mainly by bending deformations of the pre-Cenozoic basement, 3—middle-mountain plateaus and plateaus formed on the place of undifferentiated areas of the primary archaic uplift of the territory Negative morphostructures: 4—formed on the site of relatively lowered and mostly single-sided blocks, 5—formed on the site of stable or moderately elevated blocks, 6—Mesozoic deflections, 7—interdepressions low and middle-mountains; Faults: 8—regional and deep; 9—local, 10-Pliocene and Quaternary faults;, 11—Holocene seismic faults; Faults (Roman numerals in circles): I-Tsagaan Shiveet, II-Eastern Altai, III-Khovd, IV-Tolbo Lake V-Tal Lake, VI-Khovd lake, VI-Upper Khovd, VII-Turgen mountains, VIII-Bulgan. Intermountain depressions (Arabic numerals): 1—Uureg lake, 2—Achit lake, 3—Oigon lake, 4—Tsagaan lake, 5—Khovd lake, 6—Tal lake, 7—Tolbo lake, 8—middle Khovd, 9—Namir river basin, 10—Deluun, 11—Lower Khovd, 12—Mankhan, 13—Tsetseg

yet glacial traces, such as moraines, are still evident.

The Eastern Khuvsgul mountains belong to the Southern Baikal taiga region and create a distinct relief contrast from the western part. Here, there are mountains and ranges, such as Kheven Zaluu Uuriin Saridag (2,534.0 m), Ganshuuliin Saridag (2,106.8 m), Khukh Baits (2,130.74 m), Unduh Mon (2,149.9 m), Undur Degeen (1,976.0 m), Toj (2,548.2 m), Saravslag Saridag (2,358.1 m), Yalar Saridag (2,375.0 m), Zaluu Uuriin Undur (2,189.6 m), Tsagaan Uul (2,367.0 m), and Armagiin Range (2,030.8 m). All of these mountains are mid-elevation

**Fig. 3.9** Glacier circus (upper Tsagaan Gol, Mongol Altai) (own photo)

mountains with elevations between 1,900 and 2,500 m a.s.l. and have sphere-shaped summits. The mountains are separated by the Arig and Uur river valleys and their tributary rivers.

### 3.5.3 The Khangai Mountains Region

The Khangai mountains in Central Mongolia are a 500 km long mountain system, the largest of which runs from northwest to southeast. For example, the Khan Khukhii and Bulnai mountains stretch from east to west, and the Tarvagatai and Buren ranges stretch from southwest to northeast (Fig. 3.10).

Among the mountains of the Khangai region, the highest point is Otgontenger (4,021 m a.s.l.), located in the west. Otgontenger is covered with small quantities of permanent snow and has glacial rivers. The mountain ranges are Angarkhai (3,540.0 m), Erhet Khairkhan (3,535.0 m), Undur Jargalant (3,441.0 m), Untaa Jamaat (3,335.0 m), Asgat Gurvanbulag (3,240.0 m), and Suvarga Hairhan (3,179.0 m). The height of these mountains decreases from west to east.

The Khangai mountains have an ancient peneplain surface and thus rounded peaks. In this region, there is an intersection between three major water basins of the globe, which distinguishes this area from others. The central part of the mountains is highly dissected by rivers and surface flow, contributing to steep rocky slopes in the southern exposition and smooth, less-steep slopes covered by forests in the northern exposition. West of Otgontenger, the Khangai mountains system includes mountains such as Baga Bogd (3,643 m), Bayanzurkh (3,242 m), Ikh Myangan (3,369 m), Asgat (2,795 m), Tsagaan Khairhan (2,762 m), Eastern Arts (2,626 m), and Western Arts (2,355 m). The plateau is dissected by the Zavkhan and Khungui rivers, the watershed of which is composed of a series of mid-elevation mountains. Sediments from Pleistocene glaciations are widely distributed in the Bogd, Shar us, Orkhon, Uliastai, Chuluut, and South and North Tamir river valleys (Fig. 3.11). There are also many moraine lakes and large basalt covered areas near the Terkhiin tsagaan lake, Chuluut and Orkhon river basins (Fig. 3.12).

Western Khangai is occupied by the Khan Khukhii mountains, which are deeply inclined to the west in the Great Lake Depression and separate the Uvs and Khyargas lake basins. Compared to the other mountains in the Khangai region, the Khan Khukhii range is relatively low

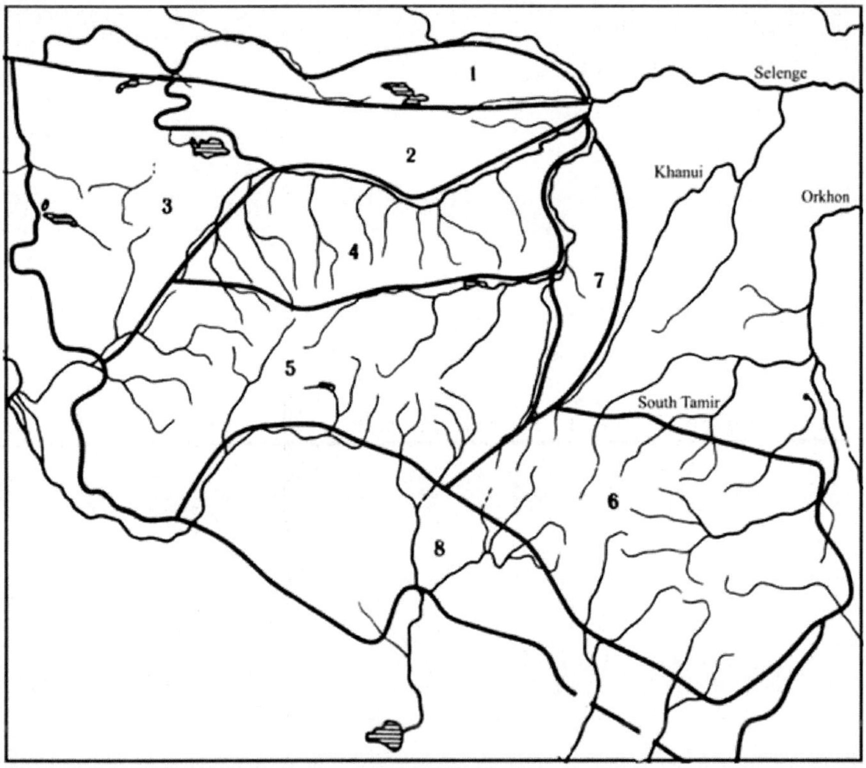


**Fig. 3.10** Orographic scheme of Khangai mountains range. 1—Northern border of range with latitudinal direction: ridges, massifs, plateaus and depressions; 2—Bulnai ridge—a series of narrow flat top ridges (about 2,500 m) and depressions, except for the eastern part where the direction of the ridges is northwestern and their height is not more than 2,300 m; 3—Northwestern Khangai of sub-latitudinal ridges (2,800–2,900 m) and wide intermountain depressions (1,600–2,000 m, sometimes almost isometric); 4—Tarvagatai-asymmetrical sub-latitudinal ridge with a long gentle northern and short southern steep slope, with a broken line of the apex bending to the northeast in its eastern part; 5—Western and central Khangai isometric massifs, different directions of ridges (some parts exceed 3,500 m) and plateaus (more than 3,000 m); 6—Eastern Khangai is a complex system of ridges with mainly northwestern and different directions; 7—Eastern border zone of Khangai, submeridional ridges (2,450–2,775 m) of the Khanui and Chuluut rivers; 8—South-Khangai plateau-plain, hills, separate massifs, ridges and depressions with clearly northwestern direction of orographic elements; average height is 1,300–2,400 m

elevation, with the highest point being Mount Duulga (2,928 m) to the east. To the west, the mountains include Bor Khairhan (2,591 m), Tsagaan Khairhan (2,551 m), Bayankhairhan (2,270 m), Togtokhiin Shiliin Khurmen[1] Ovoo (2,357 m), and Dovt (2,273 m), which have clear absolute height reductions. At their peaks, there are well-preserved traces of the ancient peneplain that created flat summits. Their back slopes are smoothly inclined to the west and reach Uvs lake depression, and the southern slopes are steep and sharply inclined toward Khyargas lake depression.

[1]Khurmen-Basalts.

### 3.5.4 The Khentii Mountains Region

The Khentii mountains are divided into two main areas, Baga and Ikh Khentii, with the highest point being Mount Asralt Khairkhan (2,800 m). Besides Asralt Khairkhan, the Baga Khentii includes Altan Ulgii (2,646.0 m), Khiidiin saridag (2,666.0 m), Shar bulagiin saridag (2,534.0 m), Abhaan (2,664.0 m), Tovshin (2,666.0 m), Asgatiin Saridag (2,320.0 m), and Gel Uul (2,551.0 m) mountains. The average elevations are low and vary from 1700 to 2500 m. The mountains have rounded tops, and the slopes are steep and highly dissected by river valleys, sourced from the mountains. The

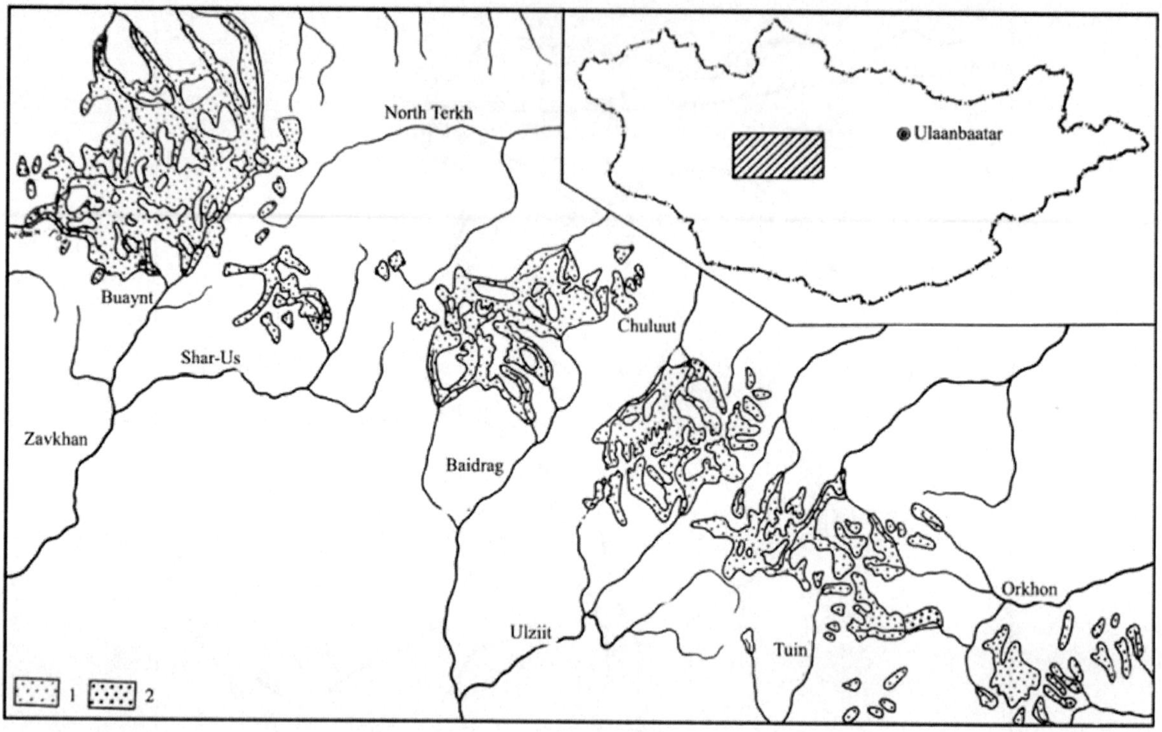


**Fig. 3.11** Scheme of the ancient glaciation of Khangai mountains: 1—the spreading area of the last glaciation, 2—the spreading area of the prehistorical

mountains are isolated from each other by intermountain depressions. The headwaters of the Kherlen, Onon, and Minj rivers are in the Ikh Khentii range, which consists of mountains such as Khentii Khan (2,362.0 m), Asralt Uul (2,452.0 m), Tsuuts (2,297.0 m), Ikh Khentii (2,283.5 m), Ikh Sukhulig (2,247.0 m), Bayan undrakh (2,236.0 m), Delger Khaan (2,111.0 m), Tarsyn Tersger (2,007.5 m), and Ikh Buural (2,125.5 m).

The mid-elevation mountains are covered by forest, above which the mountain tops are covered by stones, rocks, and loose weathering products. The mountain slopes are different, and in some mountains, north-facing slopes are steep and rocky, and south-facing slopes are less steep. This condition may be opposite in some mountains depending on their directions. Generally, the mountains are fully covered by forests and are rocky and highly dissected. The Khentii region is one of the major centers of ancient glaciation in Mongolia, and thus, there are many surface glacial landforms, including moraines, hills, glacial lakes, and karst.

Mount Bogd Khan, located south of the capital city, has an elevation of 2,257 m a.s.l. and is considered a mid-elevation mountain in Khentii. It is one large uplifted block structure that is highly dissected by surface flow, creating large and narrow ravines, gorges, and canyons. The mountain slopes are usually steep and commonly have pieces of stone and landslides. The main range runs northwest to southeast, and because the uplift was not synchronous, the slopes are different; north-facing slopes are less steep and longer than south-facing slopes. This highly impacts the formation of different landscapes.

The ancient glacial sediments are distributed in the Khag-Khongor river valley and in the headwaters of the Estii and Tuul rivers. There are many landforms reminiscent of the glacial period, including moraine terraces, traps, glacial lakes, and rivers. These landforms are mainly concentrated in the central part of the Khentii mountains at elevations between 1,700 and 1,800 m a.s.l., and thus, the relative mountain height is lower than that of the other regions. The glacial lakes in Khentii are usually located in

**Fig. 3.12** Basalt canyon in Chuluut river (own photo). Valley basalts of the Chuluut river (I) and Tagin Gorkhi (II). 1—floodplain, 2–90 m terrace, 3—basalts, 4—bedrocks: **a**—on the profiles, **b** on the plans, 5—sand with pebbles, 6—line profiles

**Fig. 3.13** Khagiin Khar lake of glacial origin in the central part of Khentii mountains

similar locations and are connected by small rivers and springs. The largest glacial lakes are Khagiin khar, Khuh, Mungun, and Mungun Khul.

The largest glacial lakes, which include Khagiin Khar (Fig. 3.13), Khukh, Mungun, and Mungun Khul, are situated at 1,790, 1,820, 1,730, and 1,720 m a.s.l., respectively. In the surrounding Estii Valley, along the Aaya River Basin, glacial landforms decrease to 1,400 m a.s.l., and in this region, 1,840 m a.s.l. is the upper limit of moraine sediments. According to our observations, the glacial landforms in Aayaa, Khag, and Khongor River basins are distributed between elevations of 1,400 and 1,840 m a.s.l.

### 3.5.5 The Middle Mountains of Eastern Khyangan

The mid-altitude mountains of eastern Khyangan form a region in the basins of the Khalkh and Numrug rivers (Fig. 3.14). The mountains primarily have gentle slopes and flat tops, and their height decreases from east to west, to the Lake Buir plateau and where broad valleys separate the mountains. Khyangan mountain runs from northeast to southwest, but formed within our country and the lower structure of these mountains falls from west to east. The highest mountains are Khalzan (1,503.2 m), Tumen (1,313.8 m), and Salkhit (1,088.3 m).

### 3.5.6 The Gobi-Altai Region

The Gobi-Altai region consists of the continuation of the Mongolian Altai and private high mountains and runs from northwest to southeast. This region is a narrow, 600 km long mountain system, and Bogd Mountain (3,957 m) is the highest point of elevation. Scientists believe that these mountains border the valley of Lake Alag Nuur. This region includes the mountains of Burkhan buudai (3,765 m), Tsahir Khaalgyn Nuruu (3,669 m), Javkhlant Nuruu (3,529 m), Gichgene (3,359 m), Khan Jargalan (3,208 m), and the Khar Azraga range (3,112 m).

The Gobi-Altai mountains cease to be a long-lasting system at the Bayantsagaan mountains and pass through the mountain systems of many individual mountains, such as Bogd (3,957 m),

**Fig. 3.14** Mid-altitude mountains in the Numrug river basin (Eastern Khyangan, own photo)

Baga Bogd (3,590 m), Arts Bogd (2,399 m), Govi Gurvan Saikhan (2,825 m), Nomgon (2,190 m), and Khurkh (1,762 m). In addition, the high mountains of this region have steep slopes and are separated by transverse deep cavities and valleys, surrounded by broken surfaces. Generally, the Gobi-Altai mountains are still very active, although research is ongoing. This situation is confirmed by the last powerful earthquake in 1957, which changed the watershed regions and resulted in cracks on the backside of the Gobi-Altai mountains. The earthquake formed lakes, such as Oyu and Nomin in the valleys of Bituut and Uliastai.

The Gurvansaikhan mountains consist of several interconnected mountain ranges, Baruun Saikhan (2,548.7 m), Dund Saikhan (2,825.0 m), Zuun Saikhan (2,815.0 m), Bayantsagaan (2,613.0 m), and Bayanbor (2,232.9 m). These are ancient and relatively young faults, and the mountains mainly run along the latitude. However, given the general elevation basis, they are located in the west and run from the northwest to the east and southeast. The mountains usually have pointed peaks and steep slopes, and Alpean surface types in the highest mountains (Fig. 3.15). The Dund and Zuun Saikhan mountains have high-elevation peaks, and the other mountains have cone-shaped structures with single peaks.

### 3.5.7 The Orkhon-Selenge Mid-Elevation Mountain Region

The Orkhon-Selenge mid-elevation mountain region is located along the deep faults of Zaamar, Bayangol, and Selenge. This region includes mountains distributed from southwest to northeast, such as the Khantai range (2,170 m), Khusht Buluu (2,093 m), Khan Jargalan (2,069 m), Buren range (1,985 m), Buteel range (1,720.9 m), Khujingiin range (1,444.0 m), and Asgat (1,444.0 m). The peaks of these mountains are flat and convex, and the main ranges do not

**Fig. 3.15** Alpine type relief in the desert region (Ikh Bogd mountain, own photo)

have peaks above the timberline and are completely covered with forest. The steep slopes have rocky formations, which rarely form rocky ridges. River valleys in the headwaters are narrow, expansive in the middle sections, and eventually become open plains.

### 3.5.8 Great Lakes Depression and Valley of Lakes

The Great Lake depression region is of tectonic origin and is located in a deep valley between the mountains of Central Asia of 46°–51°N and 92°–96°E. This region separates the Altai and Khangai mountains and extends along the longitude. In this regard, this region has a very different geographical position, type, and surface shape than the other regions (Dash 2002). Among the scholars who explored this region, Murzaev (1948) first coined the name "Great Lake Valley". The geomorphology of this valley is, by nature, erosive-accumulative and accumulative. The region includes widespread hills, sloping plains with diluvial-proluvial sediments, lakes, plains, and river and sandy sediments (Jigj 1975b). The lacustrine and proluvial sediments are situated between 750 and 1600 m a.s.l. and are divided into four parts: Lake Uvs, Khyargas, Khar Us, and Sharga Valley (Murzaeva 1982).

The majority of the valleys, located in the lake, is formed in the Mesozoic. After the Mesozoic, there was a period of relatively quiet tectonics, contributing to the modern preservation of Mesozoic valleys (Selivanov 1969) This also has caused the accumulation of sediments over a long time period, and thus, in the central part of the valley and at the river ends, there are thick layers of sand, gravel, and lacustrine sediments (Fig. 3.16). The Valley of Lakes, also of tectonic origin, is the eastern continuation of the Great Lake Depression and separates the Gobi-Altai and Khangai mountains along the longitude. Situated at the center of the Buuntsagaan, Orog, Taatsiin, and Tsagaan Lake Valleys, there are widespread sandy deposits and valley, lake,

**Fig. 3.16** Buurug Del sand dunes (own photo)

and river sediments, with inclinations of diluvium-alluvial fan sediments. Grunert et al. described that loess-like aeolian sediments on mountain slopes and sandy fields on the banks of rivers and lakes are common in the Great Lake Depression (Grunert and Lehmkuhl 2004).

### 3.5.9 Trans-Altai Gobi

The desert valley south of the Mongol Altai and Gobi-Altai mountains is called Trans-Altai Gobi. It is dominated by rocky and stony desert with dark brown rocks. A strong wind always carries the powder particles from the valley. The extremely dry desert Govi-Tien-Shan mountain system includes the mid-elevation, rocky Atas Bogd (2,695 m), Ginghis Uul (2,114 m), Shar Khulsnii Nuruu (1,754 m), Zaraa Khairkhan (1507 m), Segs Tsagaan Bogd (2,380 m), Khatan Suudliin Khyar (1,637 m), Nemegt (2,768.0 m), Gilbent (2,168.0 m), Sevrei (2,631.0 m), and Zuulun (2,422.0 m) mountains. The mountains are separated, and the low- and mid-elevation mountains have sharp peaks and steep slopes and are rocky (Fig. 3.17). These mountains run almost in a line, along the latitude. The alluvial fans and sediments are produced by weathering and mass movement processes (Owen et al. 1999).

The western part of Southern Altai Gobi is the Baruun Khuurai Valley, which stretches from west to east. However, establishing clear valley boundaries was difficult. This is due to the fact that the low mountains and hills in the central valley connect the surface and divide the interior portion into several minor valleys. Additionally, west of the desert, the Zuungar Gobi is limited to the country border, as opposed to a structural boundary. Murzaev (1952) mentioned that the Baruun Khuurai is the eastern continuation of the Zuungar Depression in Xinjiang. Baitag Bogd (3,187 m), Ikh Khavtag (2,908 m), Baga Khavtag (2,111 m), and Takhiin Shar mountain range (2,725 m) are located in the southern part of

**Fig. 3.17** Alluvial fans and sediments in the southern slope of Nemegt mountain (own photo)

Baruun Khuurai, along the mountains that form the state border.

### 3.5.10 The Dundgovi–Umnugovi Region

The Dundgovi–Umnugovi region includes the southern and eastern parts of Dundgovi, Tuv, Dornogovi, and Umnugovi *aimags* (provinces). Here, low- and mid-elevation mountains with many faults dominate. The major mountains of the region are Delgerkhangai (1,990.0 m), South Unjuul (1820.0 m), North Unjuul (1,770.0 m), Kharuult (1,700.0 m), Takhillaga (1,695.0 m), Ongon Khairhan (1,748.0 m), Ikh Khairkhan (1,666.0 m), Zorgol Khairkhan (1,668.0 m), and Choiriin Bogd (1,731.0 m), all of which are separated by ancient river valleys and depressions.

There are many places with effusive rocks proximal to large granites, such as Baga Gazriin Chuluu (1,701–1,767 m) (Fig. 3.18), Ikh Gazriin Chuluu (1,706.0 m), and Adaatsag (1,804.0 m). Hills are some of the most common surfaces in this region, the heights of which in the north and south are 1,000–1,500 and 1,000–1,200 m, respectively. The surface is very crowded with streams, ravines, and rivers, which form temporary drainages. In this region, the largest depressions are Undershil, Galba Gobi, Baishin Tsav, Ulshig, Zuunbayan, and Sainshand, which accumulate sand on their windward sides. (Namnandorj 1972)

### 3.5.11 Eastern Mongolia

Eastern Mongolia is the central element of a vast space between the Khangai and Khentii mountains, with smooth and wavy surfaces (Fig. 3.19). The East-Mongolian plain from its lowest absolute marks (about 600 m) in the areas of lake plains in the northeast and Buyr Nur Lake in the east gradually increases in the southwest direction up to 1000–1100 m, and in the south of the Dariganga plateau up to 1,300–1,400 m. This

**Fig. 3.18** Granite massif (Baga Gazriin Chuluu) (Hoyor Erhi photo)

part includes Menen plain, Badam ish, Zurkhiin Tsagaan Khooloi, Tsagaan Daravgain Khooloi, Segseegiin Khooloi, Ereen Dersnii Khooloi, and Tamsag Bulag plains with alluvial fan and lacustrine-alluvial sediments, as well as the erosion surface on the southern part of Lake Buir. The surface is 620–720 m a.s.l., and tunnels and shallow depressions have slight slopes, gradually moving toward lower elevations. Instead, old river valleys widen and have sides with different terraces. In addition to large morphological elements, there are a number of relatively small forms on the territory of the East Mongolian plain, characterized mainly by a passive reflection of the tectonic structure in the present relief. These are ridges and groups of melkosopochnik representing structural features of the Paleozoic basement. These are also circular volcanic and tectonic structures of the Mesozoic age, and among the negative forms—basins and valley-shaped depressions of erosion and deflation origin, located within the tectonic troughs.

The western part of the eastern region is dominated by the hilly plain of the Central Khalha Plateau. The total surface stands at 900–1,606 m a.s.l., and on the surface, there are single hills or knolls. The largest mountains are Ikh Khaltar Morit (1360.0 m), Zaan shiree (1,436.0 m), Kharaat Uul (1,437.1 m), Ikh Buural (1,379.5 m), Nomyn Khaan (1,414.5 m), Uyat (1,358.6 m), Baga Buural (1,294.2 m), Munkhkhaan (1,606.3 m), Asgat (1,483.8 m), Bayantsagaan (1369.8 m), Tumentsogt (1,358.3 m), and Darkhan Khan (1,055.8 m).

Dariganga plateau, which sits at 1,200–1,300 m a.s.l., occupies the southern half of the Sukhbaatar aimag territory. From the north and west, the plateau is limited by terraces ranging in height from 20–40 m to 100 m. Residual small volcanoes over 1500 m high, sand dunes and depressions of various sizes are located in the Darganganga plateau (Fig. 3.20). The distinguishing feature of this region is the accumulation of many extinct volcanoes in one place. According to Shagdar (2000), there are about 222 extinct volcanoes in Dariganga region, the highest and the largest of which are Buduun Tolgoi (1,542.2 m), Ikh Uul (1,584.2 m), Baga Uul, Shaazant (1,537.0 m), Lamiin Dush (1,574.0 m), Ganga Tsagaan ovoo (1,530.2 m),

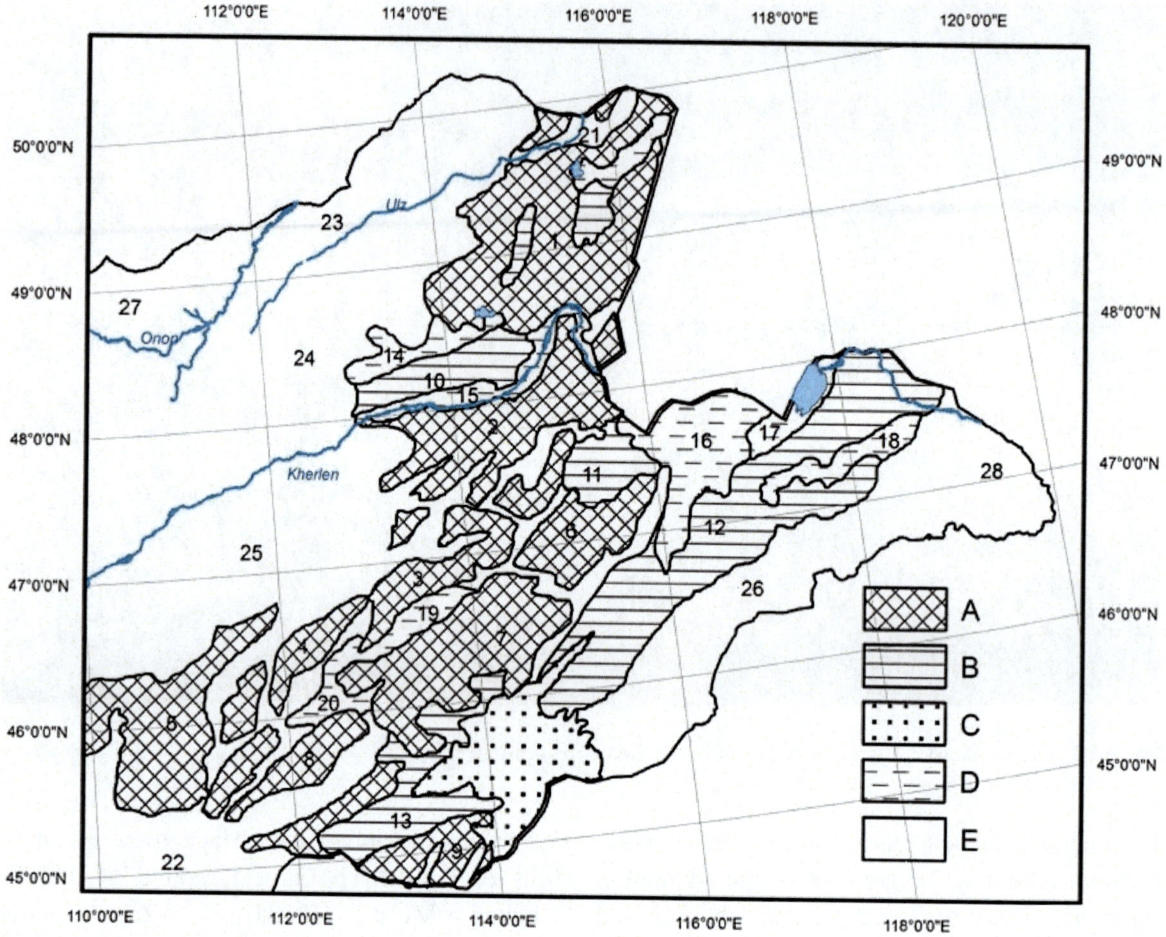


Fig. 3.19 Relief types of the Eastern Mongolia (Siernev 1982). A—Paleozoic and Early Mesozoic denudation plains: 1—North Kherlen; 2—South Kherlen; 3—Baruun-Urt; 4—Uulbayan; 5—Delgerekh; 6—Matad; 7—Asgat; 8—Bayandelger; 9—Naran; B—Upper Cretaceous and Neogene plains of the sediments: 10—Choibalsan; 11—Ulaantsereg; 12—Tamsag; 13—Ongon;C—Dariganga lava plateau; G-Accumulative plains:14—Gal river; 15—Valley of Kherlen; 16—Menen plain; 17—Buir lake; 18—Tamsagbulag; 19—Sukhbaatar; 20—Khalzan; 21—Lower Ulz river; D-hilly and sloping plains with the Mesozoic and Cenozoic saline depressions: 22—Hilly plain relief of the East Gobi depression; 23—Ereentsav low mountains; 24—Low-mountain between Kherlen and Ulz river s; 25—low mountain of the middle Khalkh; 26—Low mountains of the Nukht davaa; 27—Khentii mountains; 28—Khyangan mountains

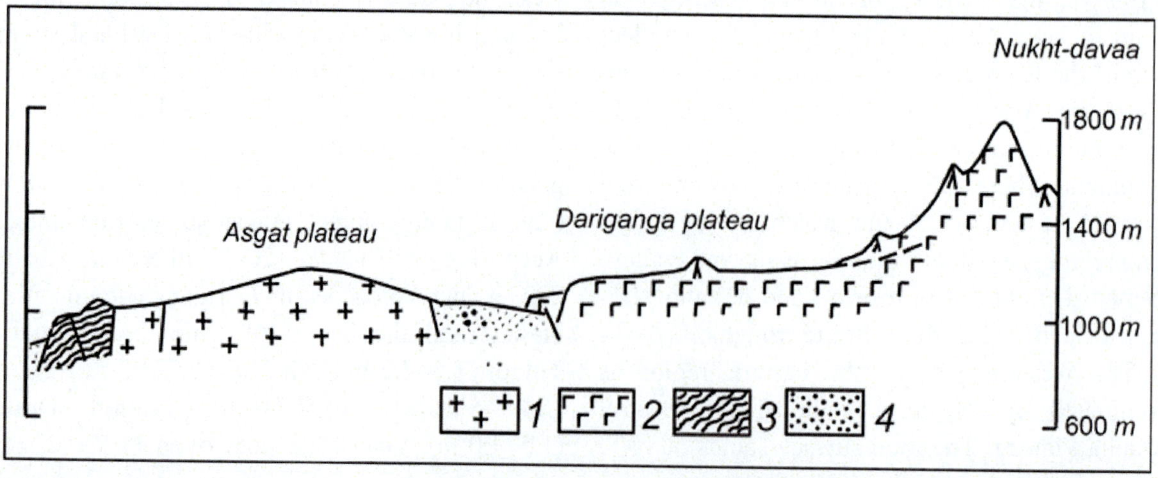


Fig. 3.20 Geological-geomorphological profile through the Eastern-Mongolian Plain Types of rocks: 1—intrusive, 2—effusive basic, 3—metamorphic, 4—sedimentary

Dush (1,547.9 m), Baruun (1,434 m) and Zuun Yargait (1,401 m), Buduun Tolgoi (1,582.3 m), Bogd Ulaan (1,749 m), Shiliin Bogd (1,777.5 m), Zotol Khan (1,425.0 m), and Bayantsagaan (1,534 m). Mountains with steep slopes highlight these extinct volcanic peaks. The Dariganga plains on the left side continue to the plains of Mount Erdenetsagaan. In the midst of the plain, the Nukht (1,650 m) and Baruun Ereen (1,090 m) mountains tower and do not form foothills. In the east, there are sand deposits of Ongon, Moltsog, and Devteer sand dunes.

The age of the main mass of the Dariganga basalts' covers is defined as a pre-Quaternary one. Thus, the leveling surface, partially interrupted by basalts, formed in the Miocene or early Pliocene, is preserved here. Apparently, in other areas in the east of Mongolia, the leveled areas of high level are the remnants of the same leveling surface. To the west of the Dariganga Plateau, flat horizontal surfaces cutting off upper Cretaceous layers are located at lower levels. This is a relatively young alignment surface.

## References


Baasan S (2006) Sand dunes of Mongolia. Munkhiin Useg publishing, Ulaanbaatar (in Mongolian)

Batsaikhan C (1999) Research on the rift zone morphostructure in Westerm Mongolia. A dissertation for philosophical doctor of geography. Institute of Geography. Mongolian Academy of Sciences, Ulaanbaatar (in Mongolian)

Blomdin R et al (2018) Timing and dynamics of glaciation in the Ikh Turgen Mountains, Altai region, High Asia. Quat Geochronol (47):54–71

Byamba J (ed) (2012). Plate tectonics of Litosphere. In: Mongolian geology and mineral series, vol 4. Ulaanbaatar, p 494 (in Mongolian)

Chistyakov KV, Seliverstov YV, Moskalenko IG et al (1994) Problems of stability of inland mountain landscapes in the changing world. Russian Geographical Society, St.-Petersburg, 94 pp (in Russian)

Corina NA (1974) Relief of valleys and troughs of Northern Mongolia and its development in Neogene-Quaternary time. In: Cand. geo-mineral. Nauka, Moscow, p 21 (in Russian)

Dash D (2002) Geographical features of the landscape of mongolia. Geogr Iss Mongolia 25–35 (in Mongolian)

Dash D (2009) The place of the Mongolian Altai in the system of physical-geographical zoning of Mongolia and the features of its differentiation. In: Proceedings of the IX international conference on natural conditions, history and culture of western Mongolia and adjacent regions, Khovd, pp 65–71 (in Russian)

Devyatkin EV (1981) Cenozoic of Inner Asia. Nauka, Moscow (in Russian)

Devyatkin EV, Murzaeva VE (1975) Pleistocene lakes of Mongolia. In: History of lakes, vol 4, Leningrad, pp 98–103 (in Russian)

Devyatkin EV, Smelov SB (1974) Tectonics of the Mongolian People's Republic. Nauka, Moscow (in Russian)

Devyatkin EV, Smelov SB (1981) Basalts and their position in the section of sedimentary Cenozoic Mongolia. USSR Academy of Science, Moscow

Dorjgotov D (ed) (2009) National Atlas of Mongolia. Institute of Geography, Mongolian Academy of Sciences, Ulaanbaatar

Enkhtaivan D (2006) Physical-geographical characteristics of the Altai region. In: Environmental security and sustainable land use-with special reference to Central Asia. Springer, Dordrecht, pp 349–351

Florensov NA (1968) Some features of the basins of large lakes in Southern Siberia and Mongolia. In: Mesozoic and Cenozoic lakes of Siberia. Nauka, Moscow (in Russian)

Florensov NA (1975) Mesozoic and Cenozoic tectonics and magmatism of Mongolia. Nauka, Moscow (in Russian)

Florensov NA, Korjuev SS (1982) Geomorphology of the Mongolian People's Republic. Nauka, Moscow (in Russian)

Grunert J, Lehmkuhl F (2004) Aeolian sedimentation in arid and semi-arid environments of Western Mongolia. In: F-HP Smykatz-Kloss W (ed) Paleoecology of quaternary drylands. Lecture notes in earth sciences, vol 102. Springer, Berlin, Heidelberg,, pp 195–218

Grunert J, Lehmkuhl F (2004) Aeolian sedimentation in arid and semi-arid environments of Western Mongolia. In: Paleoecology of quaternary drylands. Springer, Berlin, Heidelberg, pp 195–218

Hilko SD, Kurushin RA (1982) Mongol Altai. Geomorphology of the Mongolian People's Republic. Nauka, Moscow, pp 40–54 (in Russian)

Jigj S (1975a) Primary features of Mongolian landforms. Institue of Geography and Cryology. Mongolian Academy of Sciences, Ulaanbaatar [in Mongolian]

Jigj S (1975b) The characteristics of land forms of Mongolia. State Publishing House, Ulaanbaatar [in Mongolian]

Klinge M, Lehmkuhl F (2013) Geomorphology of the Tsetseg Nuur basin, Mongolian Altai–lake development, fluvial sedimentation and aeolian transport in a semi-arid environment. Journal of Maps 9(3):361–366

Kozhevnikov AV (1968) Stratigraphy of anthropogenic deposits in the Selenga river basin. In: Problems of Quaternary period studies. Khabarovsk (in Russian)

Kuznetsov NT (1962) River runoff in the Mongolian People's of Republic, vol 5. USSR Academy of Sciences, Moscow (in Russian)

Lehmkuhl F, Lang A (2001) Geomorphological investigations and luminescence dating in the southern part of the Khangay and the Valley of the Gobi Lakes (Central Mongolia). J Quat Sci (Published for the Quaternary Research Association) 16(1):69–87

Levintov ME, Kleiner UM (1978) Hypsometric characteristics of the territory of the Mongolian People's Republic. News Russ Geogr Soc 110(1):61–63

Lomborinchin R (1989) Exogenic processes of the Khangai and Khentii mountain ranges. Mongolian Academy of Sciences, Ulaanbaatar (in Mongolian)

Mikhailov NN, Ostanin OV (2002) Glaciers of the Southern Mongolian Altai and their changes in the twentieth century. Geogr Nat Manag Siberia. Barnaul. (5):3–20 (in Russian)

Munkhuu Z (1974) The structure and geomorphology of the Eastern Khangai. Institute of Geography. Mongolian Academy of Sciences, Ulaanbaatar (in Mongolian)

Murzaeva VE (1982) Depression of Great Lakes. In: Florensov NK (ed) Geomorphology of Mongolian People's of Republic, vol 28. Nauka, Moscow (in Russian)

Namnandorj O (1972) Geographical unique of Mongolia. State Publishing House, Ulaanbaatar [in Mongolian]

National Atlas of Mongolia (2009) In: Dorjgotov D (ed) Institute of Geography. The Academy Science of Mongolia. Ulaanbaatar (in Mongolian)

Nazag J, Chichagov V (1990) Depression. In: Tsegmid Sh (ed) National Atlas of People's Republic of Mongolia. Ulaanbaatar, pp 86–88

Nikolaeva NV, Shuvalov VF (1995) Landscape development in Mongolia in Mesozoic and Cenozoic. Geomorphology (2):54–65

Owen LA et al (1999) The landscape evolution of Nemegt Uul: a late Cenozoic transpressional uplift in the Gobi Altai, southern Mongolia. Geol Soc Lond: Spec Publ 162:201–218

Selivanov EI (1969) Aeolian sands of Mongolia and their origin. West, Moscow (in Russian)

Selivanov EI (1972) Neotectonics and geomorphology of the Mongolian People's Republic. State Publishing House, Nedra, 293 pp. (in Russian)

Ser-Od T (2019) Main characteristics of the landscape of the Mongol Altai Mountain Ridge and issues of its conservation and use. A Dissertation for Philosophical Doctor of Geography. Institute of Geography and Geoecology, Natioanl University of Mongolia, Ulaanbaatar, 148 pp (in Mongolian)

Siernev IP (1982) Plain of Eastern Mongolia. In: Geomorphology of the Mongolian People's Republic. Nauka, Moscow, pp 166–176 (in Russian)

Stolz C, Hülle D, Hilgers A, Grunert J, Lehmkuhl F, Dasch D (2012) Reconstructing fluvial, lacustrine and aeolian process dynamics in Western Mongolia. Zeitschrift für Geomorphologie 56(3):267–300

Timofeev DA, Chichagov VP (1974) Geomorphology of foreign countries. Nauka, Moscow (in Russian)

Tsegmid Sh, Vorovev (eds) (1990) The National Atlas of Mongolian People's of Republic. Mongolian Academy of Science, Academy of Sciences USSR, Ulaanbaatar, Moscow (in Russian and in Mongolian)

Tsegmid Sh (1951) The Physical geography and prehistorical glaciers of the Khentii mountain range. Moscow (in Russian)

Tumurtogoo O, Zaitsev NS (1990) Geology. In: Tsegmid S (ed) National Atlas of People's Republic of Mongolia. Moscow-Ulaanbaatar, pp 86–88

Vipper PB, et. al (1976) Experience of vegetation reconstruction in Western and Central Mongolia in the Holocene based on the study of fresh lake sediments. In: Structure and dynamics of key ecosystems of Mongolian People's Republic. Nauka, Moscow, vol 8 (in Russian)

Yembuu B (2014) Earth science, 5th edn. Mongolian Natioanal University of Education. Munkhiin Useg Publishing, Ulaanbaatar, pp 486 (in Mongolian)

# 4 Climate and Climate Change of Mongolia

Batchuluun Yembuu

**Abstract**

In brief, Mongolia's climate can be described as having a harsh continental character. Mongolia's climate varies considerably from region to region, not only due to latitudinal differences, but also because of altitude and vegetation cover, etc. This chapter begins with an overview of Mongolia's climate studies carried out by national and foreign scientists. It also shows the influencing factors on climatic features in Mongolia, its geographical location, remoteness from the oceans, topographical features, global circulation and their inter-relationships and the geographical distribution of the main climatic parameters, temperature and precipitation, atmospheric pressure and wind systems. Finally, it will provide an overview of climate change, its causes and consequences, as well as the future trend in the main parameters—air temperature and precipitation in Mongolia.

**Keywords**

Climate · Climate change · Annual mean temperature · Precipitation

B. Yembuu (✉)
Department of Geography, Mongolian National University of Education, Ulaanbaatar, Mongolia
e-mail: batchuluun@msue.edu.mn

## 4.1 Overview of the Climate Research

According to historical documents, the first weather station in Mongolia was installed in 1869, (Uliastai in 1879, Khovd in 1895–1897), and because of nomadic culture little information about the historical climate of Mongolia was recorded. However, a few facts about the weather in Mongolia recorded in ancient Chinese scripts and travelers' notes (Natsagdorj and Gomboludev 2018). For example, the famous Russian traveler N.M. Prejvalskii noted that "*Mongols have an extraordinary ability to predict strong winds, storms, rains and they can find their lost horses and camels with tiny signs and symptoms*" (Dash and Mandakh 2011). Nomads in Mongolia have created an entire lifestyle system based on their knowledge and tradition of weather forecasting. They would calculate weather, livestock and pasture (Shardar 1980; Khangaisaikhan 1989). The first books related to Mongolia's climate were published in 1880 by the Russian climatologist Voikov (East Asian climate) (Voikov1880), and Kamenskii (Northwest Mongolia's climate) in 1915 (Jambaajamts 1989).

Climatic studies in Mongolia have been initiated by Russian scientists since the early twentieth century, such as the temperature inversion of Shastakovich (1928), the influence of the relief on the temperature distribution by Kondratev (1929), the wind regime of Kaminskii,

© Springer Nature Switzerland AG 2021



B. Yembuu (ed.), *The Physical Geography of Mongolia*, Geography of the Physical Environment,
https://doi.org/10.1007/978-3-030-61434-8_4

etc. (Natsagdorj and Gomboluudev 2018). As noted by Douglas (1912), "*Potanin records the meteorology during 23 days from 27 June to 19 July, on the east side of the Turgen mountains. Out of 23 days, 7 were fine or clouded, and 16 were wet. There were 11 thunderstorms and twice it hailed*" (Carruthers 1912). The paleogeographic studies of Mongolia were conducted by B.M. Sinits, G.E. Grum-Grjimailo and his followers H.B. Pavlov, B.A. Smirnov, climatologist L.S. Berg, botanist A.A. Yunatov and others. According to these studies, B.M. Sinits and G.E. Grum-Grjimailo concluded that droughts had increased in Mongolia due to increased mountain elevation, and L.S. Berg and A.A. Yunatov argued that the weather in Mongolia was not dry at the time (Jambaajamts 1989).

At the national level, climate research in Mongolia was launched in 1924 at the initiative of Tseveen Jamsran. By 1926, several weather stations had been established in Ulaanbaatar, Uliastai and Zamiin-uud (Jambaajamts 1989), with 22 observation stations and 2 weather posts operating in 1947. Since then, Mongolia's meteorological posts have been merged with the international meteorological network (Natsagdorj and Gomboluudev 2018).

Systematic meteorological observations in Mongolia began in 1940, together with aerological observations. The first overview of the Mongolian climate was made by the Russian geographer E.M. Murzaev in his famous work "Mongolian People's Republic" in 1952. He compiled the first climatic map of Mongolia, described and explained the geographical distribution of meteorological parameters and temperature inversion in winter. This was the main systematic source for further climate research in Mongolia. After Murzaev, a more detailed description of the Mongolian climate made by the Russian scientist E.P. Arkhipova in 1945 (Arkhipova 1945).

After the establishment of the National Institute of Meteorology in 1966, national scientists intensified climate research, such as the distribution of elements (Badarch 1969), applied climatology, climate of natural zones, climate classification, snow cover, aeroclimatology (Gira 2004), and so on. The data on extreme climatic events and climate change in eastern Mongolia since 1740 was collected by D. Tsedevsuren, which contributed to the work on historical climate research in Mongolia. Since 1961, an actinometric network has been started, and in 1970 Mongolia received satellite information and images in the polar orbit. A digital information station had been installed in 1986–1988, and greenhouse gas monitoring was started since 1992. Since 2007, Mongolia has been receiving images from the MODIS satellite (Natsagdorj and Gomboluudev 2018).

The study of climate change in Mongolia began in 1980, following the first symposium held by the Institute of Meteorology and Hydrology (Batima 2006). The first articles on climate change in Mongolia were published in the 1980s. Since then, various studies on climate change have been initiated, such as the use of tree ring (NAMHEM 2012), snow line and moraine analyses; assessment of atmospheric-vegetation interactions; assessment of climate change risk (Dagvadorj et al. 2009); Natsagdorj 1980), and the climate trend up to 2020 (Mijiddorj 1990). Future climate change in Mongolia until 2100 was determined based on scenarios of the SRES 2000 using simulations of the Global Climate System (Gomboluudev 2007) (Natsagdorj and Gomboluudev 2018). In addition, a number of studies have been conducted on the impacts of climate change, such as drought and zud on livestock production (Natsagdorj and Sarantuya 2011); the pasture-carrying capacity for evaluating zud risks (Altanbagana 2013), the drought and zud[1] assessment (Natsagdorj 2014) and historical climate change (Mijiddorj 2012), natural disasters analysis (Munkhtseren et al. 2015) and so on. The satellite-based vegetation optical depth (VOD) in certain areas (Liu et al. 2013) and Normalized Difference Vegetation Index (NDVI) used to study climate change-induced land cover changes (Jambaljav et al. 2016).

[1]Zud (Dzud) is a Mongolian term for severe winter (white zud) and harsh spring (black zud or without snow)

## 4.2 Climate Formation Factors

The integrated factors of geographical location, incoming solar radiation, air circulation and topography influence climate formation. The climatic characteristics of the country are remarkably different due to the size, complexity of its territory and topography and global air circulation. Thus, climatic variations are characterized by a regional feature with large temperature fluctuations, and low precipitation compared to areas in the same latitudes. The climatic features of Mongolia are characterized by distinctly marked four seasons with short dry summers (from June to mid-August), long and cold winters, spring and autumn, resulting in high fluctuations in air temperature and precipitation. In addition, latitudinal and altitudinal distinctions are clearly visible in the country. Winter lasts at least four months (November–April), with average temperatures below −18 °C, and the daily maximum exceeding 35 °C in mid-summer.

According to climatologists and geographers (Murzaev 1952; Badarch 1971; Tsegmid 1969; Jambaajamts 1989; Natsagdorj and Gomboluudev 2018), the actual nature of the climate in the country is determined by the interaction of three main factors: geographical location; general atmospheric circulation; and elevation of relief and topography differences.

### 4.2.1 The Effect of Geographical Location

Mongolia's harsh continental climatic conditions are primarily due to its geographical position in the middle latitude (from 42°45′N to 52°35′N) of the Northern Hemisphere. Due to the latitudinal factor, the main climate parameter, the air temperature, is influenced by the incoming solar radiation. Low level of humidity and precipitation caused by the landlocked location, in the center of the Eurasian supercontinent with isolation from the oceans. In addition, the dry continental climate type is significantly determined by the surrounding high mountains with an average altitude of 1,500 m above the sea level. The country is isolated at a distance of 3,000 km from the Arctic Ocean, 5,000 km from the Mediterranean Sea, and 1,600 km from the Pacific Ocean.

Despite the same location of latitudes, Mongolia's climate is completely different from that of Western Europe with its oceanic type of climate. For example, in Bulgan, Mongolia is located at the same latitude as the French capital Paris 48°48′N. However, the average annual temperature in Paris is 10.6 °C; in January it is +3.4 °C, and in Bulgan −1.3 and −20.5 °C, respectively. The amount of annual precipitation in Bulgan is 279 mm, which is two times less than in Paris. Therefore, according to the Koppen-Geiger classification, Cfb is for Paris and Dwc for Bulgan, Mongolia.

### 4.2.2 The Effects of Elevation and Topography

The territory of the country is surrounded by numerous mountain ranges (Sayan in the north, Altai in the west, and Khyangan in the east) and is elevated above the sea level by an average of 1.580 m. In terms of average altitude, it is almost eight times higher than Eastern Europe (170 m a.s.l. on average) and twice as high as that of the North American plateau in the same latitude. Thus, the mean temperature in Mongolia, determined by the elevation factor, is lower than in these places. The complex of topography (altitude and surface diversity) plays an important role in the climate formation in Mongolia. In particular, the distribution of air temperature and the precipitation patterns are affected by topographical diversity; annual precipitation decreases from northwest to east and south.

The mountain ranges are natural barriers to air moisture through westerlies and change the direction of the wind. The prevailing wind direction in Mongolia is influenced by the Altai and Khangai ranges extending from northwest to southeast. Westerlies therefore change their original direction from southwest in middle latitude, with the prevailing winds blowing from the northwest in most of the territory. The slope aspect also influences the temperature, precipitation patterns, and wind conditions. The northern shadowy slopes of the mountains are colder than the southern slopes and have more precipitation covered by forests. Due to the different climatic conditions of the slopes, nomads establish a winter camp on the warmer southern and leeward slopes of Mongolia (Bazargur 2008).

The effect of mountains in the cold season, that is the temperature inversion, is created by radiation cooling. Due to the intensive inflow of cold air from the mountains to the valleys in winter, low-layer surface inversions with a thickness of 600–1,000 m and an intensity of 6–17 °C are observed. The largest inversion is observed in the basins of large rivers, such as Tes and Selenge, and depressions with low elevation between the Altai and Khangai ranges. The largest (13.0 °C) and the deepest (2,037 m) inversion observed at Ulaangom in January, near Uvs lake, is surrounded by mountains (Erdenesukh 2008). Different mean annual air temperature is observed at the same latitudes depending on the relief height: in Baruun-Urt (980 m) 0.5 °C, in Galuut (2,126 m) −4.6 °C (difference 5.6 °C), in Khujirt (1,662 m) 1.8 °C, and in Bayanbulag (2,258 m) −4.6 °C.

### 4.2.3 The Air Circulation Effects

The main air circulations playing on the Mongolian climate are the westerlies, cyclones, and anticyclones (Tsegmid 1969; Murzaev 1952). The real character of the Mongolian climate is determined by the interaction of several air circulation and, as a consequence, by a strongly varying pattern of surface pressure (Jambaajamts 1989; Natsagdorj 1980). As a rule, Mongolia is located in the middle latitude in the northern hemisphere and its climate is influenced by westerlies. Most of the moisture comes from the Atlantic Ocean through the westerlies with partial inflow from the northern seas (Natsagdorj 1980; Tuvdendorj and Myagmarjav 1986; Dagvadorj et al. 1994; Mijiddorj 2000).

In winter, due to the predominance of anticyclonic weather, air temperature is mainly dependent on radiation conditions and the air is cooled down above the underlying surface. Extremely dry air causes a strong surface radiation cooling and temperature conditions. In winter, most of Mongolia's territory is under high pressure, including the stable so-called Siberia-Mongolian High (SMH). Sometimes with a short interruption due to occasional cyclonic activity (Badarch, N, 1969; Tsegmid, Sh, 1969b; Natsagdorj and Gomboluudev 2018), 50–60 % of the air masses arriving in winter have an Arctic origin (SMH). During the SMH, in depression areas such as Murun and Ulaangom where extremely low temperatures (−40 to −50 °C) are observed without air exchange, sometimes it is recorded below −50 °C (in Tosontsengel). During the Mongolian High (by Natsagdorj, L) with the center of the Great Lakes depression in Mongolia (Badarch 1971), its frequency is 35 %. The Mongolian High period was dominated by clear, stable weather in winter throughout the country.

In the warm season, due to the conventional process, the subtropical air mass moves from south to north. In summer, cyclonic air masses prevailed for three months. The sub-tropical air masses are of continental origin, operating throughout the summer and causing hot weather. So, in summer the whole country is controlled by a low-pressure cyclonic system (Badarch 1971). Any insignificant change in the intensity of cyclonic activity trajectories will have a direct impact on precipitation patterns in most of Mongolia (Murzaev 1952; Badarch 1969). The average air pressure in summer is 770–920 hPa.

As for the monsoons, scientists have different points of view, but still questionable. Some scientists—Russian climatologist Voikov (1880), Berg (1925), national scientists Namnandorj (1963), Shagdar (1967), Badarch, (1971)—believe that the transfer of humid air with Pacific monsoons reaches the eastern part of Mongolia through the Khyangan mountains. There is evidence that precipitation in the Numrug and Khalkhgol river basins is higher than in the surrounding areas, so Mongolia's climate is affected by summer monsoons in East Asia and Indian monsoons (Sato et al. 2004; MNET 2009; Natsagdorj and Gomboluudev 2018). Sato et al. (2014) noted that "the moisture transport along the southwesterly wind of the Asian summer monsoon has difficulty reaching Mongolia as a monthly/seasonal mean perspective. However, eastern Mongolia is situated on the border area between westerly wind, moisture transport by mid latitude synoptic cyclones and southerly wind moisture transport by Asian summer monsoon" (Sato et al. 2004). However, the air moisture flow for Mongolia from the Pacific and Indian oceans has not yet been officially registered. Jambaajamts (1989) explained that this is not Pacific monsoon effect; the reason for the increase in precipitation around this area is the condensation of air moisture by creating air lifts on the mountain slope of the Khayngan ridge (Jambaajamts 1989; Murzaev 1952).

## 4.3 Spatial Characteristics of the Main Climatic Parameters

### 4.3.1 Spatial Differences in Solar Radiation and Sunshine

The vast size of the territory can significantly diversify the spatial patterns; 60–70 kcal/cm$^2$ (2,500–2,900 MJ/m$^2$) of solar radiation per year in northern part (north of 48°N) and 85–90 kcal/cm$^2$ (3,560–3,800 MJ/m$^2$) in southern part (south of 45°N). This is about 5–6 kcal/cm$^2$ (209–250 MJ/m$^2$) per one degree of latitude (Natsagdorj and Gomboluudev 2018).

The spatial difference in solar radiation within the country is caused by the vast size of the territory. According to the Jambaajamts' (1989) calculations, the total amount of global radiation in Mongolia is 120 kcal/cm$^2$ (5,028 MJ/m$^2$), 130 kcal/cm$^2$ (5,447 MJ/m$^2$), and 140 kcal/cm$^2$ (5,866 MJ/m$^2$) at 47, 44, and 43 N, respectively (Fig. 4.1). Thus, monthly temperature fluctuations between the north and south of the country are about 10 °C. For example, the average temperature in July in Rinchinlhumbe (51°7′N), which is located in the north, is about 20 °C, and in Zamiin-Uud (43°44′N) which is located in the south is 30 °C, and in January is −32 and −18 °C, respectively. The reason for this is the difference of the sun altitudes caused by latitudinal variations (Fig. 4.2).

The spatial distribution of sunshine shows that it is 10–15 % higher in the southern Mongolia than in the northern or Russian border areas. The highest values in the southern part of Mongolia can exceed 5,866 MJ, while the lower values of less than 5,000 MJ are characteristic of the western and northern regions. The amount of energy in the summer months is 5–6 times higher than that in the winter months.

The minimum solar irradiation in December is 3.0–3.4 kcal/cm$^2$ (125–129 MJ/m$^2$) in south and 0.9–2.0 kcal/cm$^2$ (37.8–83.8 MJ/m$^2$) in north (Natsagdorj and Gomboluudev 2018). The spatial distribution of sunshine shows that in June it can exceed 10 kcal/cm$^2$ (419 MJ/m$^2$) in the south and 8.0 kcal/cm$^2$ (335 MJ/m$^2$) in the north and west. However, it may differ depending on cloudiness. For instance, in December, direct radiation in Murun is 1.2 kcal/cm$^2$ (50.1 MJ/m$^2$), and in Ulaangom it is 0.9 kcal/cm$^2$ (37.8 MJ/m$^2$), where cloudiness is higher (Natsagdorj and Gomboluudev 2018), even these two stations are located at almost the same latitude.

The highest amount of horizontal solar irradiation in June is 3–7 times greater than in the winter months. Seasonal variations in direct solar radiation in the southern part of the country are lower than in the northern part. Sparsely solar radiation is 1–3 kcal/cm$^2$ (41.9–125.5 MJ/m$^2$) in winter, while 4–9 kcal/cm$^2$ (187–376 MJ/m$^2$) in summer. In southern part of the country it reaches 40–50 kcal/cm$^2$ (1,875–2,090 MJ/m$^2$).

The number of sunny days in Mongolia is high, averaging 230–260 days, and the duration of sunshine varies between 2,600 and 3,300 h per year (MEGD 2014). The highest values are in the eastern part (3,118 h in Sainshand; 2,906 h in Choibalsan), and the lowest in the northern part (2,794 h in Khovd) (Tsegmid 1969). The sunshine hours in the country increase from north to south at any time of the year, although this depends on cloudiness and seasons (Tsegmid 1969). For example, in Murun (49°38′N) it is

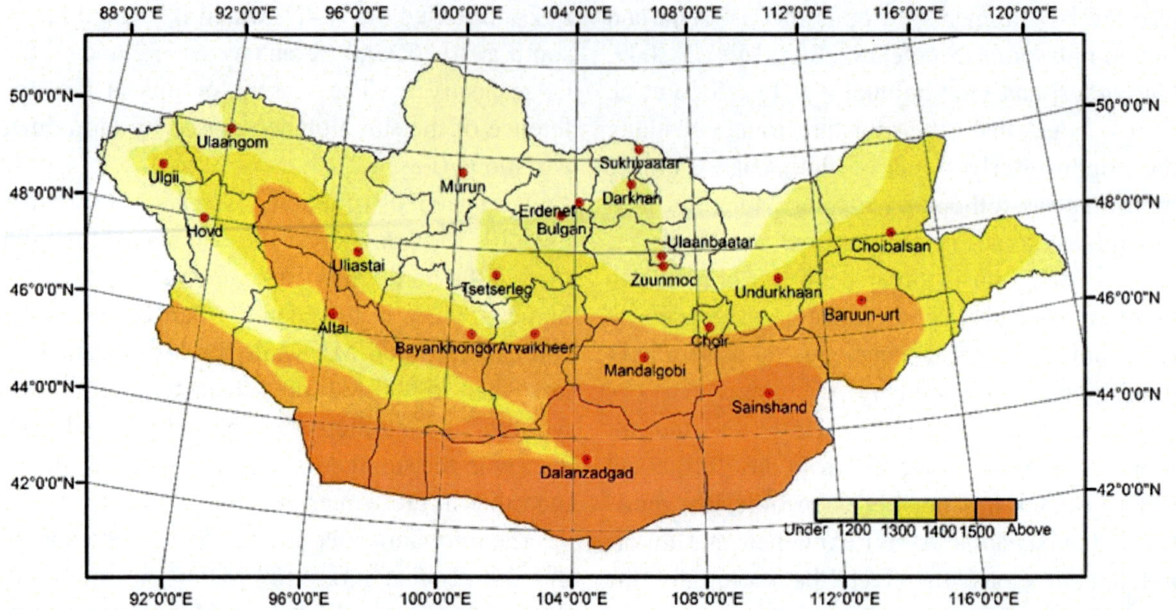


**Fig. 4.1** The spatial distribution of the yearly amount of global irradiation (kWh/m$^2$) (based on National Atlas of Mongolia 2009)

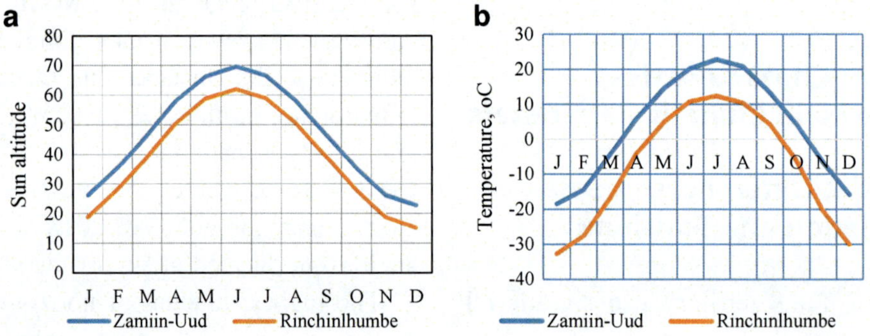


**Fig. 4.2** **a** Variations of the sun altitude at solar noon, **b** Annual mean temperature at two stations: Rinchinlhumbe (51° 7′N) and Zamiin-Uud (43°44′N)

140–280 h per year, and in Sainshand (44°54′N) it is 220–350 h (Fig. 4.3).

### 4.3.2 Regional Distribution of Air Temperature

The air temperature pattern is determined by large-scale climatic factors (e.g., incoming solar radiation, major pressure systems, terrain diversity, etc.). Regional variations in air temperature are significant in the country, determined by local orography, land cover, and so on. Long-term average annual temperature in Mongolia is almost universally negative and is characterized by huge seasonal and regional variations. As a rule, it increases with decreasing latitudes. Based on measurements (1981–2015) from 71 weather stations from various natural zones Natsagdorj et al. (2018) alculated that the average annual temperature in Mongolia is 0°5C (± 0.8). The western and northern mountainous regions of the country are characterized by lower average annual temperatures (−4 to −6 °C), while the southern and eastern regions have average annual temperatures ranging from 2 to 4 °C. The highest average annual temperatures (6 °C) are in the Gobi Desert region. The observed minimum

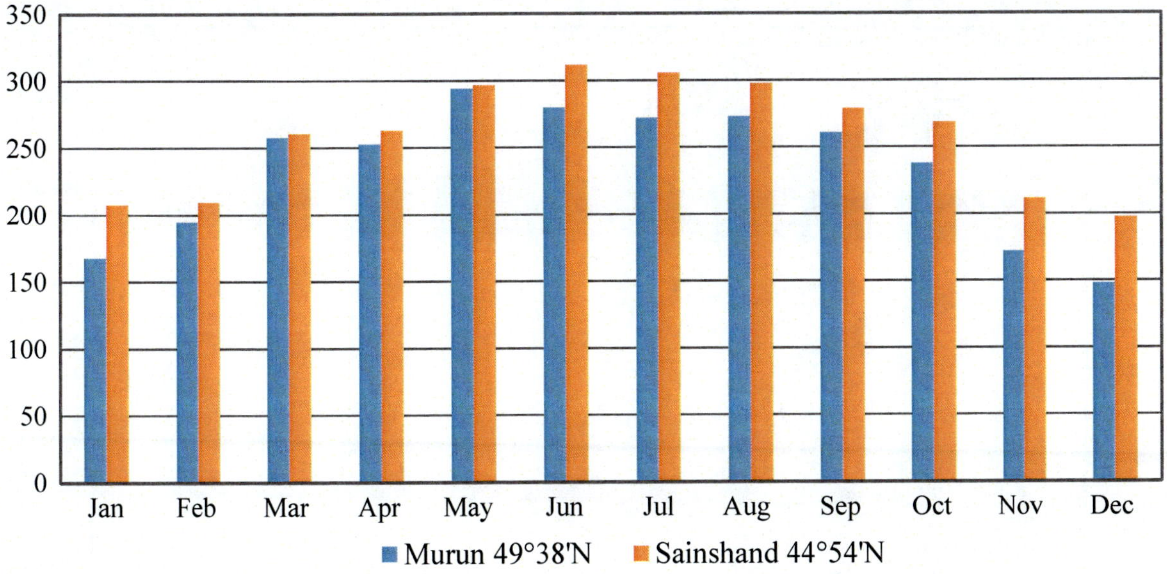


**Fig. 4.3** The variation of the sunshine hours depends on the latitudes (Murun and Sainshand)

annual mean temperatures (below −6 °C) are in the valleys of large rivers and intermountain depressions. Measurements were made at −8.4 °C in the Darkhad Depression (MEGD 2014; Natsagdorj and Gomboluudev 2018). Average annual 0 °C temperature isoline is located approximately at the northern boundary of the desert steppe, in the Gobi region of Mongolia along the northern latitude of 46°. In addition, permafrost is common in areas where the average annual air temperature is below −2 °C (Shagdar 1999).

The average air temperature in the coldest month of January (Fig. 4.4) fluctuates from −30 to −34 °C (high mountains), from −25 to −30 °C in the depressions of mountains, and from −20 to −25 °C (steppe regions). The temperature −15 to 12 °C (southern Gobi) corresponds to the 40-year average temperature of 1961–2011. In the mountains, the temperature drops sharply from −30 to −34 °C in January and rises to +15 °C in July (Dagvadorj et al. 2009; MEGD 2014; Natsagdorj and Gomboluudev 2018).

Multiyear average annual temperature in January is −26 °C in Ulaangom (939 m a.s.l) and in Dalazadgad (1,465 m a.s.l) −8 °C, which means fluctuations of 18 °C between the two stations (Fig. 4.5). The main reason for this is the latitudinal variations of these places.

The average air temperature in July (Fig. 4.6), the warmest month, ranges from +15 to +10 °C in the lower part of the mountains and from +15 to +20 °C in the basins of large rivers, and in the southern part of the steppes and the Gobi Desert between 20 and 25 °C. The average of maximum air temperature ranges from 28.5 to 44.0 °C. On the meteorological record of the 1940s, the absolute maximum temperature (44 °C) was recorded in Khongor soum, Darkhan Uul aimag on 24 July 1999 (Batima et al. 2005), the absolute minimum temperature (−55.3 °C) was in ZuunGobi soum, Uvs aimag on 22 December 1976, and −49.0 °C in Ulaanbaatar (1954) (Jambaajamts 1989; NAMHEM 2012; MNET 2009). Average annual temperature fluctuations exceed 70 °C throughout the country (NAMHEM 2012), indicating the harsh continental climate.

During the warm season, a normal vertical temperature gradient (0.65 °C/100 m) occurs, which causes the temperature to drop with height. In the cold season, almost daily temperature inversion is observed with a thickness of 600–1,000 m and intensity of 6–17 °C. As a result, air temperature increases by 0.8–0.9 °C/100 m in river valleys and intermountain lowlands (MEGD 2014). In the highland regions, the cold season ends late in spring (mid-June)

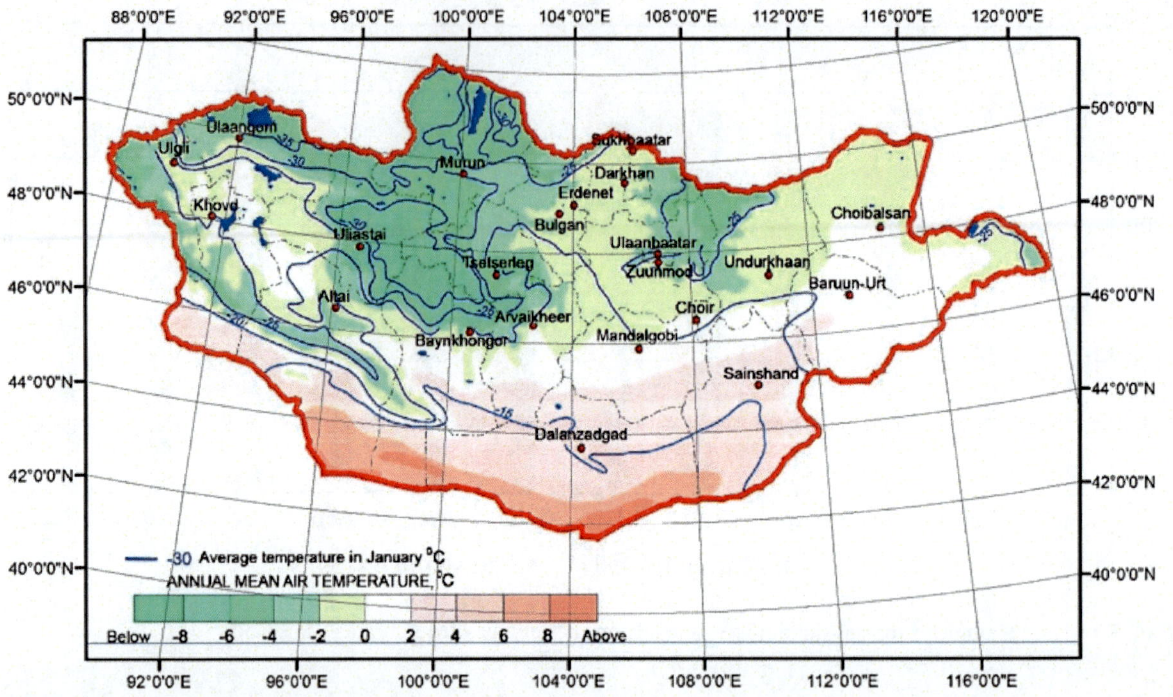


**Fig. 4.4** Geographic distribution of mean air temperature in January (modified based on National Atlas of Mongolia, 2009)

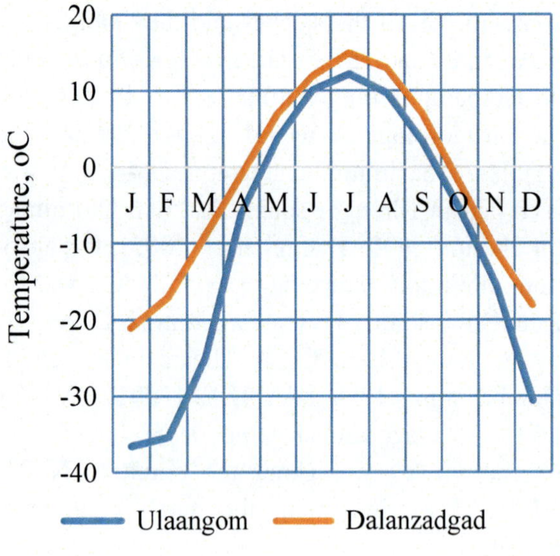


**Fig. 4.5** The regional variations of the average minimum temperature (°C): Ulaangom and Dalanzadgad

and early in the cold season (mid-August), leaving only 70 days for a warm period. In the rest of the country, warm days last 90–130 days. The number of days when the average daily temperature exceeds 10 °C is related to latitudes and topography (Fig. 4.7). These are 90 days in the mountainous areas, which rise by more than 2000 m, 100–130 days in the steppes and 130–150 days and more in the Gobi Desert (WWF Mongolia 2011). The total active temperature exceeding 10 °C is 3,000–3,500 °C and more in the desert and semi-desert areas (eastern and southern parts), 2,000–2,500 °C in steppe zone, and 1,000–500 °C and less in mountainous part of Mongolia.

### 4.3.3 Seasonal and Spatial Distribution in Precipitation

In Mongolia, frontal precipitation prevailed due to the cyclonic patterns, while orographic precipitation was occurring only in the mountainous part of the country. The interaction between atmospheric circulation (sub-tropical air mass and the period of cyclone dominance) and topographical factors in Mongolia leads to quite sharp contrasts in terms of precipitation and air humidity. The country's isolation from oceanic moisture transport has an impact on the

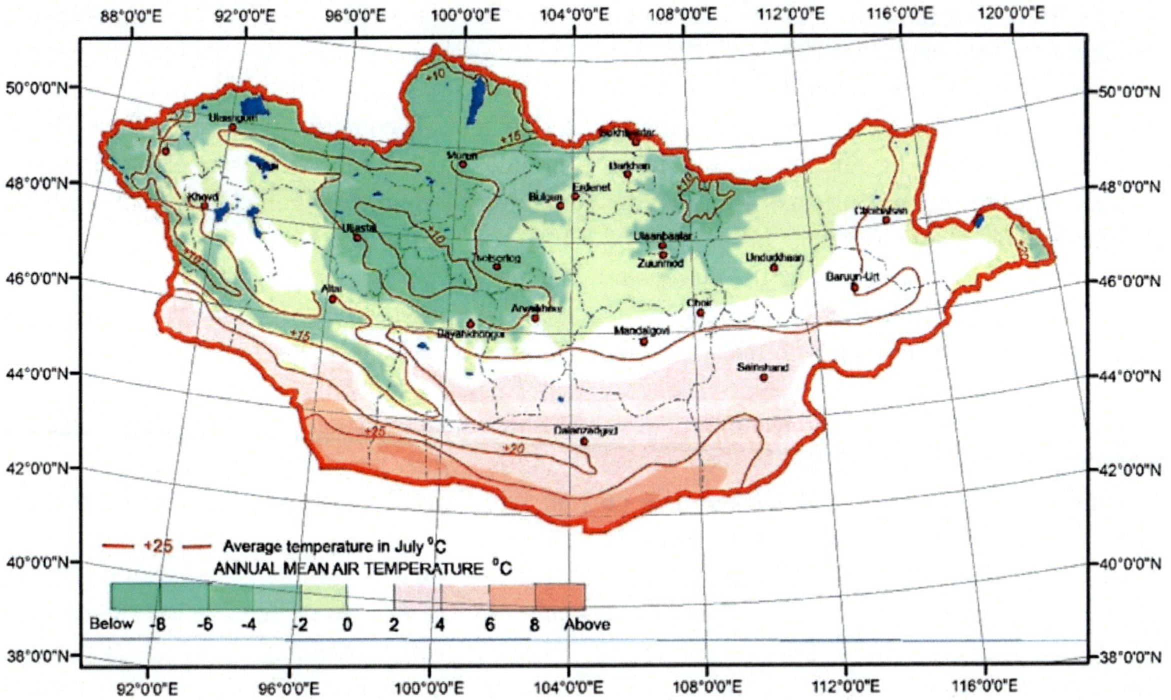


**Fig. 4.6** Geographic distribution of mean air temperature in July (modified based on National Atlas of Mongolia (2009))

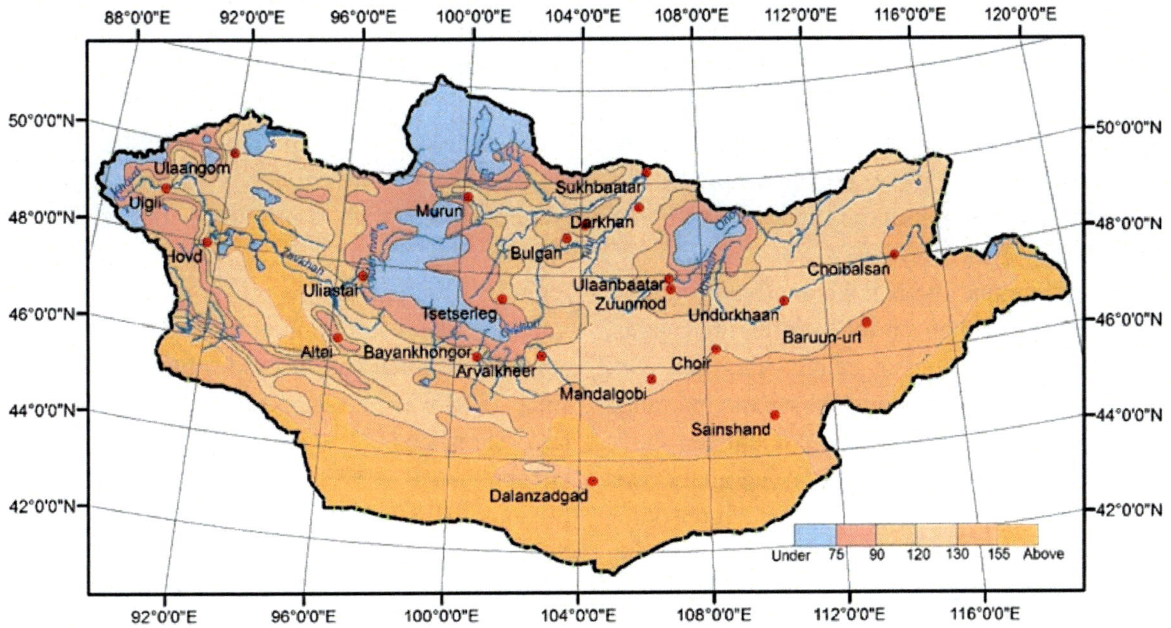


**Fig. 4.7** Dates of daily mean temperatures exceeding 10 °C (adopted from National Atlas of Mongolia 2009)

distribution of precipitation. The average annual amount of precipitation changes depending on latitude and is twice as much in the mountainous regions of the north than in the southern plains. The distribution of precipitation is extremely unbalanced both by seasons and years. The maximum number of precipitations falls in the summer months, mainly in July–August. Summer precipitation is 50–70 % per year. In winter, the high humidity and cold Arctic air mass loses moisture over Siberia and gets to Mongolia, turning into a dry continental (Badarch 1971).

Precipitation in the mountains largely depends on both the absolute height and the orientation of the mountain chains in relation to the main moisture transporting air currents. Thus, in the mountains of the Mongol Altai, the amount of annual precipitation reaches to 250–300 mm at absolute heights exceeding 3,000 m. The most humidified areas of slopes open to the west, that is, receiving some more moisture transported from the Atlantic Ocean. In the Khangai mountains, this amount of precipitation is recorded at absolute altitudes exceeding 2,000 m, and in Khuvsgul and Khentii ranges this level is reduced to heights of 1,000–1,500 m.

The diversity of topography (e.g., location and direction of mountain ranges, slopes, their absolute height, etc.) plays a significant role in the spatial distribution of precipitation. The total annual average precipitation is relatively low in the country as a whole and decreases from north to south with an increase in the continental character (Fig. 4.8); 300–400 mm in the Khangai, Khuvsgul, Khentii mountain ridges and the large river basins; 250–300 mm in Mongol Altai and forest steppe zone, 150–250 mm in the steppe region and 150–50 mm in Gobi (MNET 2009; MET 2017; Natsagdorj and Gomboluudev 2018) (Fig. 4.9).

Rainfall around the Khalkh and Numrug river basins is greater than on the surrounding plains, due to the effects of rising air cooling and condensation of the Khyangan mountain range (Jambaajamts 1989). Some scientists have explained that the precipitation anomaly was previously called the "Pacific monsoon", although this is not possible. Air may be coming from the Pacific Ocean in the form of a monsoon, but air humidity must be lost over the Khyangan ridge (Jambaajamts 1989).

The temporal distribution of precipitation in Mongolia is well differentiated by season: the majority (about 85 %) of annual precipitation falls during the warm season (April to September), due to the activity of warm air masses and the cyclonic characteristics of the low-pressure atmosphere. The annual maximum precipitation falls on July–August, or 50–60 % of total precipitation, only for these two months. In winter, during the dominant anticyclone period throughout the country, humid and cold arctic air through Eastern Siberia was transferred to dry continental air in Mongolia. Therefore, only 5–10 % of total precipitation falls in winter, with average snow cover ranging from 5 to 10 cm or less. In spring, when the air warms up, the cyclonic system develops under the influence of sub-tropical dry air coming from the south and meeting with moderate air of middle latitude. In addition, when the stable cyclonic system peaks in July, precipitation peaks at every part of Mongolia.

In Mongolia as a whole, there is a low level of humidity, as the rate of evaporation of soil (potential evapotranspiration) far exceeds the amount of precipitation throughout the country. The high evapotranspiration potential is 800–1,000 mm in southern Mongolia (Gobi Desert) and less than 500 mm in mountainous areas. It is 650–750 mm in the eastern part or steppe zone (Badarch 1971; Jambaajamts 1989; MEGD 2014). The total annual evaporation rate depends not only on rainfall but also on soil moisture.

The amount of precipitation is affected by the slopes. The windward slopes of the mountains were covered with forest and grasses due to increased precipitation in the northwestern regions. Although annual precipitation is low, the intensity of precipitation is high. The maximum amount of precipitation (138 mm/day) recorded since 1940 was in Dalanzadgad (5 August 1956) and Sainshand (121 mm/day, 11 July 1976) (Jambaajamts 1989) (Nazagdorj and Gomboludev 2018). These two records are found in desert areas of Mongolia. Rainfall days decrease from the north of the mountainous areas (60–70 days) to the east (40–60 days) and to the Gobi Desert (less than 30 days).

### 4.3.4 Snow Cover

The average depth of the snow cover is usually low in Mongolia due to the anticyclonic system and the penetration of the Arctic continental air

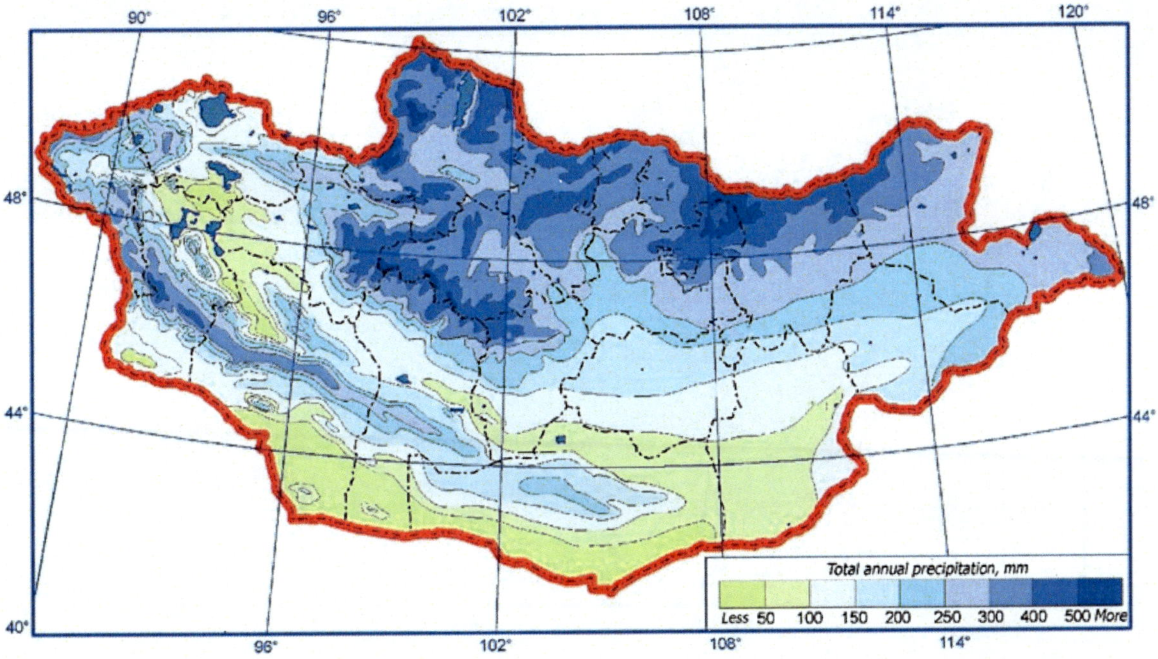


**Fig. 4.8** Spatial distribution of the mean annual precipitation between 1960 and 1990 (by D. Otgonbat)

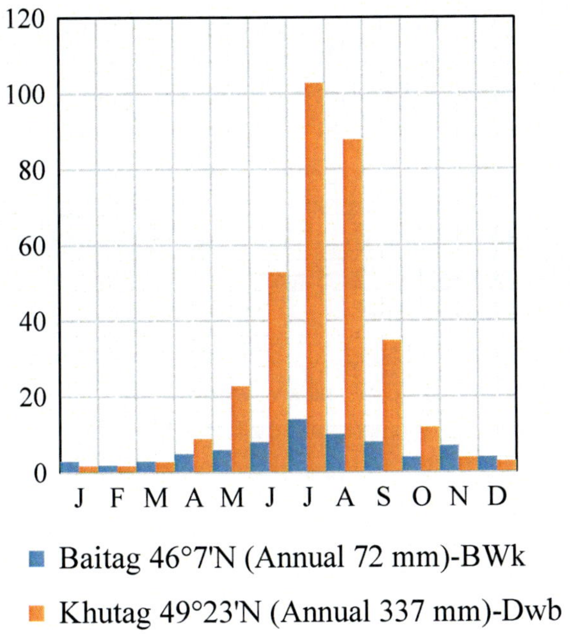


**Fig. 4.9** Spatial differences in precipitation between Khangai (Baitag) and Trans-Altai Gobi (Baitag) regions

mass in winter. However, periods of stable snow cover are characterized by low air temperatures for a long time. In the cold season, about 10 mm of snow falls in the Gobi region, in the mountainous area, and in the Uvs lake basin of 20–30 mm, and the rest of the country gets 10–20 mm of snow, respectively. The average height of snow cover in the mountains is 5 cm (maximum 30 cm) and in the steppes 2–5 cm (maximum 15–20 cm). The period of stable snow cover (Fig. 4.10) is defined as consecutive 30 days or more with snow cover and the area covered by snow is 50 % or more of total area (Jambaajamts 1989). From the northwest (more than 150 days) to the east (80 days) and south (less than 50 days in Gobi), the number of days of stable snowfall decreases (Natsagdorj and Gomboluudev 2018).

### 4.3.5 Spatial and Temporal Distribution of Air Pressure

Mongolia's atmospheric pressure regime was influenced by general air circulation, Arctic air front shifts, cyclones and anticyclone systems and seasons. In winter, Mongolia is dominated by a stable anticyclonic system [Asian maximum or Siberian-Mongolian High (SMH)], which is due to the penetration of cold air into the Arctic continental part of the Arctic. Its maximum development occurred in January, and the

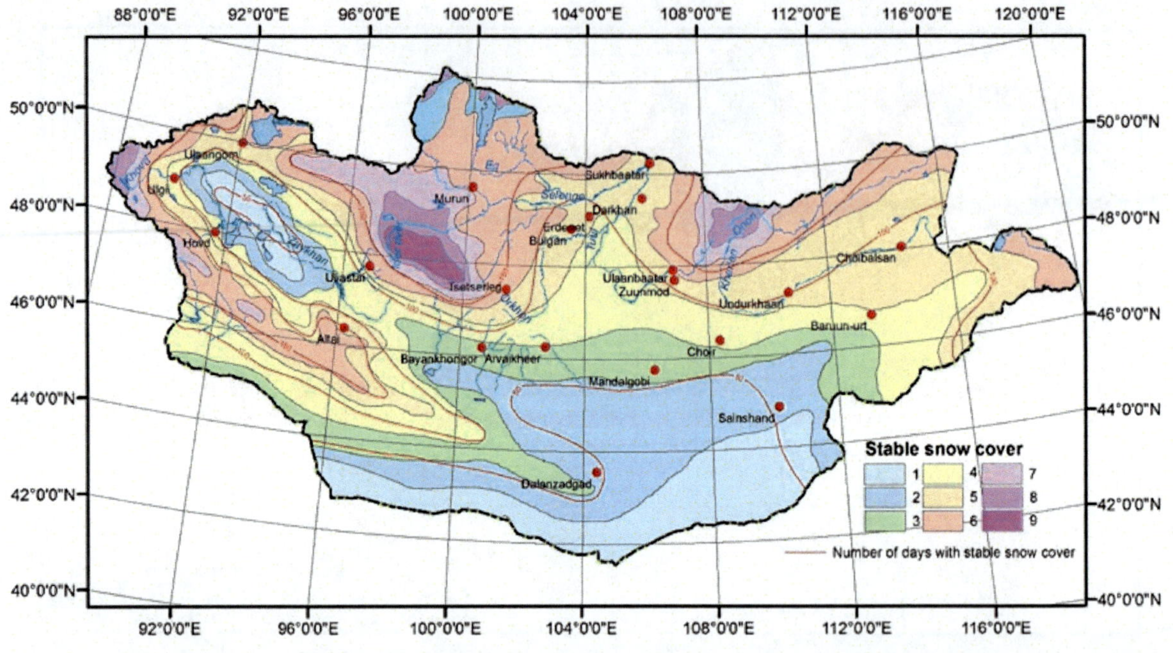


**Fig. 4.10** Stable snow covered days (adopted from National Atlas of Mongolia, 1990, 2009): **1**—late than Dec 20 to earlier than Feb 10; **2**—Dec 10-Feb 20; **3**—Dec 01-Feb 20; **4**—Nov 20-Mar 10; **5**—Nov 15-Mar 20; **6**—Nov 10-Apr 01; **7**—Nov 05-Apr 10; **8**—Nov 01-Apr 20; **9**—earlier than Nov 01 and late than Apr 20

pressure reaches to more than 1,050 hPa, where it concentrated in the northwest of Mongolia (Fig. 4.11). In spring, when air temperature rises, the effect of Asian maximum weakens. This transition of Asian maximum to cyclone characterizes the spring instability of weather conditions in Mongolia. During this period, especially in April and May, very variable and windy weather prevails throughout the country. In addition, the cyclonic system in summer is under the influence of hot air mass from the southern deserts (Fig. 4.12). Thus, Mongolia has two different atmospheric pressure systems, which vary considerably due to physical geography, such as surface height and topographic diversity with high amplitudes.

According to long-term meteorological measurements, annual air pressure ranges from 821.7 hPa (Arvaikheer 1,846 m a.s.l.) to 935 hPa (eastern plain 683 m a.s.l.). The range 1,018 (Choibalsan) to 1,027.4 hPa (Ulaangom), however, has a large variety of spatial distribution; 1,018.3–1,023.5 hPa in the northern and mountainous regions, and 1,013 hPa in the southern part because of the flat surface and relatively warm climate. The highest pressure observed in the Depression of Great Lakes (Ulaangom) was 1,054.7 hPa due to the closed depression in weakly windy weather and latitudinal factors. As evaluated by the WMO, the record sea-level pressure (SLP) of 1089.4 hPa as of 30 December 2004 was measured in Tosontsengel (1724.6 m) (Purevjav et al. 2015), the coldest soum center (48°44′N, 98°16′E) in north-central Mongolia. Synoptic analysis shows that in cold-deep snowy winters, the 500 hPa trough is located in eastern Mongolia and extends to the west (Koike et al. 2010)

### 4.3.6 Wind Regimes

Wind conditions in Mongolia are mainly determined by topographical variations as well as seasonal shifts of cyclones and anticyclones. Seasonal variation in annual wind speeds is determined by the pressure system. Mongolia generally has two maxima and two minima of wind speed. The main wind maximum is in spring, May–June and in autumn September–November. The minimum wind speed in summer (July–August) and in January is due to the

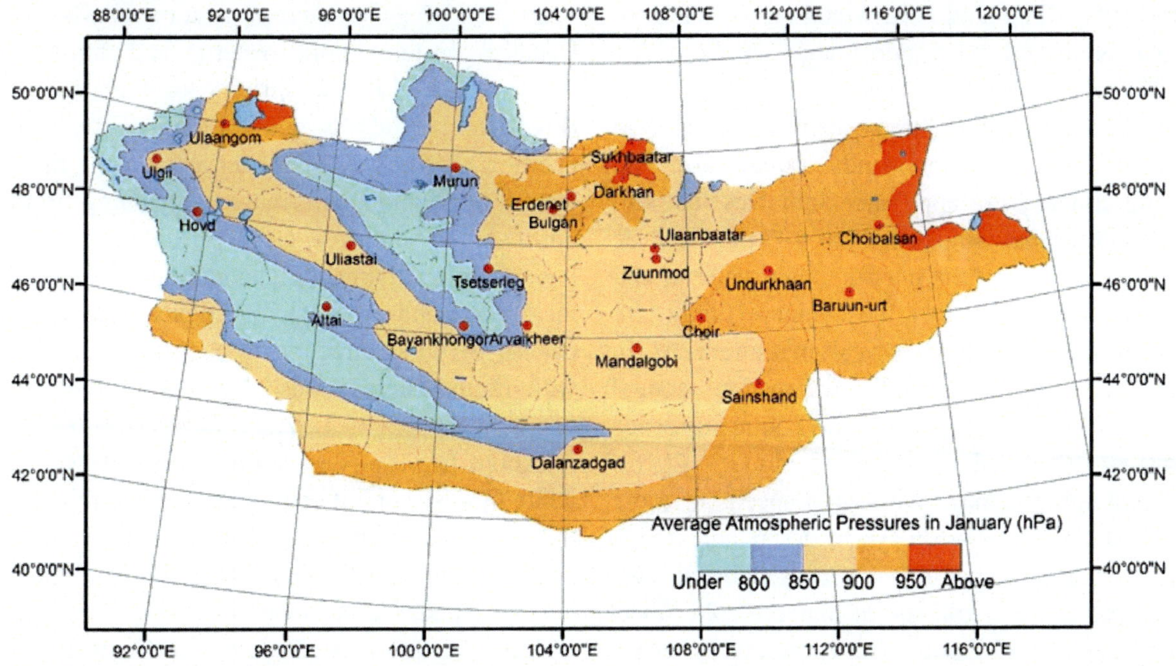


**Fig. 4.11** Spatial distribution of air pressure in January (adopted from National Atlas of Mongolia 1990, 2009)

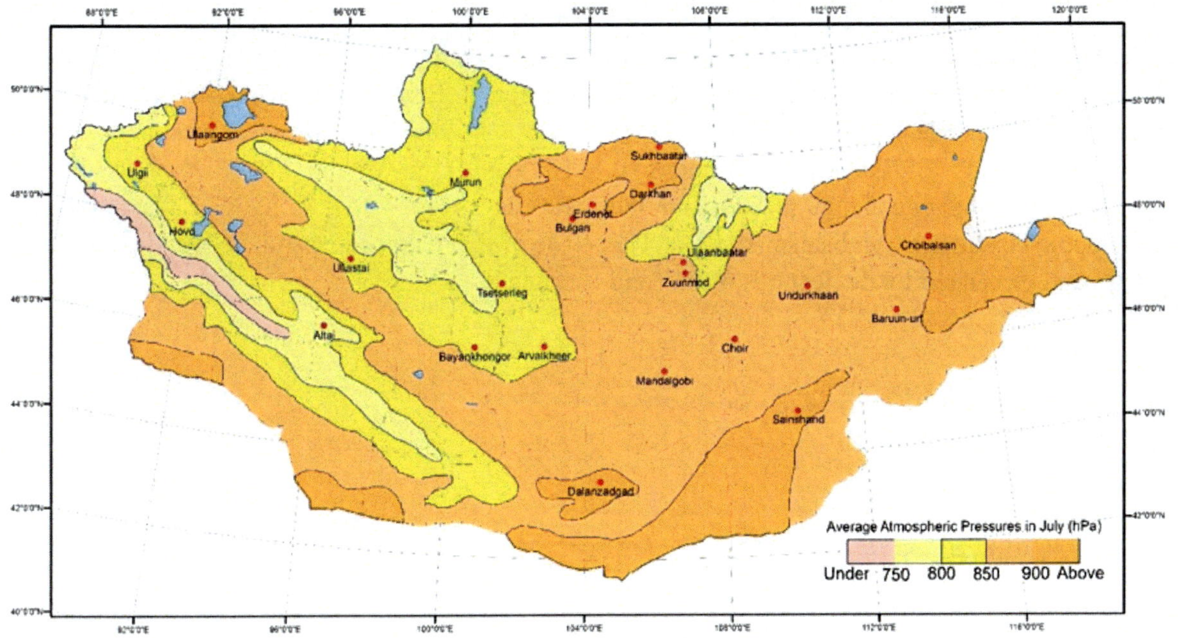


**Fig. 4.12** Spatial distribution of air pressure in July (modified from National Atlas of Mongolia 1990, 2009)

anticyclonic system in the cold season, as well as the snow-covered surface of the land. The maximum wind speed in spring, caused by the transition period from high air pressure to low pressure, continuously triggers cyclone type weather in late spring. Therefore, the wind speed and direction are very variable in the spring/autumn period, which is determined by unstable atmospheric pressure and weather conditions in different parts of the land surface. In Mongolia, there are some types of local winds, such as foehn winds on the windward slope of the Altai Ridge and bora winds on the shore of lake Khuvsgul (Natsagdorj and Gomboluudev 2018).

There are spatial differences in wind speed in the country; the most windy are the arid and semi-arid areas (steppe and Gobi Desert), with average annual wind speeds of 4–6 m/s in these areas, 1–2 m/s in the intermountain valleys, and 2–3 m/s in the rest of the country (Fig. 4.13). According to meteorological measurements in 250 localities in Mongolia, the average annual wind speed in a quarter of the country exceeds 4.0 m/s (Natsagdorj and Gomboluudev 2018). The number of days with wind speeds of more than 15 m/s is 30–40 days in desert and steppe areas. Large sandstorms are caused by air turbulence and the collision of massive cold Siberian and hot fronts from Southeast Asia (Jambaajamts 1989). The number of days with dust storms per year is less than 10 days in mountainous areas, more than 50 days in the Gobi Desert and Great Lakes depression, and more than 90 days in the southern part of the Trans-Altai Gobi and Mongol Els sand dunes (Chung et al. 2004; Dorjgotov et al. 2009).

The complexity of the wind directions is determined by the circulation phases and the topographic conditions (Jambaajamts 1989). The winds can be changed by the features of the land surface and vegetation cover on a regional scale. Usually western, northwestern, and northern winds prevail on the territory of the country (Fig. 4.14), which is due to the geographical position on the way of westerlies and the main direction of mountain ranges stretching from northwest to southeast. However, mountain valley winds are particularly common due to the local relief and orography. A typical temporary pattern of wind direction is east of Lake Uvs, with northwestern winds changing to southern winds in winter.

About 61 % of dust storms occur in March, while 7 % occur differently in various regions. 80 % of dust storms occur during the day. Dust storms occur in less than 10 days in the Altai, Khangai, and Khentii mountains and more than 50 days in the Gobi Desert and Great Lakes depression (Dorjgotov et al. 2009). According to meteorological data, dust storms occur in the Gobi Desert zone for 300–600 h per year. The largest number of dust storms is observed in the area of Trans-Altai Gobi and in the southern part of Mongol Els (in the west), where the number of days of sand and dust storms per year is 120 days, and in the Gobi Desert it is 30–100 days (Fig. 4.15). Mongolian dust storms are one of the main sources of "yellow dust of Asian origin".

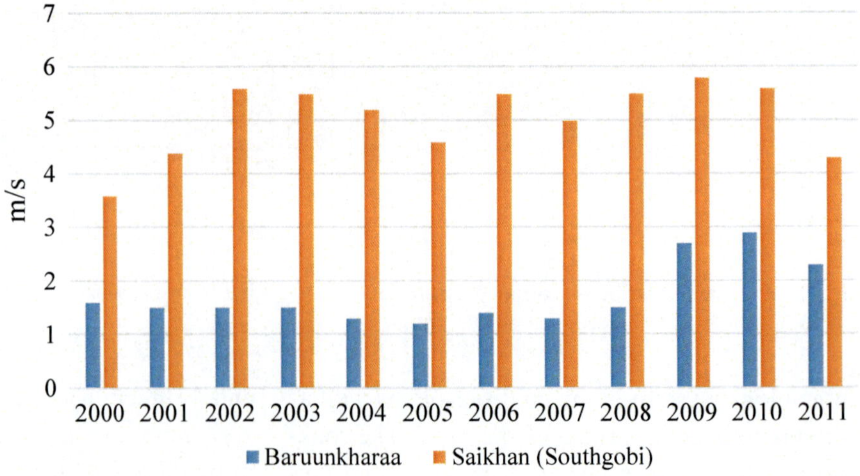


**Fig. 4.13** Average wind speed (m/s) in foreststeppe and desert regions (2000–2011) (based on climate data of the Ministry of Environment, 2018)

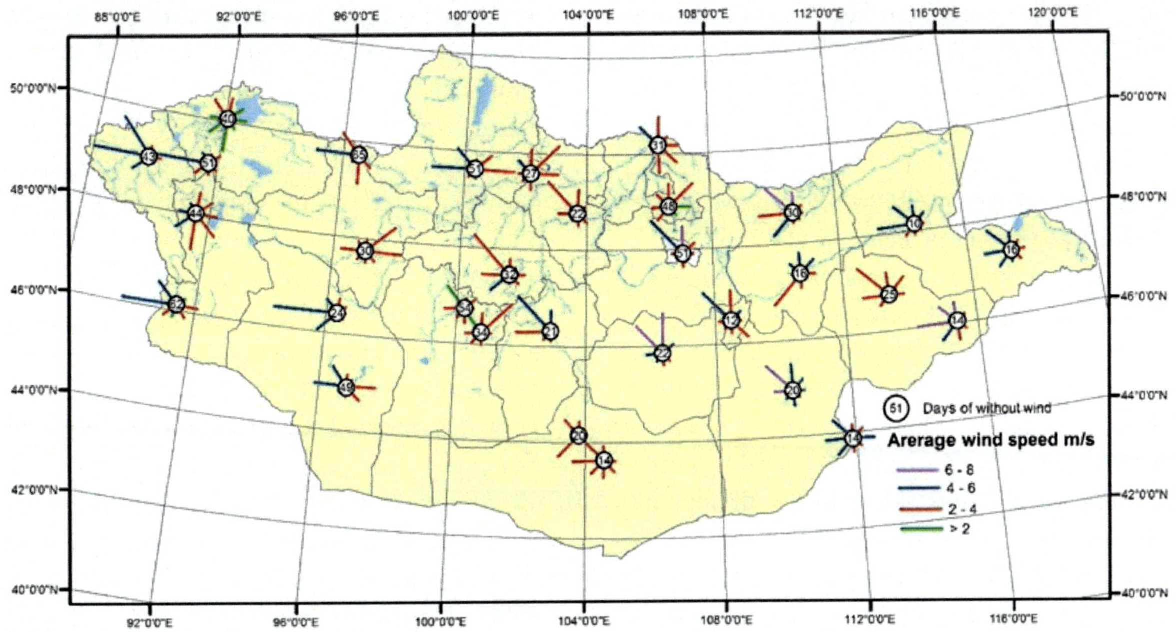


**Fig. 4.14** Wind directions and speed (NIHM 1985)

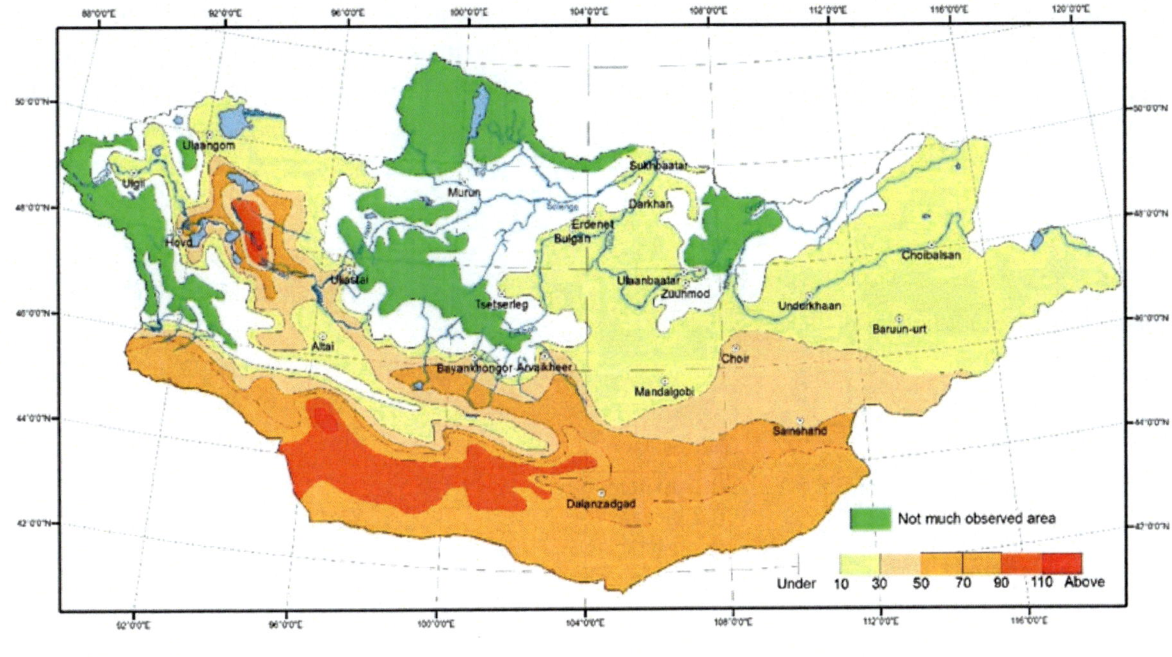


**Fig. 4.15** Distribution of dust storms/1960–2008/ (modified from National Atlas of Mongolia 1990, 2009)

## 4.4 Climatic Regions and Main Characteristics

The first climatic regionalization of Mongolia was developed by N. Badarch and B. Gungaadash (1960) based on geographical feature, soil and vegetation covers characteristics. Badarch (1967) has defined four main types of climate: (1) Humid climate of Khangai and Khentii, (2) moderately dry climate of Altai region, (3) moderate climate of steppe and plain, and (4) continental dry and warm climate in Gobi region. According to the regionalization by Natsagdorj (2000), based on Budiko Aridity Index, Mongolia is divided into four climatic zones: (1) Desert zone (aridness index or K > 8); (2) Gobi and semi-desert zone (K = 5–8); (3) dry steppe zone (K = 2–5); (4) forest steppe zone (K = 0.5–2.0); and (5) Alpine zone in high mountains (K < 0.5). There are also some scientists who have used the Ivanov-Mezentsev humidity coefficient, which is common in Russia. The common climatic regions in Mongolia (Table 4.1; Fig. 4.16) were included in the main atlases and sources, "*Atlas of climate and water resource*" *(1985), National Atlas of MPR (1990), National Atlas of Mongolia (2009), Atlas of Desertification in Mongolia (*2011*).* In terms of heat and water supply, Mongolia is divided into five climatic regions and 15 sub-regions (Badarch 1971; Jambaajamts 1989).

The main characteristics of each climatic region have been described on the basis of moisture and heat accumulation as well as winter conditions (Jambaajamts 1989) as follows:

I. ***Humid cold region***: This climate region includes parts of the Alpine belts with elevations of more than 1,800 m a.s.l. in the Mongol Altai, Khuvsgul, Khangai, and Khentii ranges, which are characterized by large seasonal temperature differences, with warm and humid summers and cold (sometimes very cold) winters. The total active temperature in summer is lower than +15 °C, and in winter is lower than −25 °C, and the air humidity coefficient is higher than 2.6–4.0. The amount of precipitation exceeds 350 mm per year, especially 400–500 mm in mountain ranges. Due to the deficit of temperature in the soil cannot accumulate a sufficient amount of nutrients. The average height of snow cover is more than 20 cm, and in winter snowfall occurs with a frequency of 6–7 times.

II. ***Moderately humid and moderately cold region***: This climate zone includes part of the mountainous areas that rose to a height of 1500–3000 m (the middle part of the Mongol Altai and Khangai). The total summer active temperature ranges from 1,500 to 2,000 °C and the humidity coefficient is 2.2. In winter, the air temperature reaches −25 to −30 °C (sometimes colder) and in summer +15 to +20 °C. The annual rainfall is 250–350 mm. The average height of snow cover is 10–15 cm and snowstorm is 10–16 days a year. The period without frost is 80–90 days a year.

**Table 4.1** Main characteristics of climatic regions

| Regions (summer) | Sub-regions (winter) | Aridness[a] $\sum d/p$ | Radiant heat[b] $\sum t > 10$ °C |
|---|---|---|---|
| I. Humid, cold | I.2. Severe | Less than 1.0 | Less than 1,500 |
| II. Moderately humid and moderately cold | II.1. Extreme severe<br>II.2. Severe<br>II.3. Moderate | 1.1–2.5 | 1,500–2,000 |
| III. Semi-arid and cool | III.1. Extreme severe<br>III.2. Severe<br>III.3. Moderate severe | 2.6–4.0 | 2,000–2,500 |
| IV. Arid and cool | IV.1. Extreme severe<br>IV.2. Severe<br>IV.3. Moderate severe<br>IV.4. Cold | 4.1–5.5 | 2,500–3,000 |
| V. Dry warm | V.1. Extreme severe<br>V.2. Severe<br>V.3. Moderate severe<br>V.4. Cold | More than 5.6 | More than 3,000 |

*Source* National Atlas of Mongolia (1990, 2009)

[a]$\sum d$ –The amount of average daily values of moisture deficit for one year; p–Annual mean rainfall

[b]$\sum t > 10$ °C–The average sum of daily temperatures more than 10 °C on surface

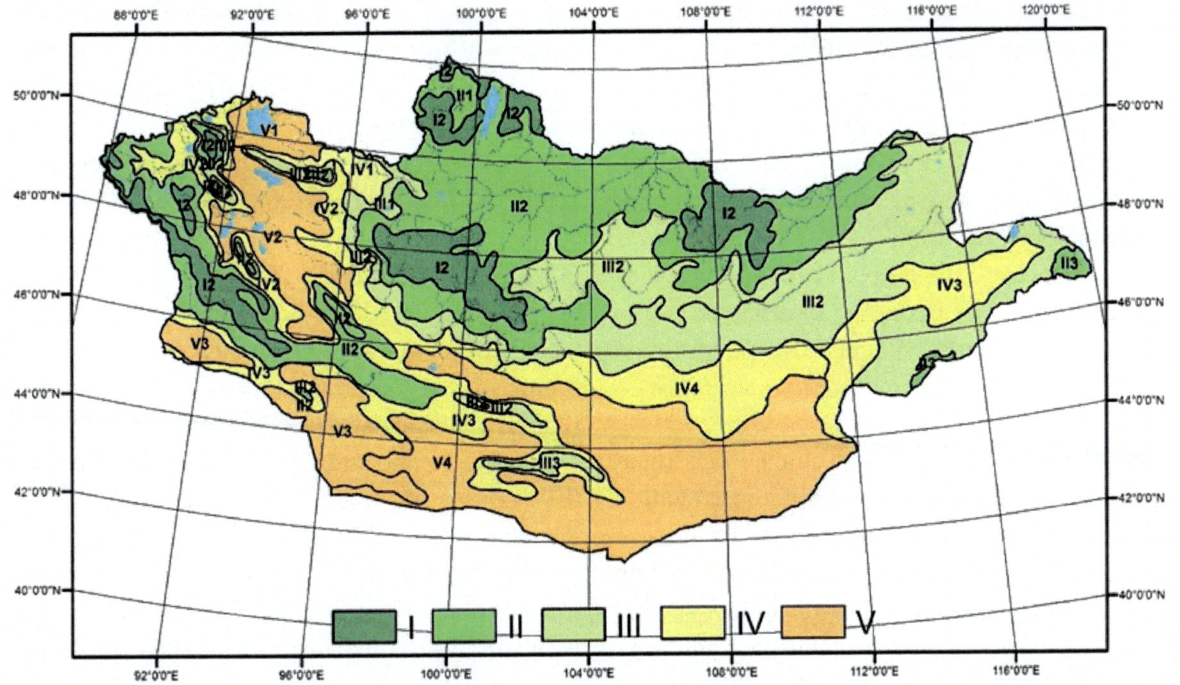


**Fig. 4.16** Climatic regions in Mongolia (based on (Tsegmid and Vorovev 1990)). I: Humid cold; II: Moderate humid and moderate cold; III: Semi-arid and cool; IV: Arid and cool; V: Dry warm; Sub-zones: 1: extreme severe winter; 2: severe winter; 3: moderate severe winter; 4: cold winter

The frequency of drought is 40–50 %, once every 2–2.5 years. This climatic region is divided into three sub-regions. The Darkhad and Khuvsgul lake depressions and basins of the rivers Tes and Ider belong to the extremely cold sub-region, and other parts of this area belong to the cold sub-region.

III. ***Semi-arid and cool region***: Most of the steppe area of the country belongs to this climatic region, which is divided into three sub-regions: extremely severe winter (Tes river basin), severe winter (Tuul river basin) and moderately severe winter (rest areas of this zone). The total summer active temperature is 2,000–2,500 °C and the humidity coefficient is 1.8–2.2. Most of this climate zone has a moderately severe winter with little snow, but the average snow depth in the southwestern part of the Khangai mountain is more than 15 cm, and the winter temperature is below −30 °C. Temperature ranges from −15 to −30 °C in winter and +15 to +25 °C in summer. Precipitation is 150–250 mm per year. The frequency of windstorms is 30–40 %, and spring droughts 50–60 %. The period without frost is 90–110 days per year.

IV. ***Arid and cool region***: This region includes a transition zone between steppe and desert steppe (Gobi Desert). The total active summer temperature is 2500–3000 °C and the humidity coefficient is 0.5–1.0. Temperatures range from −15 to −30 °C in winter and +20 to +25 °C in summer. The annual rainfall is 100–150 mm. The frequency of strong wind is 30–40 %. The frequency of droughts is 60–70 %. The period without frost is 100–140 days per year. The region of arid and cool climate is divided into four sub-regions: extremely cold (around the Uvs lake depression), severe cold (the western part of the Khangai ridge), cold (the intermountainous area between Altai and Khangai), and mild cold (Great Dornod plain).

V. ***Dry warm region***: This climatic region includes the Gobi Desert and the extremely arid parts of the southern desert of the country

(Trans-Altai Gobi Desert). The total temperature of active summer is higher than 3,000 °C and the humidity coefficient is 0.1–0.3. Temperatures range from −15 to −30 °C in winter and +20 to +25 °C and higher in summer. The annual rainfall is 100 mm or less. Winter is mild, so stable snow cover does not exist. The frequency of strong winds is 50–60 %. The frequency of droughts is 70–80 %. The period without frost is more than 140 days a year (Tsogtbaatar et al. 2014).

On the basis of the global climate classification systems, regional climatic differences within the country have been identified due to the size and surface diversity of the country. According to the Koppen-Geiger classification, several types of climate can be distinguished in Mongolia: BSk (cold desert climate), BWs (cold semi-arid climate) and Dwc, Dfc (cool continental or subarctic climate) are most common, while Dsc (cool continental climate) and Dsb (temperate continental climate) are the most common types of climate in Mongolia (Fig. 4.17).

## 4.5 Climate Change and Its Impact

### 4.5.1 Climate Change and Tendency

Mongolia is extremely vulnerable to climate change due to its geographic location and the fragile of natural ecosystems. The country's diverse climate, high dependence on coal-based energy, and growing urban population contribute to Mongolia's ranking on the 2014 climate risk index as the eighth most vulnerable country to the impacts of extreme weather (Germanwatch 2014). Long-term trends in air temperature show that Mongolia has already observed warming caused by the effects of global warming (Dagvadorj et al. 2010; MNET 2009).

Compared to the global average annual temperature, which increased by 0.85 °C between 1880 and 2012, climate change is occurring rapidly in Mongolia (MEGD 2014). The increase in air temperature is found in all seasons of the year, but there are seasonal variations that cold seasons of the year had an increase in temperature by 3.6 °C, spring and autumn seasons had an increase in temperature by 1.8–1.9 °C. In the summer season, the temperature increase was 1.1 °C (Gomboluudev 2007) (Fig. 4.18).

Analyzing data from 70 meteorological stations between 1961 and 2011 (MNET 2018), air temperatures rise at all locations with average values of 1.4–2.4 °C. The highest values are found in Baruunturuun (2.9 °C), Tosontsengel (2.5 °C), Ulaanbaatar (2.4 °C), and Khovd (2.5 °C), and in steppe and Gobi regions (Ekhiin Gol, the values of Undurkhaan, Dashbalbar, and Khanbogd) are relatively low (Table 4.2).

The number of extremely hot days (with a maximum temperature of 260 °C and above) and their duration is increasing by 5–28 days, especially in eastern Mongolia (MEGD 2014; MNET 2009), while the number of extremely cold days is decreasing by 13–14 days (Mijiddorj et al. 2011). The determined multi-year trend (Gomboluudev 2011) shows that in Mongolia's coldest place (Ulaangom), the number of extreme cold days (below −25 °C) is clearly decreasing (Fig. 4.19).

In the 1960s and 1970s, deep snowy winters coincided with extreme cold weather and now have changed with warmer conditions (Koike et al. 2010). Since 1961, the warm season precipitation has increased slightly in the western region (Altai mountain and Trans-Altai Gobi), while in the central and eastern regions there has been a rapid decline (Table 4.3), in some places with a 95 % value (MEGD 2014). In the rest of the country, precipitation decreases at a rate of 0.1–2.0 mm per year (Natsagdorj and Gomboluudev 2018) (Fig. 4.20).

According to trends of some extreme climate indices, frost days have decreased by nearly 15 days, while summer days have increased by 19 days between 1971 and 2015 (GCF 2019).

According to the long-term tendency, the annual mean temperature in Mongolia is expected to become warmer by 2.1–3.0 °C by 2050, while eastern Mongolia will be warmer in winter than other regions (Dagvadorj et al. 2009), and precipitation in the central and eastern parts of

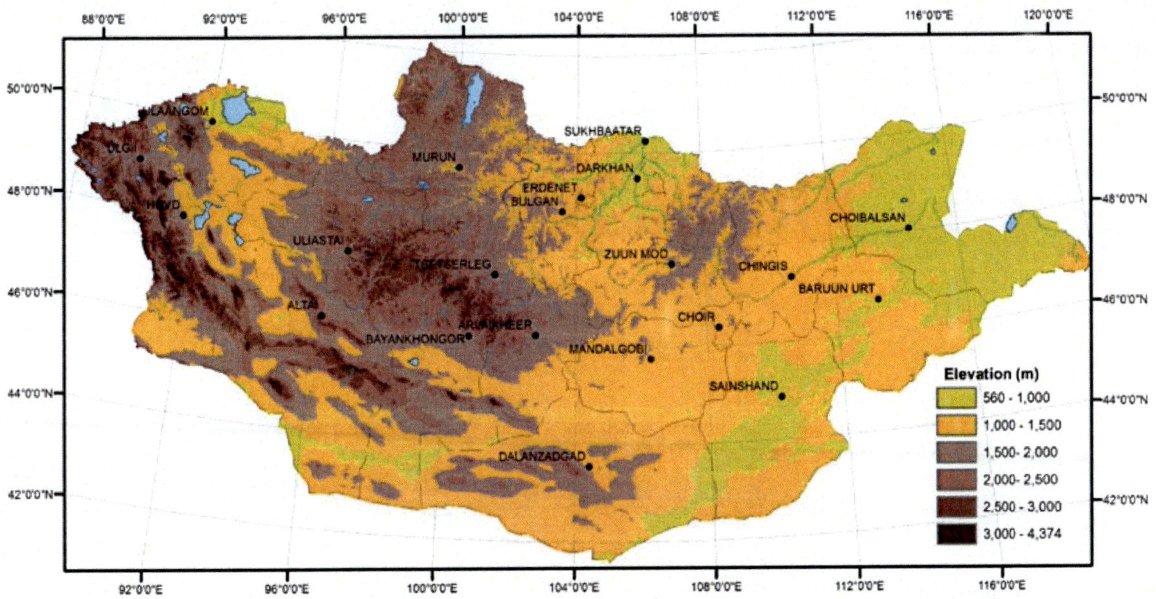


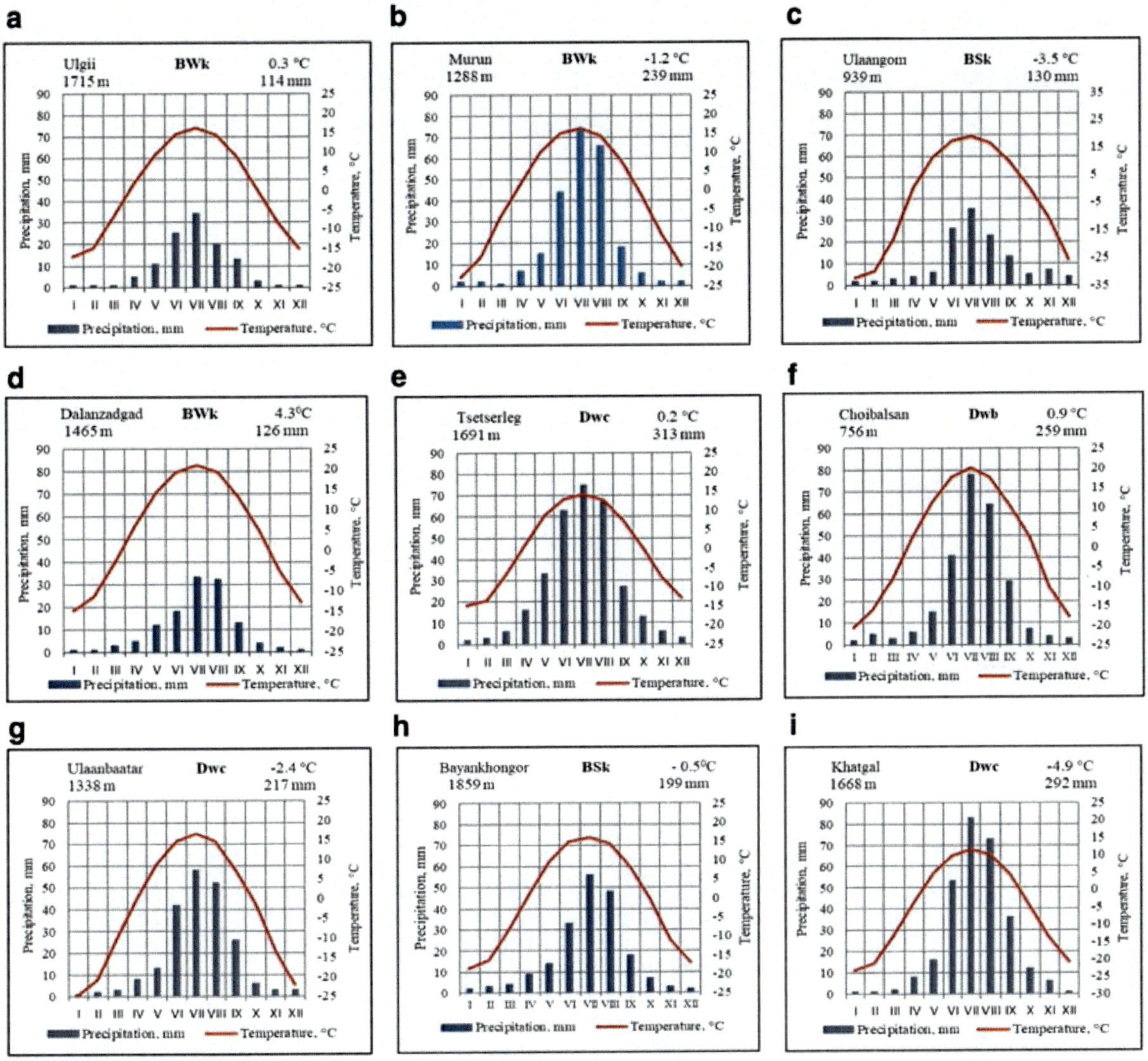


**Fig. 4.17** Regional differences of the climate (by Batchuluun). **a** Bayan-Ulgii aimag, Ulgii., **b** Khuvsugul aimag, Murun, **c** Uws aimag, Ulaangom, **d** UmnuGobi aimag, Dalanzadgad, **e** Arkhangai aimag, Tsetserleg, **f** Dornod aimag, Choibalsan, **g** Ulaanbaatar, **h** Bayankhongor aimag, Bayankhongor, I) Khuvsugul aimag, Khatgal sum

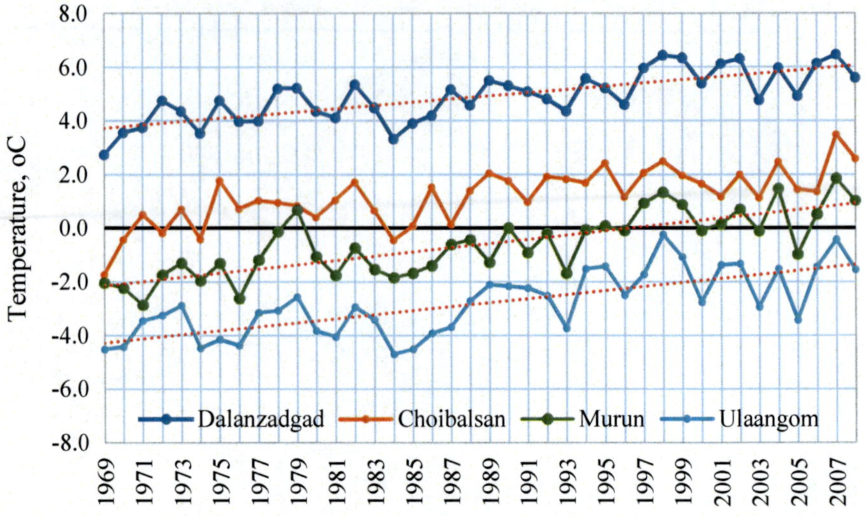


**Fig. 4.18** Mean temperature trend in different distinct ecological zones: West (Ulaangom);North (Murun); East (Choibalsan); and South (Dalanzadgad) in Mongolia (adopted from MARCC-2014 (MEGD 2014)

**Table 4.2** The annual air temperature increase rates, between 1961–2011 (ten stations in each with highest and lowest rates)

| Stations | Highest rate (°C) | Stations | Lowest rate (°C) |
|---|---|---|---|
| Baruunturuun | 2.5 | Ekhiin gol | 0.4 |
| Murun | 2.5 | Undurkhaan | 0.8 |
| Khovd | 2.5 | Duchinjil | 0.9 |
| Tosontsengel | 2.5 | Tsetsen-Uul | 1.0 |
| Ulaanbaatar | 2.4 | Bayan-Ovoo | 1.2 |
| Ulnaagom | 2.2 | Tooroi | 1.2 |
| MandalGobi | 2.1 | Bulgan | 1.2 |
| Choir | 2.1 | Tarialan | 1.3 |
| Baruun-Urt | 2.1 | Khutag | 1.4 |
| Erdenesant | 2.1 | Khalkhgol | 1.4 |

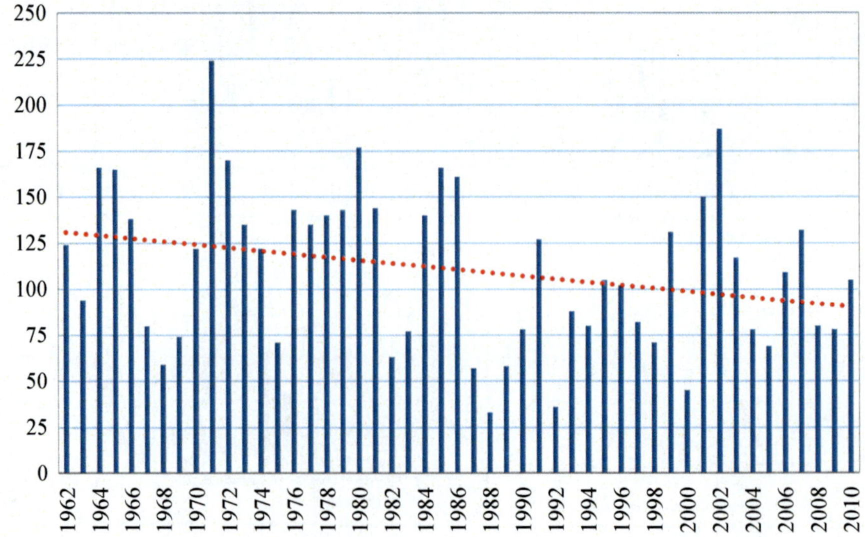


**Fig. 4.19** Multi-year trend of number of days when the daily highest temperature is below 25 °C (Ulaangom station) (Adopted from MARCC-2014)

**Table 4.3** The precipitation changes between 1961 and 2011 (mm) (selected weather stations with highest and lowest rate of changes)

| Stations | Increased rate (mm) | Stations | Decreased rate (mm) |
|---|---|---|---|
| Tooroi | 40.3 | Dashbalbar | −123 |
| Zavhan | 36.5 | Erdenemandal | −106.9 |
| Bayanbulag | 35.7 | Erdenesant | −101.5 |
| Duchinjil | 33.7 | Choir | −89.6 |
| Khanbogd | 31.2 | Khutag | −82.2 |
| Rinchinlhumbe | 26.4 | Yeroo | −78.7 |
| Baitag | 26.1 | Khujirt | −71.1 |
| Yalalt | 19.2 | MandalGobi | −65.7 |
| Khovd | 16.6 | Arvaikheer | −64.8 |
| Huvsgul | 15.7 | Choibalsan | −60.7 |

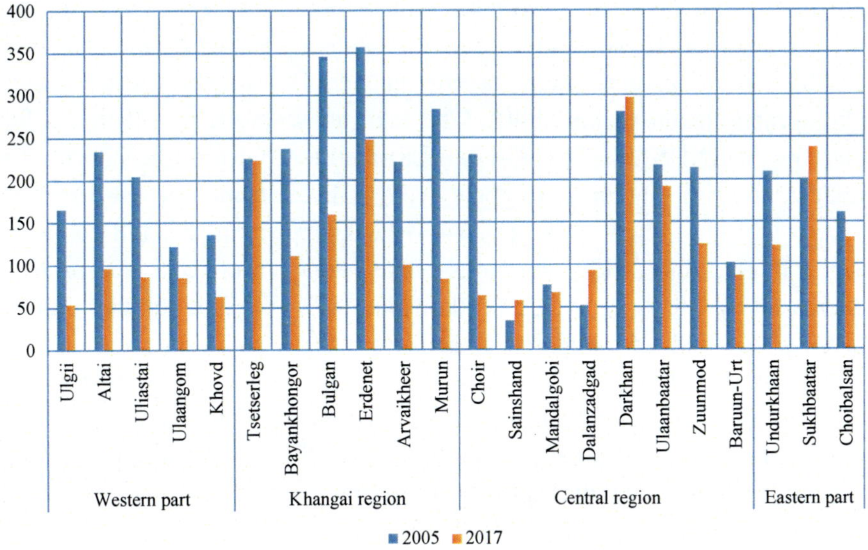


**Fig. 4.20** Precipitation change by regions (2005 and 2017) (data from NAMHEM)

the country will increase in winter and during summer in southeastern part. Decrease in precipitation is projected to be 10 % in Western Mongolia. According to this study, by the middle of the century (at the 10-year level of 2040–2050) the total annual precipitation in Mongolia could decrease by 20–50 mm, especially if the largest decrease in precipitation in the Great Lakes depression area is predicted (Gomboludev 2007). The estimated growth in potential evapotranspiration is 7–12 % (Tsogtbaatar 2013), which significantly increases water scarcity and causes the drying up of many rivers and lakes, desertification and land degradation, and increased frequency of dust and sandstorms.

### 4.5.2 Climate Change Impact

Climate change-related heat waves, air pollution, floods, drought, water scarcity and pollution, as well as the impact of climate on agriculture, can have direct or indirect effects on the health of population. The consequences of climate change have been confirmed in Mongolia, so ecosystems have changed significantly (Batima et al. 2005; MET 2018), and about 70 % of the territory is under the influence of desertification (Mandakh et al. 2017); depletion of water resources due to changes in the water regime; decline in the level of groundwater; intensive degradation of pastures and soil; extreme climatic events; growth in the number of grazing insects (Dagvadorj et al. 1994); and so on. Since the 1960s, the area of forests has decreased by 26 %, the biomass of pastures has decreased by 20–30 % droughts, and the average annual snow height is decreasing in the northern mountainous areas (Batjargal and Enkhjargal 2013; Dagvadorj et al. 1994), as well as affecting the well-being of local population (MNET and UNDP 2012). Surface water evaporation has increased by 118.1 mm since 1961, and precipitation has decreased by 33.0 mm, leading to aridity and being a major cause of desertification (MEGD 2014) and negative impact of extreme warming on crops occurs during its growth period (Mijiddorj et al. 2011). For instance, there were 4,296 lakes with a total surface area of 15,514.7 km$^2$, derived from a 1:100,000 topographic map based on aerial photographs taken in the 1940s. The lake area data from LANDSAT ETM, TM and L8 satellite images showed that in 2015 there were 3,464 lakes with a total area of 14,312.6 km$^2$. Accordingly, in 2015, the total lake area decreased by 7.8 %, or by 1,201.9 km$^2$, and 832 lakes were dried up compared to data from the 1940s (MET 2018).

The results of climate change impact assessment shows that the average river flow in the Arctic Ocean basin decreases by 4–20 %, and in the Pacific basin decreases by 15–30 %. Due to increased evaporation potential, surface waters cannot be provided with soil or ground moisture. In the internal basin of Central Asia, the river flow will decrease by 6.7 % with an increase in air temperature of one degree due to increased air temperature, melting of high mountain glaciers and permafrost in the north of Mongolia, deepening of summer melting, and disappearance of shallow permafrost at the southern fringe (MEGD 2014). Since the 1940s, Altai glaciers have decreased by 30 % (Davaa et al. 2015). The minimum concentration of carbon dioxide in the atmosphere increases in summer due to vegetation degradation and desertification in Mongolia. These changes have affected desertification, water supply, and soil organic matters (MEGD 2014) (Table 4.4).

The results of the model experiments show that if the land surface temperature is raised by 1.50 °C and total precipitation is reduced by 14 % in Mongolia, the steppe region will turn into a desert steppe, and the desert steppe into the desert region. However, global warming could bring some benefits to Mongolia. For example, lower temperatures will lead to a reduction in energy consumption and improvement of living conditions for people (Dagvadorj et al. 2010).

The climate condition of the country has influenced the traditional culture, civilization, housing, and economy style of Mongolians. Nomadic traditional cultures of Mongolia developed warm durable, fur, and animal skin clothing which is necessary for survival in the extreme cold in winter. The traditional housing the ger (yurt) is a portable, circular dwelling made of a lattice of flexible poles and covered in felt or other fabric. It is a type of original shelter from ancient time, has an ideal structure for nomadic and semi-nomadic herding cultures in Mongolia, and protect from fierce winds.

Due to the nomadic life, arid climate, and poor soil, the agriculture in Mongolia is not sufficiently developed, except the Orkhon and Selenge river basins, where it is more humid and flat topography. However, the traditional way of life of nomads is changing due to climate change (Yembuu 2016).

### 4.5.3 Climate-Related Natural Disaster

With regard to climate-related disasters, extreme events of atmospheric convection (thunderstorms, flash floods, hail, etc.) have increased significantly since 2000, resulting in loss of life and economic loss. The frequency and intensity of natural disasters, including drought and zud, are increasing due to the climate change (Natsagdorj 2014). In 2011, 70 dangerous meteorological warnings were reported, and their numbers multiplied in 2012 (MEGD 2014).

Drought is one of the most serious weather-related challenges in Mongolia, caused by lack of

**Table 4.4** Current content of soil organic matters ($g/m^2$) and its future changes (%) (Adopted from MARCC-2014) (MEGD 2014)

| Natural zones | Soil organic carbon | | | Soil organic nitrogen | | |
|---|---|---|---|---|---|---|
| | 1961–2010 | Change, % | | 1961–2010 | Change, % | |
| | | 2020 | 2050 | | 2020 | 2050 |
| High mountains | 2921.9 | −1.02 | −1.78 | 165.8 | 0.62 | 0.76 |
| Forest steppe | 5445.1 | −6.30 | −9.47 | 448.7 | −0.73 | −1.77 |
| Steppe | 3328.6 | −5.76 | −8.40 | 249.4 | −1.06 | −1.84 |
| Desert, desert/steppe | 1857.5 | −0.32 | −1.06 | 90.7 | 0.24 | 0.31 |

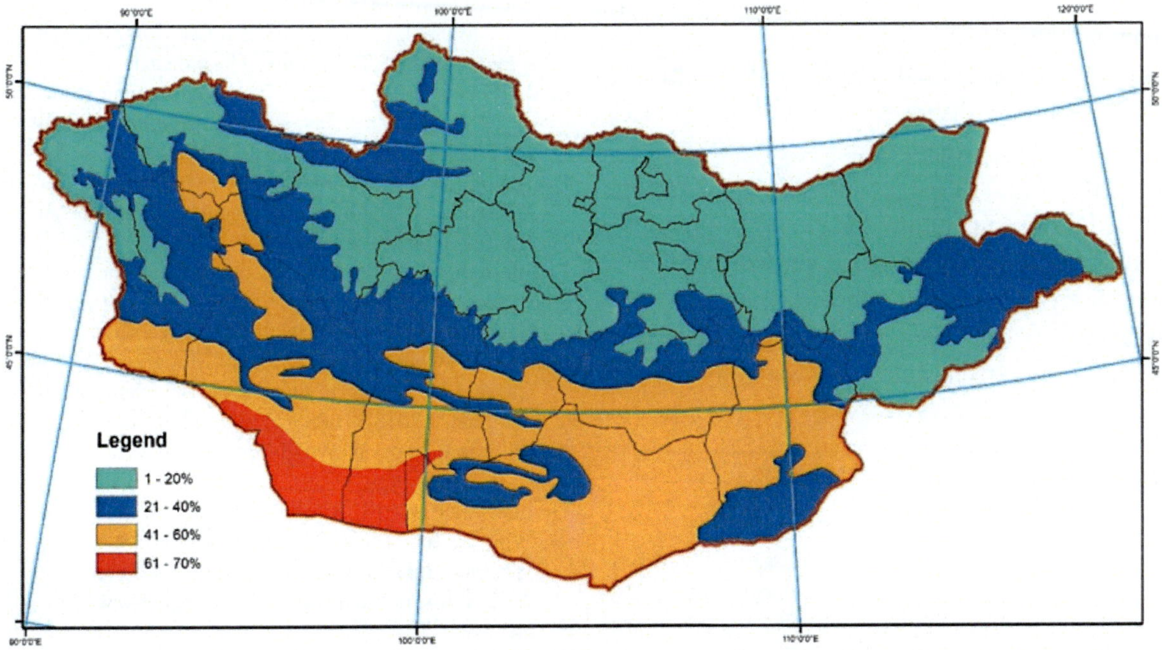


**Fig. 4.21** Drought frequency in Mongolia (modified from: MARCC, 2014, (MEGD 2014))

precipitation in the growing season. In some years, more than 50-day-long dry period was recorded without any precipitation. The spatial pattern of drought occurrence shows high variability. Most part of the country is affected by drought that can occur in every third year on the arid and semi-arid part of the country (Jambaajamts 1989). Drought can occur during periods of reduced precipitation, reduced air humidity or higher temperatures, which is often the case in south-eastern Mongolia. In spring, mild to moderate droughts are prevalent throughout the country, and desert areas of Mongolia tend to experience severe and extreme droughts (Fig. 4.21).

In Mongolia, the most commonly used index to describe drought is the Selyaninov hydrothermal coefficient (SHC). The SHC is suitable to describe the yearly changes of drought rate, which is the ratio of total precipitation to total mean daily air temperature over the growing period ($T > 10$ °C) (i.e., $SHC = 10\sum P/\sum T$, where P is mean daily precipitation (mm) and T is mean daily air temperature). Drought creates a SHC value of less than 0.6 (Natsagdorj and Gomboluudev 2018). Drought occurs in spring and summer once every 6–10 years, which corresponds to an annual occurrence probability (AOP) of drought of 10–15 % in most areas of mountain forests, and in the steppe once in

4 years (AOP = 25 %). A higher prevalence of drought is observed in deserts, or AOP > 50 % (Gomboluudev 2007).

A special type of weather-related disaster in Mongolia is the zud, which is the result of a combination of harsh winter conditions with extremely low temperatures and heavy snowfalls (Batima 2006). There are two different types of zud in Mongolia, depending on the snow cover in winter and spring. White zud is caused by heavy snowfalls and thick snow cover, while the black zud is common in the Gobi Desert and steppes, where snow cover is more temporary and without snow (Batjargal and Enkhjargal 2013). The intensity of zud can be expressed by zud index ($S_{sum}$, where $S_{sum} = \{[(T_i - T_{ave})/\sigma_T - (P_i - P_{ave})/\sigma_P]/n\}$). $T_i$ and $P_i$ are average monthly air temperature and monthly precipitation for i = May to August, respectively. $T_{ave}$ and $\sigma_T$ are average and standard deviation of the average monthly air temperature over a long period of time (Natsagdorj and Sarantuya 2011).

## References


Altanbagana M (2013) Vulnerability assessment of pasture socio-economic system to climate change. Dissertation. 272. Joint publication: Risk assessment and assumptions on disasters of natural and technogenic origin

Arkhipova EP (1945) Climatic description of the Mongolian People's Republic. NIU 16(1):1

Badarch N (1969) The climate of Mongolia. In: Tsegmid S (ed) The physical geography of Mongolia. The State Publishing House, Ulaanbaatar, pp 78–90

Badarch N (1971) The climate of Mongolia. State Publishing House, Ulaanbaatar (in mongolian)

Batima P (2006) Climate change vulnerability and adaptation in the livestock sector of Mongolia. Final report of the AIACC (Assessments of Impacts and Adaptations to Climate Change) project

Batima P et al (2005) Observed climate change in Mongolia. AIACC (Assessments of Impacts and Adaptations to Climate Change) working paper No. 15

Batjargal Z, Enkhjargal B (2013). Interference impact of global warming and globalization on the Society and Ecosystem in Mongolia. In: Y. E (eds) The Mongolian Ecosystem Network, Environmnetal issues under climate and social changes. Springer, pp 295–313

Bazargur G (2008) The pastural animal husbandary of Mongolia. Admon publishing, Ulaanbaatar

Carruthers D (1912) Exploration in North-West Mongolia and Dzungaria. Geogr J 39(6):521–551. https://www.jstor.org/stable/1778190?read-now=1&seq=1#page_scan_tab_contents. Accessed 20 Jan 2019

Chung YS, Kim HS, Park KH, Gao T (2004). Dust-storm observations in China, Mongolia and Korea from March to May 2003. Asia Pac J Atmosp Sci 40(3): 241–257

Dagvadorj D et al (2010) The Climate Change Adaptation. UB

Dagvadorj D et al (2009) Mongolia Assessment Report on Climate Change. Ministry of Environment, Nature and Tourism, Mongolia, Ulaanbaatar, Mongolia

Dagvadorj D et al (1994) Climate change in Mongolia. Pap Meteorol 17:3–14

Dash D, Mandakh N (2011) The history of geographical science. Admon Publishing, Ulaanbaatar

Davaa G et al (2015) Surface water regime and resources of Mongolia. In: Davaa G (ed) National Institute Meteorology and Hydrology, Ulaanbaatar

Dorjgotov D et al (eds) (2009) National Atlas of Mongolia. Institute of Geography. Mongolian Academy of Sciences, Ulaanbaatar

Erdenesukh, S. (2008). Studies on inversion near ground surface layer in Mongolia. *Ph.D. dissertation.* National University of Mongolia, Ulaabaatar

GCF (2019) Country programme of Mongolia. The Green Climate Fund. https://www.greenclimate.fund/documents/20182/1688867/Mongolia_Country_Programme.pdf/da4b9c33-75ce-3ed7-69bf-7cdadd84697e

Germanwatch (2014) Global climate risk index 2014: Who Suffers Most from Extreme Weather Events? Weather Related Loss Events in 2012 and 1993 to 2012. https://germanwatch.org/en/download/8551.pdf. Accessed 15 May 2017

Gomboluudev P (2011). Studies of interaction between atmosphere and land cover using regional climate model. Ulaanbaatar (in mongolian)

Gomboluudev P (2007). Future climate change in Mongolia—Detailed assessment results of the regional from the global scale, using dynamic downscaling method. In: Scientific conference modern issues of water resources and sustainable agricultural development, Ecology and Sustainable Development, vol 8, pp 56–59. Ulaanbaatar

Jambaajamts S (1989) The climate of Mongolia. State Publishing House, Ulaanbaatar (in mongolian)

Jambaljav Ya et al (2016) Long-term Research monitoring on permafrost of Mongolia, Research Report of thematic Work. Institute of Geography, Academy of Sciences, Ulaanbaatar

Khangaisaikhan A (1989) Ajiglaj chadval medej bolno (If you can observe the sky, you can know it). State Publishing House, Ulaanbaatar (in mongolian)

Koike T et al (2010) Cold- and warm-deep-snow winters in Mongolia. J Agric Meteorol 66(2):103–110

Liu YY, Evans JP, McCabe MF, de Jeu RAM, van Dijk AIJM, et al (2013) Changing climate and overgrazing are decimating Mongolian Steppes. PLoS ONE 8(2). doi:e57599. https://doi.org/10.1371/journal.pone.0057599

Mandakh N et al (2017) Desertification in Mongolia. In: MET, The environment of Mongolia, vol 1. Munkhiin Useg Publishing, Ulaanbaatar, pp 97–170

Mongolia WWF (2011) Assessment of climate change and anthropogenic impact into hydrological systems of Onon, Kherlen and Khalkh River basins, Mongolia. Munkiin Useg Publishing, Ulaanbaatar

MEGD (2014a) Convention on biological diversity: the 5th National report of Mongolia. Ministry of Environment and Green Development. Mongolica Publishing, Ulaanbaatar

MEGD (2014) Mongolia second assessment report on climate change-2014. In: Dagvadorj D, Batjargal Z, Natsagdorj L (eds) Ministry of Environment and Green Development, Ulaanbaatar

MET (2017) Mongolia's initial biennial update report under United Nations framework convention on climate change. In: Batjargal Z (ed) Ministry of Environment and Tourism, Ulaanbaatar. https://unfccc.int/files/national_reports/non-annex_

MET (2018) Mongolia Third National Communication (TNC) Under the United Nations framework convention on climate change, Batjargal Z ed. Ministry of Environment and Tourism, Ulaanbaatar

Mijiddorj R (2000) Climate change and sustainable development. Ulaanbaatar

Mijiddorj R (2012) Climate change (Issue that deserves attention, will be too late if we delay). UB 111 pp (in mongolian)

Mijiddorj R et al (eds) (2011) Assessment of climate change and anthropogenic impacts into hydrological systems of Onon, Kherlen and Khalkh river basins, Mongolia. WWF Mongolia Programme Office, Ulaanbaatar, pp 10–15

MNET & UNDP (2012). Mongolia Second National communication under the United Nations framework convention on climate change. UB

MNET (2009) Mongolia assessment report on climate change 2009. Ministry of Environment, Nature and Tourism of Mongolia (MNET), Ulaanbaatar

MNET (2018, 12 24) Retrieved from data base of the climate change. http://eic.mn/climate/

Munkhtseren Sh et al (2015) Research report on a case study on -National data collection and the state implemented follow-up disaster measures and compensations for the disaster||, NGO. World Bank, Center of Human development, research and training, Ulaanbaatar

Murzaev EM (1952) Mongolian People's of Republic. State Publishing House, Ulaanbaatar (in Mongolian)

NAMHEM (2012) Information note on weather condition in Mongolia for 2010–2011. Ulaanbaatar

Natsagdorj L (1980) A problem of classification of the atmospheric circulation patterns over the Central Asia. In: Proceedings of the IMH (4), pp 15–39 (in Mongolian)

Natsagdorj L (2014) Estimation of climate change impact in drought and dzud assessment. Pap Meteorol (34):133–138

Natsagdorj L, Gomboluudev P (2018) The climate change, desertification and permafrost of Mongolia. Munkhiin Useg publishing, Ulaanbaatar (in Mongolian)

Natsagdorj L, Sarantuya G (2011). Extreme cold may result in loss of livestock,. Pap Meteorol Hydrol Ulaanbaatar 32/8:7–17

Natsagdorj L, Gomboluudev P (2009). Pasture degradation (desertification) of Mongolia and assessment issue of its climate factor. In: International symposium on "Mongolian ecosystems and Desertification", Ulaanbaatar

NIHM (1985) The atlas of the climate and water resources in the Mongolian Peoples Republic. National Institute Hydrology and Meteorology

Purevjav G, Balling RC Jr, Cerveny RS, Allan R, Compo GP, Jones P, Svoma BM (2015) The Tosontsengel Mongolia world record sea-level pressure extreme: spatial analysis of elevation bias in adjustment-to-sea-level pressures. Int J Climatol 35 (10):2968–2977

Sato T et al (2004) Cloud frequency in eastern Mongolia and its relation to the orography. In: Proceedings of the 3rd international workshop on terrestrial change in Mongolia, Tsukuba, Japan

Shagdar Sh (1999) Geography of Mongolia. Ministry of Education and Science, Ulaanbaatar

Shardar Sh (1980) Tenger yamar baih be? (How to know the weather). State Publishing House, Ulaanbaatar (in mongolian)

Tsegmid S, Vorovev (eds) (1990) The National Atlas of Mongolian People's of Republic. Mongolian Academy of Science, Academy of Sciences USSR, Ulaanbaatar, Moscow (in Russian and in Mongolian)

Tsegmid S (1969) Physical geography of Mongolia. Institute of Geography and Permafrost. Mongolian Academy of Sciences. State Publishing House, Ulaanbaatar 405 pp (in Mongolian)

Tsegmid Sh (1969b) The physical geography of Mongolia. State Publishing House, Ulaanbaatar (in Mongolian)

Tsogtbaatar et al (eds) (2014) The desertification Atlas of Mongolia. Institue of Geoecology, Mongolian Academy of Science, Ulaanbaatar

Tsogtbaatar J (2013) Deforestation and reforestation of degraded forestland in Mongolia. In: Yamamura NFN (eds) The Mongolian ecosystem network: environmental issues under climate and social changes. Springer, Utrecht, p 83

Tuvdendorj D, Myagmarjav B (1986) Atlas of the climate and water resources in the Mongolian People's Republic. IMH, Ulaanbaatar

UNDP & Government of Mongolia (n.d.) Human development report 2011: from vulnerability to sustainability, environment and sustainable development, Ulaanbaatar

Voikov AN (1880) Climate of monsoon region of East Asia, Amur region. Transbaikalia, Manchuria, Eastern Mongolia, China and Japan, 5th edn, vol 15. Press of the RGS (in Russian)
Yembuu B (2016) Mongolian nomads: effects of globalization and social change. In: Robertson MT (ed) Everyday knowledge, education and sustainable futures. transdisciplinary approaches in the Asia-Pacific region. Springer, pp 89–105

# 5 Hydrography of Mongolia

Navchaa Tugjamba

**Abstract**

In the first section of the chapter, the overview of Mongolian hydrological research is introduced. The geographical background of the surface water is introduced in the second part, describing the morphological, hydrological and ecological characteristics of rivers and lakes, including the description of the genetic lake types. Detailed analysis is given on the Tuul and Ulz River, Lake Khuvsgul, and Shargaljuut mineral spa. The surface water body, such as rivers, streams and lakes cover 10,560 sq. km or 0.67% of the total territory. Mongolia is divided into three hydrological basins, such as Northern Arctic Ocean Basin, Pacific Ocean Basin and Central Asian Internal Drainage Basin, in the Central and Eastern Asia. The total surface water resource of Mongolia is estimated at 599 km$^3$/year. The main water resource is stored in lakes (500 km$^3$/year) and glaciers (62.9 km$^3$/year). Water resource in the rivers is shared only 5.8% of the total surface water resources, that is, 34.6 km$^3$/year. The third section introduces groundwater resources and its geographical distribution. The amount of water resources in the renewable groundwater (i.e., groundwater with a smaller residence time that can be replenished relatively quickly) is estimated at 10.8 km$^3$/year. Groundwater is the main source for drinking, agricultural and industrial usage in Mongolia.

**Keywords**

Surface water · Groundwater · River basin · Water resources · Water management

## 5.1 Overview of Hydrographical Research of Mongolia

In this chapter we followed the main content of “Rivers and lakes of Mongolia” (Tsegmid 1969) and developed analyzing today’s scientific documents on “Hydrography of Mongolia” (Baldandorj 2017; N Jadambaa 2002; Myagmarjav and Davaa 1999; Namnandorj 1960; Tserensodnom 1971, 2000; Tuvdendorj 1985).

The earliest document of Mongolian hydrography could be “The Secret History of the Mongols. *A Mongolian Epic Chronicle of the Thirteenth Century*” which was written in the thirteenth century by an unknown author(s) (Damdinsuren 1957). Geographical description of Mongolian main rivers, such as Onon, Tuul, Orkhon, Kherlen (“*The source of Three rivers”*—as noted in the book), was described and

N. Tugjamba (✉)
Department of Geography and Planning, Macquare University, Sydney, NSW, Australia
e-mail: navchaa.tugjamba@mq.edu.au

© Springer Nature Switzerland AG 2021



B. Yembuu (ed.), *The Physical Geography of Mongolia*, Geography of the Physical Environment,
https://doi.org/10.1007/978-3-030-61434-8_5

issues of exploring and digging wells along the dryland area, using the medical spas, for instance, Gurvan nuur, Avraga toson lakes have water with high concentration of mineral elements and yellow mud, were presented in the book.

Because of Russian and Western countries interests on Central Asia, numerous Russian scientists and explorers traveled to Mongolia throughout the nineteenth and twentieth centuries and published their notes as the books, monographs and reports, some of those are by Bichurin (1828); Przhevalsky and Morgan (1876); Pozdneev (1892). Most of these publications included a general and narrative description of Mongolian surface water resources, rivers, lakes and glaciers.

Documented history of Mongolian limnology starts from 1883. Russian ethnographer Grigory Potanin (1883) had traveled around northwestern Mongolia and Tuva in 1876–1877 and 1879–1880, Central Mongolia in 1884–1886 and 1892–1893, and to the Greater Khingan in 1899. During the trips, he had conducted a study of natural history and a study of ethnography of Mongolia. Moreover, he studied and documented some of the lakes' chemistry of lakes, such as Uvs, Khyargas, Durgun, Uureg, Sangiin Dalai etc. The longest glacier river in Mongol Altai is named by after him, Potanin Glacier in the Mongolian Altai Mountains.

However, the history of Mongolian modern hydrography research started in 1921, and many research sources had illustrated Mongolian hydrography and water resources in different periods. The Government of Mongolia issued a resolution declaring the establishment of "The Institute of Literature and Scripts" (present "The Mongolian Academy of Sciences) in 1921. At the beginning of the history, Russian scientists had been contributed well to establish our scientific organizations and document the research materials. Since 1925, because of the joint efforts of the Mongolian and Russian scientists on hydrology and limnology, research studies began to be carried out alongside the country.

"*Telmen lake and West of Khangai plateau*" (Kondratiev 1929) and "*Account about hydrochemical works of Mongolian expedition of 1926*" (Smirnov 1932) were the first special publications of the joint expedition. Obviously, the main research publication of Mongolian hydrography is "*The Mongolian People's Republic. A physical geographical description*" (Murzaev 1948) that describes three hydrological basins of Mongolia. Following the book, research materials on Mongolian large lake hollows were published (N. T. Kuznetsov 1951; Tsegmid 1955). In 1959–1960, "Lake Khuvsgul" Joint Expedition on Limnology of Mongolian and Irkutsk State universities initiated to work under the guidance of A. Dashdorj and A. Tomilov. As a result of the joint expedition, the books "*Hydrobiology of Darkhad lake hollow*" (1964) and "*Khuvsgul lake and the opportunities its fish supply utilization*" (Kozhov M.M. 1965) were published (Sevastyanov and Dorofeyk 2005).

V. Smirnov (1932) conducted a research study on water quality of saline lakes, such as Uvs, Khyargas, Durgun, Sangiin Dalai and Tsookhor and springs in the Valley of the Great Lakes. He concluded that the Valley of the Great Lakes in the Gobi region was covered large bodies of fresh water, covering 50,000 km$^2$ area in the past, and modern fresh and saline lakes of the Valley are the remnants.

Institute of Geography has played a key role in the limnology research study in Mongolia since the 1960s. Scientific documents of Tsegmid (1955), Avirmed (1959), Luvsandorj (1959), Davaasuren (1961) and Tsend (1966) presented the paleolimnology and evolution of lakes in the Western Mongolia (Valley of the Great Lakes), the chemical composition, and location of Lake Devter (Western Mongolia, Gobi region). Especially, Luvsandorj (1959) investigated the lake deposits in the Valley of the Great Lakes and clearly explained the chemical composition and economic value of Mongolian salt lakes. He studied the salt lakes and their deposits in the Valley of the Great Lakes (Egorov 1993).

Medical importance of salt lakes with mineral mud such as lakes Toson, Ikh-Tukhum, Dolon, and Davst was described in the book of "Mineral water of Mongolia" (Namnandorj, Sh, and Nyamdorj 1966).

In 1971, "Lakes of Mongolia" (Tserensodnom 1971), the first general description of Mongolian limnology was published in Mongolian. It described the physical, chemical and morphometric characterization of the largest lakes in Mongolia. The publications of Dulmaa (1979), Williams (1991), Egorov (1993) contributed well on a Mongolian salt lakes research study. One of the modern research results is "Catalog of Mongolian lakes" (Tserensodnom 2000), which is based on the 1970s research on limnology and lake studies. As presented in the catalog, the total water resources of Mongolia is 500 $km^3$, and among them 410 $km^3$ is fresh water. The primary research publication on springs of Mongolia (Namnandorj 1960) was published in 1960.

The first map on average runoff volume of the Mongolian rivers was divided into several regions, as illustrated in 1955 (N. Kuznetsov 1955) and J.Tserensodnom updated in 1969. B. Myagmarjav (1975) and B.Bat, B.Myagmarjav (1985) classified 13 regions of the rivers' average runoff distribution. A new map with 14 regions of runoff distribution was produced by G. Davaa and D.Batkhuu in 1990. In 1973, Russian hydrologist Krashnikov (1973) estimated that total water resources of Mongolian rivers is 40.1 $km^3$, and among them 29.2 $km^3$ is surface water and 10.9 $km^3$ is groundwater resources. Because of less monitoring cases and calculation of hydrology balances in a wide area, that assessment was a bit higher than potential water resources. As described in "Mongolian surface water" (Myagmarjav and Davaa 1999, 2014), total surface water resource is 599 $km^3$, and in that 34.6 $km^3$ is from rivers.

The Mongolian-Polish Physical Geographical Expedition to the Khangai mountain range in 1974 was organized within the framework of scientific cooperation between the Institute of Geography, Polish Academy of Sciences and the Institute of Geography and Geocryology, Academy of Sciences of Mongolia, as the first stage of two years' expeditionary investigations. The main purpose of the expedition was to learn about the resources of the geographical environment on the southern slope of the Khangai. Studies of water circulation in the mountainous part of the Tsagaan-Turuut river basin were carried out by W. Froehlich, J. Słupik and Ts. Sugar and those in the Bayan nuuriin khotgor basin by Z. Babiński and J. Grześ; their aim is to learn water resources and establish the mode of alimentation of rivers draining the Khangai (Klimek et al. 1976).

The initial groundwater resources assessment was conducted by Russian scientist Ivanov (1958). As he assessed, the total groundwater resources are 5.58 $km^3$/year and the potential consumable resources are 0.6 $km^3$/year. Russian hydrogeologist N.A. Marinov made a significant contribution to the research study on hydrogeology and groundwater exploration in Mongolia. "Hydrogeology of the Mongolian People's Republic" (Marinov and Popov 1963) was published based on their research results until 1962.

Nowadays, IMHE[1] is a leading organization for the national network of hydrological observations including various components of river hydrological processes. The first hydrological monitoring station was established in 1942. The network of hydrological monitoring consists of 126 permanent gauging stations at 110 rivers and 16 lakes. IMHE presented that the total groundwater resource is 12.93 $km^3$/year, in 1973. Later, the organization documented that the potential consumable groundwater is 6.28 $km^3$/year. The Hydrogeological Map of Mongolia (scale 1:1,000,000) was produced cooperatively by Mongolian and German hydrogeologists in 1993–1996 (N Jadambaa et al. 2003).

Surface water quality samples are taken at 140 sites by IMHE. History of groundwater monitoring is quite short. However, many monitoring sites are increased year by year, and a few sites have records longer than ten years. Since 1945, river ice information has been collected by the

[1]IMHE- Institute of Meteorology, Hydrology and Environment, Mongolia.

IMHE as hydrometric observations and measurements at gauging stations.

## 5.2 Hydrological Characteristics of Surface Water

Due to the geographical location of the country, Mongolia is divided into three hydrological basins (Fig. 5.1) in the Central and Eastern Asia, such as The Northern Arctic Ocean Basin, The Pacific Ocean Basin and The Central Asian Internal Drainage Basin (Murzaev 1948; Myagmarjav and Davaa 2014; Tsegmid 1969).

The northern and eastern part of the country's rivers drain to the oceans, and western and southern rivers drain at Endorheic basins in the Gobi Desert and the depressions of Inner Asia. The river network is highly densified in the Arctic Ocean basin and the major river system of the basin is The Selenge River. The Selenge River drains into the Arctic Ocean via Lake Baikal, Russia with its main tributaries of Orkhon (its main tributary is Tuul), Eg, Ider and Delgermurun. However, northern Arctic Ocean basin occupies 20% of the total area of the country, where 49% of the water resources belong to the basin (Table 5.1). Also, the basin shares 50% of the total length of Mongolian rivers. It is estimated that the total length of the rivers in the basin is 35,000 km.

One of the main rivers in the Arctic Ocean basin is the Tuul river. The Tuul river basin, which is in the central part of Mongolia, collects water of rivers originating from the southwestern slopes of the Khentii mountain range and drains into the Orkhon river, the main tributary of the Selenge river (largest river basin in Mongolia) and which is part of the Arctic Ocean basin. The Tuul river watershed area is 49774.3 km$^2$ and the length is 717 km. Many tributaries of the Tuul river originate from the Baga Khentii mountain. The river flows at the beginning through mountain taiga and forest steppe region, then down from Ulaanbaatar, the river flows through the steppe region which occupies 80% of the river basin area. The valley becomes wider downstream of Ulaanbaatar and it reaches a width of 8–10 km at Ulaanbaatar city. The river basin covers the territories of 7 districts of Ulaanbaatar city, 37 soums of 5 aimags and occupies a total area of 49774.3 km$^2$.

The Pacific Ocean basin occupies only 12% of the country. The rivers mainly originate in the Khentii and Khyangan mountain ranges and drain into Russia and China. Onon, Kherlen, Ulz and Khalkh, which are the main rivers in north-eastern Mongolia, drain into the Pacific Ocean.

The Ulz river is an example of a lowland river in the Pacific Ocean basin (Fig. 5.2). Ulz river originates from the spring named Ikh and Baga Burd in Norovlin soum of Khentii aimag that flows through Khentii and Dornod provinces to the northeast, crossing the state border it flows in Baruun Tari located in Tari Lake concavity in Russia. The total length of Ulz river is 510 km (from what 495 km on our territory) and average fall for all length is 500 m or 1.0 m per 1 km. The catchment area of Ulz river is 36,456 km$^2$ at the Ereentsav hydrological gauging station, and the average density of the network is 0.015 km/km$^2$ (MEGD 2014). Regarding the river-sorting classification of Horton stream lengths with Strahler ordering, Ulz river (Fig. 5.3) by its priority is a river of fifth category and there are 498 rivers of first category, 98 rivers of second category, 21 rivers of third category and 3 rivers of fourth category in this river basin (Adiyabadam 2013).

The Central Asian Internal Drainage Basin occupies 68% of the total area and generates 40% of total water resources in the country. The main rivers of the basin are Khovd, Bulgan, Uench, Bodonch, Buyant, Zavkhan, Baidrag, Tui and Ongi rivers of the basin. Khovd river drains into Khar-Us lake, Zavkhan river drains into Khyargas in the Valley of the Great Lake, Ongi river drains into Ulaan lake in the Gobi Desert.

Surface water resources are unevenly distributed across the country. According to the estimation of IMHE, the total volume of fresh surface water is approximately 535 km$^3$. As presented in the "*Catalog of Mongolian lakes*" (Tserensodnom 2000), the main surface water resources are stored in lakes. The volume of water, which is stored in lakes is 500 km$^3$/year. Khuvsgul lake contains 63% of total surface

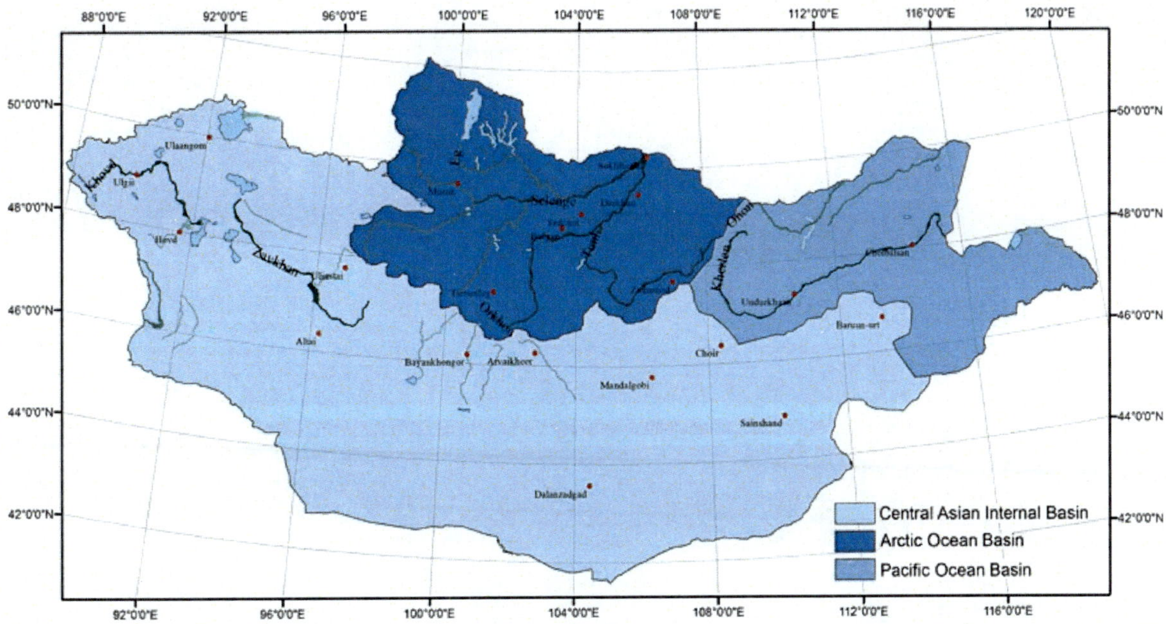


**Fig. 5.1** Hydrological drainage basins of Mongolia Modified for the educational purpose (Batchuluun 2012)

**Fig. 5.2** Ulz river valley (own photo)

water resources and 75% of freshwater. Davaa G. et al. (2012) estimates that fresh water resource in glaciers is 19.4 km$^3$/year. Rest of the total surface water resources, only 6.2%, that is 34.6 km$^3$/year, are from rivers. The water resources in the salt lakes are 90 km$^3$.

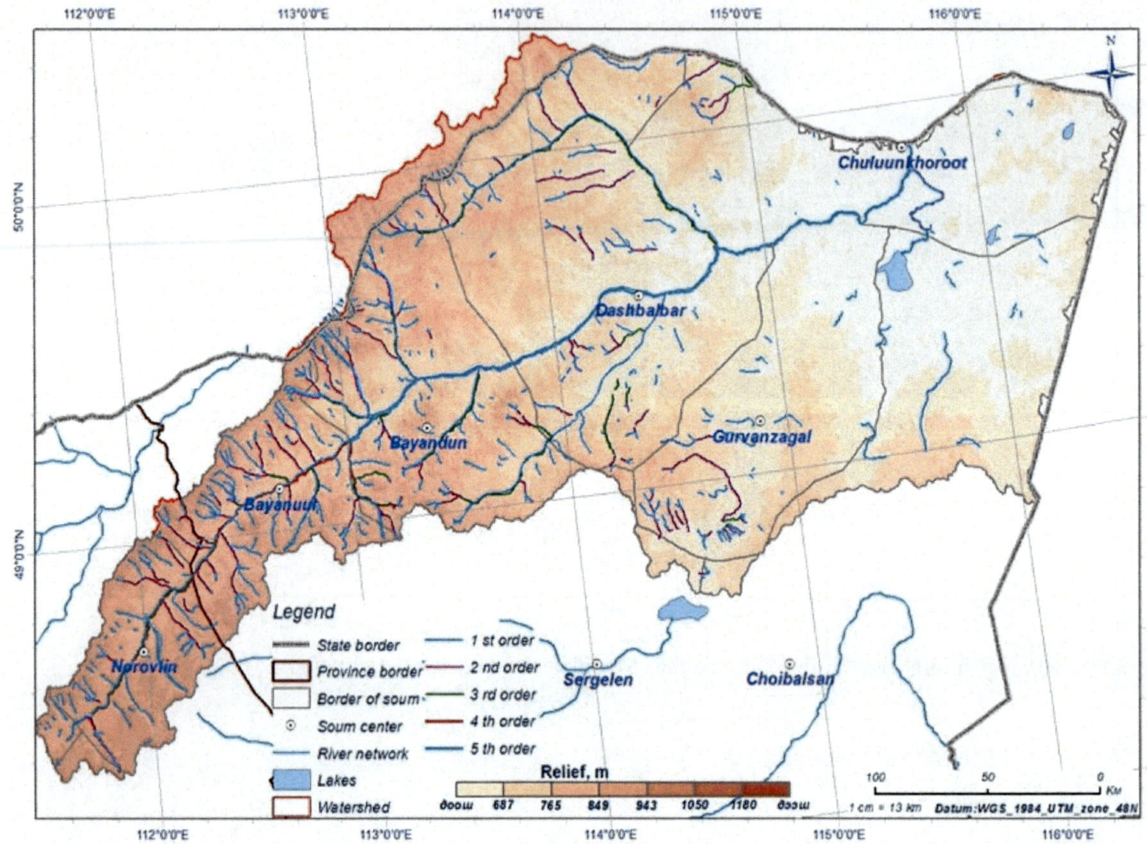


**Fig. 5.3** Stream order of Ulz river basin. *Source* "Ecological and socio-economic baseline studies conducted in Ulz river basin" report. MNET

The Water Authority, Ministry of Nature and Environment started to conduct surface water inventory in 2007. Regarding the order of the Ministry of Nature and Environment, surface water inventories must be carried out by the River Basin Organizations, every 4 years. Up today, we have four inventories statistics (Table 5.1). However, it is not so much reliable data (e.g., depending on the methodology of the inventories, the capacity of the organization, and as well as the timing of the inventory execution); inventory results show how the surface water bodies are changed in the country. Table 5.2 illustrates that the year 2011 was wetter than 2007, which is indicated by the lower percentage of dry rivers and lakes. And 2014 was quite drier than in 2011. A substantial number of dry lakes, which were counted in 2007, reflects the following years' inventories and gradual decrease in number of lakes indicates how surface water resources are reduced.

## 5.3 Rivers

### 5.3.1 River Runoff and Discharge

Myagmarjav and Davaa (2014) assessed that the total runoff of the rivers is 34.6 $km^3$/year. 88.4% (30.6 $km^3$) of the river runoff formed within Mongolia and surface inflow water of 4 $km^3$/year from neighbor countries of Russia and China. As the estimation of IMHE, which is based on the long-term observed runoff, the usable volume from this runoff for water supply and other requirements is only 15%. However, the river's runoff is not much stable every year and seasonal reflection is quite high within a year. Snow and ice melting contribute a lot in the spring flow. It is occurred from the end of March to the end of April, depending on the rivers' source altitude. And it is followed by a low flow period in June.

**Table 5.1** Percentage of the area and water resources of the three main hydrological basins in Mongolia

| Hydrological basin | Area (%) | Water resources (%) |
|---|---|---|
| Northern Arctic Ocean Basin | 20 | 49 |
| Central Asian Internal Drainage Basin | 68 | 40 |
| Pacific Ocean Basin | 12 | 11 |

**Table 5.2** Total numbers of rivers, lakes and mineral waters in Mongolia

| Assessed year | Rivers | | Streams | | Mineral waters and springs | | Lakes | |
|---|---|---|---|---|---|---|---|---|
| | Total | Dried | Total | Dried | Total | Dried | Total | Dried |
| 2007 | 5128 | 852 | 9306 | 2277 | 429 | 60 | 3747 | 1181 |
| 2011 | 6646 | 607 | 10557 | 1587 | 265 | | 3613 | 486 |
| 2014 | 5800 | 387 | 10809 | 1091 | 450 | | 2738 | 399 |
| 2016 | 5585 | 263 | 11420 | 1163 | 490 | 19 | 2214 | 346 |

*Source* Annual report of the National office of Statistics (2017)

Geographical differences of the river regimes in the three main river basins are quite significant (Table 5.3). The northern Arctic Ocean basin shares 51.4% of the total annual runoff in the country. Pacific Ocean basin flow accounts approximately 15 and 33.6% of the total flow that belongs to the Central Asian Internal Drainage Basin. Sources of rivers' runoff are also varied by the regions. For the rivers of the Khuvsgul, Khangai and Khentii mountains, rainfall is the main source (56–75%) of rivers runoff.

The seasonal difference in the river flows is high for the entire country (Fig. 5.4). For instance, the river network density ranges between 0 and 998 m/km$^2$ with the highest value in the upper part of the catchment and the lowest in the steppe and low land areas in the Tuul river basin. Due to the differences in morphology, the distribution of long-term runoff varies from 0.2 to 7.0 l/s/km$^2$. About 69% of the annual runoff forms from rainfall, 6% from snow melting and 25% from a groundwater source (Table 5.3). Depending on flow condition, 62–64% of annual runoff forms within June-August. The long-term means runoff of the Tuul river at Ulaanbaatar station is 21.6 m$^3$/s. During high flow years, the mean runoff with 5% probability of occurrence can reach 59.1 m$^3$/s and in case of low flow years with 97% probability of occurrence, the mean runoff of the Tuul river is 6.0 m$^3$/s (MEGD 2012).

The spring flows continue until June in the Central Asian Internal Drainage Basin rivers, specifically in the Mongol Altai mountains, because of the snow cover that remains until the end of June (Fig. 5.5). The peak precipitation period is summer (July and August) in Mongolia. It contributes to occur the largest river flow in these months to nearly entire country and continues until the end of September in some areas. From September the monthly rainfall decreases and it follows the river flows recede. It continues until November when the water starts to freeze. Obviously, the lowest river flow occurs in the winter.

As of Ulz river regime, it is a river with the flood that has spring water flow from melting snow and rainwater in the summer season. Spring flood is observed at the beginning of April and continues about 40 days in average along the Ulz river. Long-term average of the spring flood is 70 m$^3$/s, and rainwater flood reaches to 272 m$^3$/s on average. In most cases summer flood is higher than spring flood, but in some years it is observed that spring flood is dominated (Fig. 5.6). During observation period maximum summer flood discharge during the flooding that occurred on 24 July 1990 was 575 m$^3$/s. The run-off variation coefficient of Ulz

**Table 5.3** Rivers runoff components of Mongolia

| № | River-station | Basin area ($km^2$) | The average height of the runoff surface (m) | Annual runoff components % | | |
|---|---|---|---|---|---|---|
| | | | | Groundwater | Snow and ice melting water | Precipitation |
| *Rivers in the Mongol Altai* | | | | | | |
| **1** | Khovd river-Ulgii | 22,057 | 2,820 | 32 | 63 | 5 |
| **2** | Khovd river-Myangad soum | 59,939 | 2,300 | 40 | 57 | 3 |
| **3** | Sagsai river-Sagsai soum | 4,665 | 2,525 | 27 | 69 | 4 |
| **4** | Buyant river-Khovd | 7,230 | 2,500 | 33 | 63 | 4 |
| **5** | Kharkhiraa river-Tarialan soum | | | 31 | 57 | 12 |
| *Khangai mountain area* | | | | | | |
| **6** | Delgermurun-Murun soum | 16,300 | 2,005 | 30 | 17 | 53 |
| **7** | Ider-Zurkh soum | 19,800 | 2,040 | 30 | 25 | 45 |
| **8** | Eg- Khantai soum | 41,500 | 1,650 | 29 | 10 | 61 |
| **9** | Orkhon-Orkhon soum | 23,600 | 1,640 | 39 | 11 | 50 |
| **10** | Orkhon-Sukhbaatar soum | 132,000 | | 36 | 18 | 46 |
| **11** | Ongi-Saikhan Ovoo | 6,894 | 1,950 | 36 | 5 | 59 |
| **12** | Tui-Bayankhongor | | | 36 | 18 | 46 |
| *Rivers in the Khentii mountain area* | | | | | | |
| **13** | Yuruu-Yuruu soum | 8,975 | 1,260 | 34 | 20 | 46 |
| **14** | Kharaa-Baruunkharaa soum | 9,580 | 1,325 | 43 | 15 | 42 |
| **15** | Tuul-Ulaanbaatar | 6,300 | 1,820 | 25 | 6 | 69 |
| **16** | Kherlen-Kherlen bridge | 7,350 | 2,200 | 28 | 8 | 64 |
| **17** | Kherlen-Undurkhaan | 39,400 | 1,490 | 34 | 8 | 58 |
| **18** | Kherlen-Choibalsan | 71,500 | 1,280 | 31 | 11 | 58 |

*Source* Adopted from IMHE documents

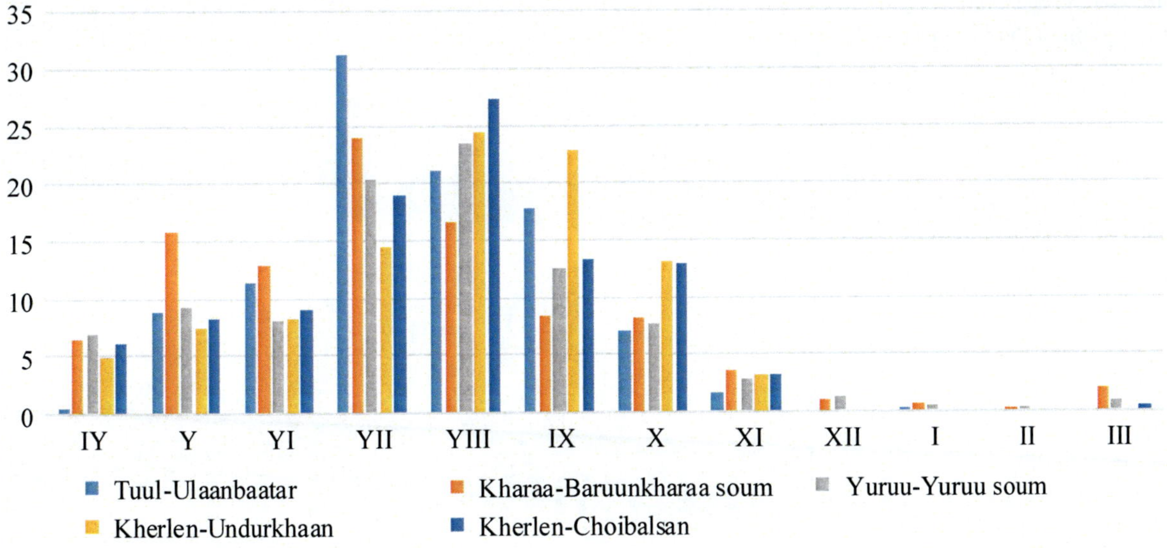


**Fig. 5.4** Annual runoff distribution of the Khentii Mountain Rivers, by the percentage

**Fig. 5.5** The Mongol Altai Rivers' annual runoff distribution (percentage from annual mean)

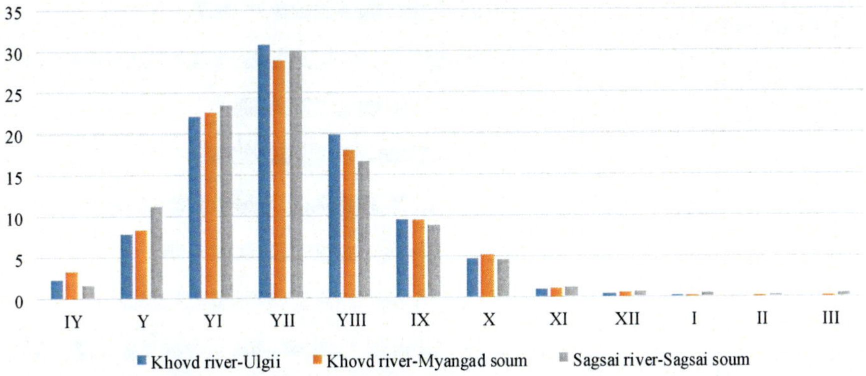


river is 0.66, which shows its high declination from a norm and sharp changes from year to year. River and streams of Ulz river basin are frozen for a cold season; due to shedding precipitation in the warm season it is observed a period with minimum flow between summer floods (MEGD 2014).

Most of the rivers in the Altai mountain area are snow and ice melting water fed (50–70%). For example, more than 60% of water fed is from snow and ice melting water in Khovd river, while the maximum flow of the Tuul river is observed during summer rainfall floods. Mixed sources, such as snow and ice melting water, rainfall and groundwater, contribute to water runoff of the other rivers (Fig. 5.7). The maximum discharge observed at Ulaanbaatar will be reduced after the flood routing. The magnitude of the spring flood maximum due to snow and ice melting is less compared to maximum rainfall flood. The total surface water resources of the Tuul river basin are 1.49 km$^3$/year.

As estimated by Myagmarjav and Davaa (1999), groundwater-fed contributes to 36.1% of the total annual runoff and the base flow component fed by groundwater is 15–40% of the entire country. The specific runoff is highest in the Khentii, Khuvsgul and Altai mountain regions and lowest in the dry southern area of the country. The variability of the annual surface water flow is high.

Based on the long-term river flow data, which was collected at 130 hydrological monitoring stations, the specific discharge (Q) was derived.

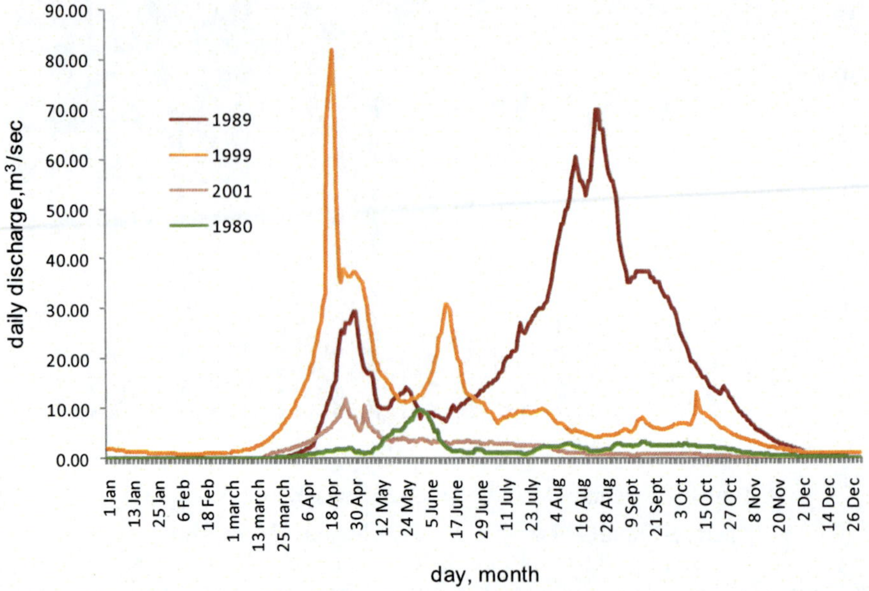


**Fig. 5.6** Typical annual hydrograph of the Ulz river

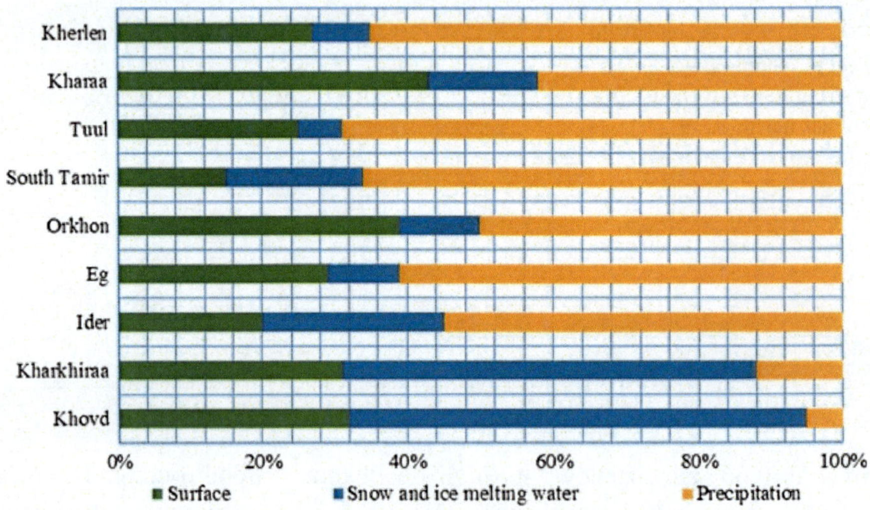


**Fig. 5.7** Some rivers' water fed, by percentage

Using the data, Davaa, Oyunbaatar, and Sugita (2006) classified the country into four regions of surface runoff, such as "*(i) high flow region with* $2 < Q < 16$ *l/(sec/km*$^2$*), (ii) medium flow region with* $0.5 < Q < 2$*, (iii) low flow region with* $0.02 < Q < 0.5$*, and iv) very low flow region with* $Q < 0.01$". Since 1978, the annual surface water flow had been increasing until 1993 (78.4 km$^3$) and decreased back, and minimum flow was observed in 2002 (16.7 km$^2$). According to the latest studies, the surface water flow in the country depends on annual precipitation. However, its overall trend is continuously decreasing due to climate change (Myagmarjav and Davaa 2014).

### 5.3.2 River Thermal Regime

The latitude impact has been observed on water temperature for the rivers which flow to the north from the south and the rivers which flow to the south; the temperature increases along the river length. Daily maximum temperature occurs in the afternoon, around 3.00–5.00 pm, and minimum temperature occurs in the early morning around the sunshine. The daily temperature amplitude is low in the big rivers, around 1–2 ° C, and higher in the small rivers.

As mentioned in "*MARCC-2014: Mongolia Second Assessment Report on Climate Change—2014*" (D. Dagvadorj et al. 2014), the water temperature of the rivers has increased by 1–4 °C, in the last 30–60 years. The river water temperature of Mongolia has increased from the mid of spring and reached a maximum in July. In May, the average water temperature of Selenge river is 9.4 °C near the Ikh-Uul soum, 10.0 °C along the Zuunburen soum and 6.3–9.3 °C in the Eg river basin. Average temperature between two is 0.2 °C for spring, and for autumn it is 9.0 °C in the Mongolian rivers and has fluctuated by 5–15 °C. Rivers form Mongol Altai mountain ranges and Khuvsgul mountains are covered by ice at the beginning of November, and rivers of the Khangai and Khentii mountains are by mid of November (Table 5.4).

The rivers and lakes are covered by thick ice layers of 0.8–3.2 m for 5–6 months in the winter in Mongolia. Thus, ice is one of the essential elements of the surface water hydrological regime in the country. Small rivers are even frozen to the bed. "More than 40 forms (border ice, ice pan, ice boom, frazil ice, bottom ice, hanging ice dam etc.) of ice occur on Mongolian rivers during the cold season" (Punsalmaa et al. 2004). Dates of first ice and water clearing of ice, freeze-up, break-up and occurrence of another form of ice are recorded at the same hydrological stations. The average date of first ice occurrence on rivers is the third week of October. The rivers freeze-up take place from the end of October through the third and last week of November. Ice occurrence is continued approximately 190 days in the rivers. The ice cover duration averages 140 days annually. The longest duration of ice cover is recorded by the river Tes (165–180 days). The spring ice break-up occurs in third and last weeks of April.

The long-term annual average of maximum ice thickness is 100 cm in the northern Arctic Ocean drainage basin and maximum thickness (2.0–3.0 m) occurs in Chuluut, Ider, Suman, Tuul and upstream part of Selenge river. Ice phenomena begin in the Tuul river in the second week of October, and by the mid of November an ice cover is established along the Tuul river. The river has on average 149 days with ice cover until the end of April. The mean depth of the ice cover of the river is 43 cm in November and it increases up to 66 cm in December. The maximum ice depth is observed in mid-February reaching 116 cm. Spring ice phenomena begin from mid-April and by the third week of April ice drifting is observed. Spring ice phenomena end by late April.

In the rivers of Pacific Ocean drainage basin, an annual average of the maximum ice thickness is also 100 cm and maximum 1.9–2.7 m in the Onon, Kherlen and Khalkh river. For instance, water of Ulz river and Khukh lake has intensively become warmer in July–August months, and maximum warming is observed in July. Since August, following to air cooling, it reduces water temperature and at the end of October it begins ice phenomena in the river. In this period, the water temperature is near 0° (Table 5.5).

From long-term observation data of water temperature fluctuation, the maximum temperature of water in Ulz river basin was increased by 0.9 °C in the last 15 years. Ice thickness has fluctuated between 20 and 65 cm and the maximum is observed at the end of January and at the beginning of February which is increased to 50–65 cm. In the middle of October in autumn, ice phenomenon begins and stable ice cover in Ulz river is formed at the beginning of November. Spring ice phenomena are observed since the beginning of April, and in the middle of this month ice is fully released. Ulz river has ice phenomena averagely for 110–190 days annually (MEGD 2014).

The annual average of maximum ice thickness is 80 cm in the rivers of Central Asian Internal Drainage Basin. "*MARCC-2014: Mongolia Second Assessment Report on Climate Change—2014*" (D. Dagvadorj et al. 2014) states that the temperature has increased most rapidly in March, May, September and November in the country, "the time when river ice processes take place" (Batima and Dagvadorj 2000). It impacts on the timing and duration of the occurrence of rivers and lakes ice phenomena, for example, autumn and spring ice occurrence, ice cover duration, and maximum ice thickness mean.

**Table 5.4** The average water temperature of the rivers in Mongolia (°C)

| Season | The northern arctic ocean Basin | Pacific ocean basin | Central asian internal drainage basin | Average |
|---|---|---|---|---|
| Warm season | 13.5 | 14.4 | 12.5 | 14.4 |
| Cold season | 4.9 | 7.2 | 5.6 | 4.9 |
| Average | 9.3 | 10.4 | 8.2 | 8.9 |

*Source* Adopted from "*Water Resource and its appropriate usage*" (Baldandorj 2017)

**Table 5.5** Water temperature of Ulz river and Khukh lake

| River/Lake-Station | An observation period | The period of the water t° exceed of 0.2 °C, in the spring | Monthly average t°C | | | | | | The period of the water t° exceed of 0.2 °C, in the autumn | Max t° of water | |
|---|---|---|---|---|---|---|---|---|---|---|---|
| | | | IV | V | VI | VII | IX | X | | Period | T°C |
| Ulz - Bayan-Uul soum | 2008–2011 | IV/13 | 4.0 | 11.3 | 16.8 | 18.3 | 7.7 | 1.5 | X/26 | VII/22 | 25.9 |
| Ulz - Ereentsav soum | 1967–2011 | IV/10 | 2.1 | 10.3 | 17.6 | 20.5 | 11.2 | 2.9 | X/22 | VII/20 | 27.1 |
| Khukh lake-37th railway junction | 2003–2011 | IV/14 | 1.3 | 4.5 | 13.3 | 19.5 | 9.4 | 2.4 | X/29 | VII/24 | 26.5 |

*Source* Adopted from "Ecological and socio-economic baseline studies conducted in Ulz river basin" report (Adiyabadam 2013)

## 5.4 Lakes

The lakes occupy only 0.4% of the total area in Mongolia. But, approximately 75% of all the freshwater resources are stored in the lakes in the country. The total volume of the freshwater lakes is 500 $km^3$ (Baldandorj 2017). Therefore, Mongolia is the country of lake-water resources. Based upon water chemistry, the lakes are classified into freshwater and saline water lakes. The freshwater lakes have an outlet and the water is refreshed by inflow. Mongolia's biggest lake in terms of surface area is Uvs nuur[2] ($A = 3{,}518.3$ $km^2$). By the volume and depth, Khuvsgul lake is the biggest and contains 74.0% (380 $km^3$) of the total freshwater resources of Mongolia, drains via the Selenge river to the Arctic Ocean. Other large freshwater lakes are Buir, Khar and Khar-Us (Table 5.6). The easternmost lake, Khukh nuur (556 m a.s.l.), the lowest point of Mongolia, is situated the Ulz river Basin.

83.7% of the total lakes are small lakes with the area $(A) < 1.0$ $km^2$ (Table 5.6) but the surface area of these small lakes shares only 5.6% of the total lake area of 16,003 $km^2$ (Tserensodnom 2000) (Table 5.7).

Mongolian lakes are classified into various types based on their formation by which their basins were shaped (Fig. 5.8). The descriptions of these different types of lakes are as follows (Tserensodnom 1971, 2000):

- *Tectonic*. Tectonic lakes often result in the formation of some of the deepest and largest lakes in the world. As the name suggests, such lakes are formed by the tectonic movements

[2]Nuur (Mongolian: нуур)-lake.

**Table 5.6** Mean values of morphologic elements of lakes

| Lake | Water level. m | Surface area. km$^2$ | Volume. km$^3$ | Average depth | A/h km$^2$/m |
|---|---|---|---|---|---|
| *Arctic ocean basin* | | | | | |
| Khuvsugul | 1,647.60 | 2,770.0 | 383.7 | 138.5 | 20.0 |
| Terkhiin tsagaan | 2,069.21 | 54.9 | 0.828 | 6.1 | 9.0 |
| *Central asian internal drainage basin* | | | | | |
| Uvs | 759.94 | 3,518.3 | 66.7 | 10.1 | 96.6 |
| Khyargas | 1,035.29 | 1,481.1 | 75.2 | 50.7 | 19.7 |
| Khar-Us | 1,160.08 | 1,496.6 | 3.12 | 2.1 | 479.4 |
| Khar | 1,134.08 | 565.2 | 2.34 | 4.1 | 137.8 |
| Boon tsagaan | 1,321.0 | 252 | 2.355 | 10 | 25.2 |
| Orog | 1,217 | 140 | 0.42 | 3 | 46.7 |
| Ulaan | 1,008 | 20.3 | 0.158 | 0.9 | |
| *Pacific Ocean drainage basin* | | | | | |
| Buir | 583.02 | 615.0 | 3.75 | 6.1 | 100.8 |
| Khukh | 562 | 51.6 | 0.258 | 3.2 | |

**Table 5.7** Statistics of Mongolian lakes (Based on Catalog of Mongolian Lakes 2000)

| | Area. km$^2$ | Number | Name |
|---|---|---|---|
| 1 | 0.003–1.0 | 47901 | |
| 3 | 1.0–50.0 | 400 | |
| 4 | 50.0–100 | 11 | Ulaagchinii Khar, Khoton, Khurgan, Tolbo, Khukh (Dornod), Yahki, Dayan, Oigon, Khar-Us (Uvs.Unmugobi),Bayan, Terkhiin tsagaan |
| 5 | 100–500 | 8 | Achit, Durgun, Buun tsagaan, Uureg, Telmen, Sangiin dalai, Airag, Orog |
| 6 | 500–1000 | 2 | Buir, Khar nuur |
| 7 | 1000< | 4 | Uvs, Khuvsgul, Khar-Us, Khyargas |

of the Earth's crust like tilting, folding, faulting, etc. "Such lakes are deep and have steeply sloping banks. Generally, they are elongated along the fault line or depression" (Dulmaa 1979). Khuvsgul, Uvs, Khyargas, Achit, Sangiin dalai and Orog lakes are some of the examples of tectonic lakes.

- *Volcanic*. Lakes with a volcanic origin are known as volcanic lakes. These lakes are usually formed in volcanic calderas or craters or when lava flows interrupt the flow of a river or stream. Volcanic lakes are formed in volcanic craters or calderas when the rate of precipitation is higher than the rate of loss of water via evaporation or drainage through an outlet. Two lakes in the upstream part of Khovd river, Khukh lake and Terkhiin tsagaan lake in the Khangai mountain range are example volcanic lakes that were formed by the damming of a river and by lava flow. The

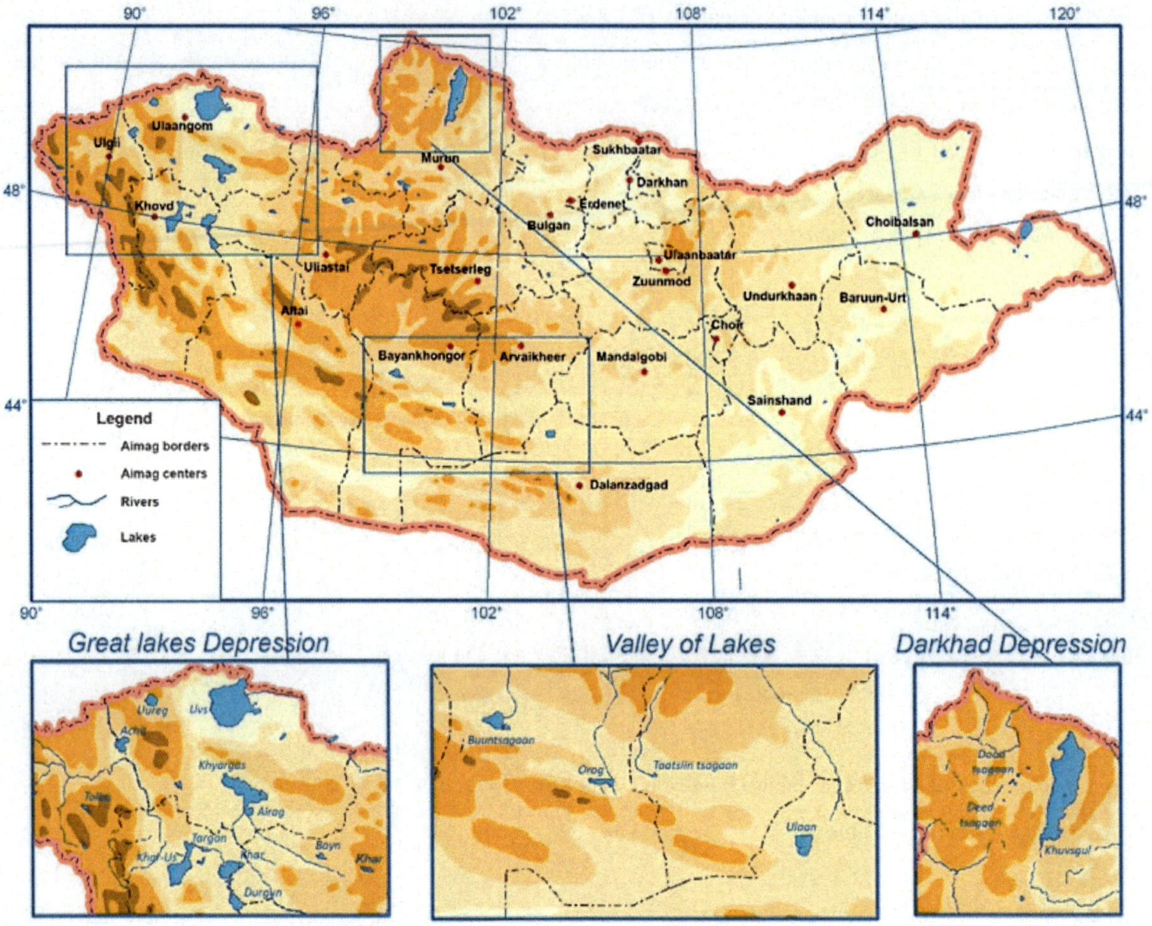


**Fig. 5.8** Lakes of Mongolia Modified for the educational purpose (Batchuluun 2012)

water-filled craters left behind after periods of volcanic activity, such as lakes Terkhiin tsagaan and Sharga belong to the group.

- *Glacial*. This type of lakes are located in the deepest parts of glacial valleys and bordered on the edges by high steep mountain sides. The shape of those lakes is various and is much shallower than tectonic lakes. "The relief of the lake beds is uneven, and the deposition of moraine material on the bottom is different in each lake depression. Such lakes are usually arranged in short chains, in which the uppermost lake is also the largest and deepest (up to 80 m)". Khoton, Khorgon and Dayan lakes in the Altai Mountain range and Khar zurkhnii khukh lake in the Khentii mountain belong to this type.
- *Fluvial.* Valley and subterranean lakes, which originated from the dislocation or at the point of disappearance of rivers. In most cases, they are elongated in shape. Examples include Ugii lake in the Orkhon Valley, lake Erkhel in the Eg river valley, as well as the lakes of the Tamsagbulag Steppes of Eastern Mongolia, Khar-Us lake in the Kherlen river valley.
- *Lakes originating from the thawing and sinking of the upper soil layers* in regions of permafrost. These are especially widespread in the northern parts of the country. This type of lakes is distributed in the Darkhadiin khotgor (Sharga lake, Khujirt and Rashaan lake) and Eastern Mongolia (Tsagaan, Sumiin lake).
- *Deflation lakes* which originated from wind activity on the sand basin of the plains. Such lakes are located in the Gobi region. Their overflow water drains into the sandy ground.

Four geographical regions can be distinguished based on the limnological features, such as *the Altai mountain region, Khangai-Khentii mountain region, the Gobi region, and the Eastern steppe region.* Most of the lakes are situated in the Gobi region; this is an endorheic region with many terminal lakes. 34% of all lakes are located in the Altai, Khangai and Khentii mountainous regions, and 66% in plain steppe desert regions (Table 5.8). 36.5% of all lakes are located in the Gobi region despite its low precipitation (150 mm/year), great evaporation (900–1,000 mm/year), scarcity of rivers and low relief (Egorov 1993).

More than 80% on Mongolian lakes (± 3,500) of area > 0.1 km$^2$ have salinities > 1 g/l. Most saline lakes occur within the Central Asian endorheic drainage basin (Western Mongolia), where they are found in three subregions referred to by Dulmaa (1979) as the Valley of the Great Lakes, the Khingan Plateau and the Gobi Valley. "Smaller numbers of salt lakes occur within the generally exorheic regions of the Arctic Ocean drainage basin (northwestern Mongolia) and the Pacific Ocean drainage basin (far-eastern Mongolia)" (Williams 1991).

The salt lakes have no outlet and concentrations of dissolved solids increase over time due to evaporation of the water. The total volume of the Mongolian saltwater lakes is 90 km$^3$. The largest salt lakes are Uvs lake and Khyargas lake.

The Valley of the Great Lakes in Western Mongolia can be sub-divided as follows (Dulmaa 1979):

- The Basin of the Great Lakes, including Khyargas, Airag, Khar, Durgun and Khar-Us lake s, Lakes Khar and Khar-Us with their outlets, the freshwater Airag lake and the remaining salty lakes without outlets are the final members of one entire hydrographical net. The elevation of the lake levels ranges from 1,132 to 1,157 m a.s.l.
- The Altai Mountain lakes, including Lakes Orog, Achit, Tolbo, Khoton, Khorgon, Dayan and others are in the western border region of Mongolia and are situated in the highest mountain system of the Mongol Altai and Gobi-Altai. They are freshwater lakes with flow-through. The water level ranges from 759 to 2,232 m in elevation. The outlet streams connect these lakes with the Valley of the Great Lakes, with which they share a common fauna.
- The lake system of the Khingan Plateau: Lakes Khinganiin khar, (Bayan), Ulaagchinii khar, Khukh, Oigon, Telmen, and Sangiin dalai. The first two of these are deep and contain fresh water, presumably having originated through karst formation. The remaining lakes contain salty water and either has no outlets at all or empty only periodically into the Valley of the Great Lakes. The water levels in these lakes are between 1,491 and 1,888 m in elevation.

Due to the continental origin, all lakes have ionic compositions dominated by Na', K', Cl$^{-}$' and $SO_4$. Especially, the lakes originated from the ancient more extensive internally drained water bodies in the arid Gobi region (Murzaev 1948; Tserensodnom 1971). Thus, the wetter regions of northern Mongolia (Khuvsgul, Khangai, Khentii, and two northwestern regions of the Mongol Altai) contain mainly fresh or only slightly saline water, while in the more arid regions of the Gobi and east steppe, lakes are more saline. The least saline lakes (0.08–0.2 g/l) lie high in the mountains of the Mongol Altai and Khangai, and the most saline ones lie in arid regions of the Gobi and eastern steppe regions where salinity can reach 308 g/l (Davsan lake), 321 g/l (Suuzh lake), and even higher. High salinities in these regions result largely from high evaporation (ca. 1000 mm p.a.) and widely distributed soil salinization; here are quite a few salt lakes situated in the generally exorheic regions of the Arctic Ocean drainage basin (northwestern Mongolia) and the Pacific Ocean drainage basin (far-eastern Mongolia). The major ions in Mongolian salt lakes are predominantly Na for cations, and usually Cl for anions. Even so, many departures from this pattern occurs, with Mg, Ca,

SO, and HCO, +CO, sometimes important in particular lakes (Dulmaa 1979; Tserensodnom 2000).

As Egorov (1993) states, approximately 74% of all lakes are in this semi-arid region, many being closed basins with unstable water regimes and most highly saline. The seasonal pattern of the lakes is driven by the continental climate. This leads to rapid and intense warming of lakes in spring and summer, and in winter to the formation of an ice cover up to 22 m thick (with some shallow lakes frozen to the bottom). The period of ice cover is from 7 to 8 months in lowland lakes. In the north, ice formation begins at the end of October, and in the south about mid-November. The annual range of temperature in salt lakes is characteristically wide: in highland lakes it is 20–25 °C, and in lowland ones, 30 °C or more. Warming begins in March when, after the snow has disappeared from lake ice, the temperature of surface water layers begins slowly to rice. After the ice melts in May, the water temperature quickly rises above 4 °C, whereupon isothermal conditions develop in shallow lakes and thermal stratification develops in deep lakes (e.g. Khyargas).

Long-term observation shows that the water levels of lakes were decreasing during the 1996–2011 period. However, the water levels of the following lakes increased in 2013: Terkhiin tsagaan lake located in the forest steppe zone increased by 12 cm, Ugii and Buir lakes located in the steppe zone increased by 45 and 81 cm, respectively, and Khar-Us, Ganga and Duut lakes located in the Gobi Desert zone increased by 18, 24 and 36 cm, respectively (D. N. Dagvadorj, L; Dorjpurev, J; Namkhainyam 2009).

The response of a lake to climate change depends on the geographical location and the bathymetry of the lake. The potential evaporation exceeds annual precipitation in all areas (Davaa et al. 2006). The water level of the lakes in the Valley of the Great Lakes and the Gobi may decrease in future, as evaporation from the lake surface increases more than the river inflow (Table 5.8). This effect will be amplified by an increase in upstream water use. On the other hand, lakes with a large drainage basin and substantial inflow, located in the mountains, including Lake Khuvsgul, will be more sensitive to changes in river inflow than to changes in evaporation.

**Khuvsgul Lake**

The main representative of the Arctic Ocean basin lakes is the Khuvsgul lake. Khuvsgul lake is one of only 17 ancient lakes on earth and is thought to be approximately 2 to 5 million years old.[3] The lake locates in the northwest of Mongolia near the Russian border, at the end of the eastern Sayan mountains. It is 1,645 m above the sea level. The oval-shaped lake is 36.5 km wide, 136 km long and 262 m deep (Table 5.9). The volume of water is 381 $km^3$. Because of soft and rich oxygen water, Khuvsgul is one of the most freshwater lakes in the world. Khuvsgul lake contains 63% of total surface water resources and 75% (381 $km^3$) of freshwater. In terms of water volume, it is the second biggest freshwater lake in Asia. The lake contains approximately 70% of the country's freshwater and 0.4% of the world freshwater. The lake watershed is small and narrow, and the tributaries are small (Fig. 5.9). Ninety-six rivers and streams drain into the Khuvsgul, and only one river drains from the lake. Egiin gol origins from the lake and it is one of the main tributaries of the Selenge river. Selenge river is also one of the biggest tributaries of Lake Baikal. Thus, between the two lakes, the water travels a distance of more than 1,000 km, and falls 1,169 m, although the line-of-sight distance is only about 200 km. Its location in northern Mongolia forms one part of the southern border of the great Siberian taiga forest, of which the dominant tree is the Siberian larch (*Larix sibirica*).

The lake is surrounded by several mountain ranges. The highest mountain is the

[3]http://www.worldlakes.org/lakedetails.asp?lakeid=8663.

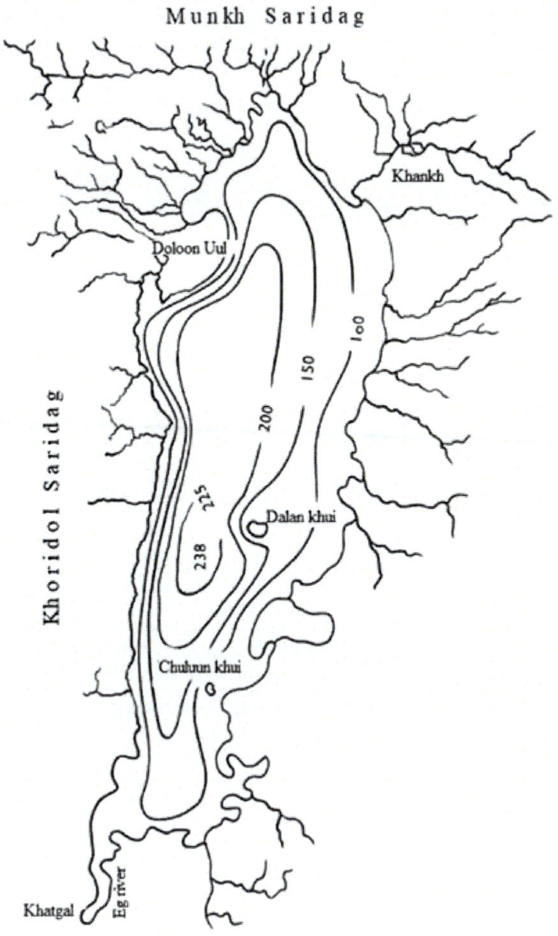


**Fig. 5.9** Bathymetric map of Khuvsgul Lake. *Source* Modified from the "The Atlas of Khuvsgul Lake" Editor Bogoyavlenskii.B.A (1989)

Munkh saridag (3,492 meters), which has its peak north of the lake exactly on the Russian–Mongolian border. Thus, the area of Khuvsgul lake represents a forest steppe with the high-mountain depression terrain and is characterized by a lower degree of channelization ranging from 0.32 km/km$^2$ for the Delgermurun river basin to 0.34 km/km$^2$ for the Egiin gol river basin.

## 5.5 Mineral Springs

The first main research study on the mineral springs of Mongolia was conducted by Namnandorj et al. (1966) in 1964–1965. As a result of the study, a small-scale scheme of mineral waters regionalization was developed. Then, a schematic map of the natural mineral waters and a brief description of Mongolian mineral lakes was published in 1974 (Marinov et al. 1974). As a result of these geological and hydrogeological investigations in Mongolia, more than 400 mineral springs are known, along with several dozen mineral lakes. Regarding the studies, there are approximately 9,600 springs in the country.

A schematic map gives (Marinov et al. 1974) a more accurate picture of the regions of occurrence of the three underground mineral water provinces (Fig. 5.10): (i) *cold carbonated; (ii) thermal nitric and (iii) highly mineralized sodium chloride methane*. It also shows the regions of occurrence of cold radon and fluoride mineral waters within the provinces.

Mongolia has 43 hot springs, with measured surface temperatures ranging from 20 to 92 °C and flow rate ranging from 1.2 to 50 l/s, mainly distributed in the central and western provinces. At that time, intense tectonic development gave the Mongolian mountains their present appearance. A geophysical survey on the crystal structure has established that accumulative thermal sources (magma lumps) are located near the surface under the Khangai, Khentii mountain area (Table 5.9). Geothermal resources in Mongolia are mainly distributed in Khangai, Khentii, around the Khuvsgul, Mongol Altai plate forms (Gendenjamts 1984).

One of the main representatives of the thermal nitric mineral spa is Shargaljuut (Fig. 5.11). The Shargaljuut mineral water deposit is located in front of the Khangai mountain range, Bayankhongor province. It is the biggest complex of natural spring water spa in Mongolia which consists of over 160 mineral water with 40–96 °C temperature springs, which cover an area of 10 hectares.

The mineral water deposit has the highest indicators of surface and artesian water reserve. The underground reserve covers an area of 270 hectares. Hot springs with 40–50 °C temperature is used for bath treatments, and hot springs with 90–96 °C is used for heating and consumption and usage of hot water in the spa resort's facilities. The

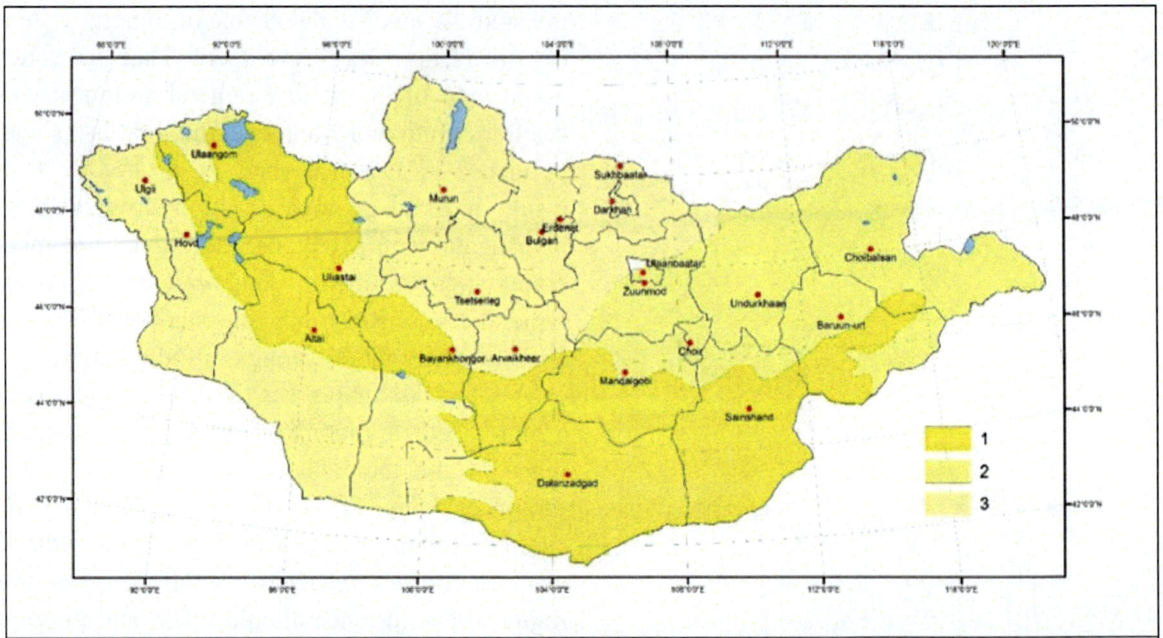


| Province | Representative | Flow of the mineral springs | The temperature of the mineral waters | The principal gas |
|---|---|---|---|---|
| *1. Highly mineralized sodium chloride methane* | Occur in intermountain areas in southern eastern and western Mongolia, where oil prospecting has led to the drilling of deep boreholes | | | sodium chloride water |
| *2.Thermal nitric* | Khujirt, Shargaljuut, Yeruu | 8-10 lit/sec | 20 to 91.5° C | Nitrogen hydrogen carbonate (94%) |
| *3.Cold carbonated* | Ulan Bator spa | 0. 01 to 7-10 lit/sec | less than 4-5°C | Carbon dioxide (92. 9-99.5 %) |

**Fig. 5.10** Mineral springs of Mongolia Modified for the educational purpose (Batchuluun 2012)

**Table 5.8** Water balance of the lakes, mm/year

| Lakes | Surface input | | Surface output | |
|---|---|---|---|---|
| | Precipitation | Inflow | Evaporation | Outflow |
| Khuvsgul | 269 | 408.1 | 665 | 187 |
| Uvs | 96.6 | 395.4 | 689 | 0 |
| Khyargas | 55.9 | 652.4 | 937.1 | 0 |
| Khar-Us | 55.4 | 1,979.2 | 942.7 | 675.3 |
| Khar | 54 | 1,786.9 | 1,117.8 | 1,287.8 |
| Terkhiin Tsagaan | 237.2 | 7,574.3 | 504.2 | 7,307.8 |
| Buir | 250 | 1,394.7 | 860 | 1,472.7 |

*Source* "Surface water of Mongolia" (Myagmarjav and Davaa 2014)

flow of total headwater of hot springs is 78 l/s and the daily natural reserve is 6,739 t/day. It is 10.2% of abyssal reserve and 9.4% of energy reserve. There is 29,250 t/day potential reserve in the depth of 500 m from the surface. When the above natural reserve was transmitted into energy it is calculated as 51123 kcal/s or 214 mW or 184 kcal/s potential reserve.

The Shargaljuut group of mineral waters is valued for its therapeutic feature. Structure and composition of the Shargaljuut group of mineral waters contain various minerals, biologically

**Table 5.9** Some hot springs of Mongolia

| | Name of spring | Elevation. m (a.s.l.) | T°C | Longitude. C | Latitude. C | Mineral (mg/l) |
|---|---|---|---|---|---|---|
| | *Arkhangai aimag* | | | | | |
| 1 | Tsenher | 1,860 | 86 | 101°39″ | 47°20″ | 239.9 |
| 2 | Tsagaan sum | 1,840 | 69 | 102°00″ | 47°40″ | |
| 3 | Gyalgar | 1,900 | 52 | 101°33″ | 47°12″ | 221.6 |
| 4 | Shivert | 1,710 | 55 | 101°31″ | 47°38″ | 24.8 |
| 5 | Bor tal | 1,880 | 46 | 101°36″ | 47°12″ | 258.9 |
| 6 | Chuluut | 2,190 | 45 | 101°15″ | 47°55″ | 250.5 |
| 7 | Noyon | 2,370 | 38 | 101°25″ | 47°20″ | |
| | *Bayankhongor aimag* | | | | | |
| 8 | Baga Shargaljuut | 1,900 | 58 | 101°15″ | 46°13″ | 37.3 |
| 9 | Shargaljuut | 1,335 | 92 | 101°15″ | 46°13″ | 259 |
| | *Zavkhan aimag* | | | | | |
| 10 | Tsetsuuh | 2,050 | 36 | 98°30″ | 48°21″ | 231.2 |
| 11 | Zaart | 2,080 | 44 | 98°46″ | 48°20″ | 247 |
| 12 | Khaluun Us | 2,170 | 35 | 98°24″ | 48°15″ | 236 |
| 13 | Khojuul | 2,170 | 45 | 98°08″ | 48°12″ | 181 |
| 14 | Otgontenger | 2,510 | 56 | 97°30″ | 48°45″ | 223 |
| 15 | Ulaan khaalga | 2,280 | 37 | 97°20″ | 48°55″ | 340.6 |
| | *Khuvsgul aimag* | | | | | |
| 16 | Bulnai | 1,660 | 47 | 100°48″ | 50°48″ | 251 |
| 17 | Salbart | – | 44 | 99°37″ | 48°35″ | 226 |
| 18 | Urtrag | – | 21 | 99°42″ | 51°45″ | – |
| 19 | Tsuvraa | – | 35 | 99°39″ | 48°30″ | – |
| 20 | Khunjil | 1,950 | 62 | 99°22″ | 48°34″ | – |
| 21 | Jalga | – | 40 | 98°01″ | 51°20″ | 105 |
| | *Khentii aimag* | | | | | |
| 22 | Baga Onon | 1,450 | 73 | 108°00″ | 48°56″ | – |
| 23 | Ikh Onon | 1,470 | 88 | 108°49″ | 48°56″ | – |
| | *Uvurkhangai aimag* | | | | | |
| 24 | Khujirt | 1,748 | 55 | 102°46″ | 46°54″ | – |
| 25 | Gyatruun | 2500 | 36 | 102°00″ | 46°36″ | 200.6 |

*Source* Adopted from Namnandorj et al. (1966); (Tseesuren 2001)

active microelements, exclusive features of gas and radon. One of the special features of the mineral water is it contains thermophilic bacteria that can survive at 90 °C which has an antagonistic feature to resist pathogenic microorganisms. Mongolians have been using the Shargaljuut mineral water for thousands of years to cure their diseases.

**Fig. 5.11** Shargaljuut hot spring

## 5.6 Groundwater

### 5.6.1 Groundwater Resources and Geographical Distribution

Because of few monitoring stations having records longer than 10 years, information on groundwater is quite limited in Mongolia (Table 5.10). As stated in the "*Integrated Water Management Plan of Mongolia*" (MEGD 2013), the potential exploitable renewable groundwater resources are 13 $km^3$/year. The officially approved groundwater resources for exploitation, however, are only 0.6 $km^3$/year, or about 6% of the groundwater potential.

The Hydrogeological Map of Mongolia distinguishes between (i) basement complexes consisting of metamorphic and crystalline rocks of the Precambrian, oceanic and volcanic rocks of the Paleozoic, and (ii) platform covers with "surrounding or covering layers" of Perm, Mesozoic or Cenozoic age. Hydrogeological studies focused on areas of high demand for groundwater near urban areas, mines or industrial sites. As a result, descriptions of the so-called "groundwater deposits" appeared. Groundwater is taken from two types of aquifers: fractured and granular. The most productive aquifers are the granular alluvial quaternary aquifers along river valleys. Where these sediments do not exist or are insufficient for recharge, groundwater is taken from fractured aquifers, the most important of which is the Cretaceous sediments.

The hydrogeology of Mongolia is described in few reports and publications, including the explanatory note to the Hydrogeological Map of Mongolia, scale 1:1,000,000 (Nanjiliin Jadambaa et al. 2004) and the report on groundwater prepared for the inception phase of Strengthening Integrated Water Resources Management in Mongolia project (MNE 2007). Specific information on the geology and hydrogeology of the South Gobi Region is available in JICA (2003) and in the background document for this study,

**Table 5.10** Groundwater resources assessment results

| | Groundwater resources. km$^3$ | | Source |
|---|---|---|---|
| | Total | Availability | |
| **1** | 5.58 | 0.6 | A.T.Ivanov 1958 |
| **2** | 12.9 | – | IWER[41] 1973 |
| **3** | – | 6.07 | IWER, 1975 |
| **4** | 6.88 | 6.28 | N.A. Marinov, 1977 |
| **5** | 12.0 | 5.6 | G. Davaa, B. Myagmarjav, 1999 |
| **6** | | 10.79 | N. Jadambaa, G. Tserenjav, 2003 |

*Source* Adapted from "Groundwater Assessment of the Southern Gobi Region" (Tuinhof and Buyanhisnig 2010)

Groundwater Assessment of the Southern Gobi (Tuinhof and Buyanhisnig 2010). Hydrogeological conditions and present status of groundwater resources of Selenge River basin is described in the "Groundwater Resources in Shallow Transboundary Aquifers in the Baikal Basin: Current Knowledge, Protection and Management" report (Vrba 2015).

Groundwater in geological unit of northern Mongolia occurs in fluvial deposits of Quaternary age, sedimentary rocks of Mesozoic age and fractured rocks of pre-Mesozoic age. Shallow aquifers in highly permeable fluvial deposits contain significant groundwater resources widely used for water supplies of municipalities and rural settlements. Deep aquifers in sediments of Cenozoic and Mesozoic age occur in the medium elevated area of the Orkhon-Selenge Basin. It consists of conglomerates, sandstones, argillites, siltstones, clays, and sands. The groundwater level in the sediments of the Cretaceous age significantly differs (from 4 to 80 m) well and yield varies from 0.15 to 10.4 l/s.

Continuous permafrost rocks of 200–500 m or even more thick are widely developed in high massifs of the Khuvsgul, Khangai, and Khentii mountain ranges. Non-continuous permafrost islands of 15–25 and 50–100 m average thickness are spread in the small river basins and valleys. Aquifers in permafrost have not been studied yet in Mongolia. However, groundwater in the permafrost is a valuable source of drinking water for several small rural settlements and for pasture livestock. Renewable groundwater resources amount to 5.08 billion cubic meters (13,931 thousand m$^3$/day), and the total potential exploitable groundwater resources amount to 2.36 billion m$^3$ (6,473 thousand m$^3$/day) in the northern Mongolia groundwater unit (Vrba 2015).

The main source of water is groundwater in southern Mongolia. Most of the territory groundwater aquifers are fossil, so they are hardly recharged. The only source of recharge in the Gobi region is limited precipitation, averaging 115–150 mm/year, and is estimated at only 1 mm/year. Most of this water circulates in the upstream aquifers (0–20 m) and a small part leaches into shallower aquifers (20–50 m) and possibly deeper aquifers (below 50 m). Therefore, the potential of groundwater in these aquifers is directly related to the selected period of time (number of years) and the acceptable reduction in groundwater levels. It is estimated that groundwater potential varies between 200 and 500 million m$^3$/year, based on the assumption of a 25–40-year period and a 50–100 m reduction in groundwater table. The lower limit of 200 million m$^3$/year is equal to 550,000 m$^3$/day and it can be concluded that the groundwater potential for the Gobi region as a whole is sufficient to cover the water needs in the next 10–12 years (Tuinhof and Buyanhisnig 2010).

### 5.6.2 Groundwater Mineralization and Usage

Groundwater is the main source of drinking and industrial water in Mongolia. It can be assumed that 99% of the population uses groundwater for drinking water supply. Groundwater from wells in remote areas is used for drinking water supply of livestock. Most irrigation systems use surface water, but groundwater use is increasing. In addition, most mines and enterprises extract groundwater. Mines also pump groundwater for dewatering the quarry. Industrial enterprises in cities use water from either the central system or their own wells. Groundwater is currently being monitored by MEGD, IMHE, water companies and private mining companies at an increasing number of sites. Typically, the amount of available observation data is too small to be used at present to assess groundwater resources.

The total dissolved solids (TDS) content of groundwater increases with decreasing precipitation and flow. This means that the concentration of soluble in the groundwater increases in a southern direction and away from rivers. Groundwater with high total dissolved solids contents is found in areas with high evaporation and low groundwater flow velocities, in confined aquifers, which receive little recharge and in areas with easily soluble mineral deposits, such as rock salt, gypsum or soda. The groundwater quality can be defined by four geographical zones (MEGD 2013):

*The Khangai–Khentii mountainous region*, the forest steppe area (30% of Mongolia) that covers the northern and central part of the country. Groundwater in this region generally has a TDS between 100 and 800 mg/l, with rare exceptions of > 1,000 mg/l. The hardness of water in the region is about 4.5 mg-eqv/l.

*The Altai mountainous region, the mountains of Western Mongolia.* Average TDS of the groundwater ranges between 510 and 810 mg/l and hardness is 4.5–5.2 mg eqv/l, with the highest concentrations found in Gobi-Altai.

[4]IWER—Institute of Water Exploratory Research, Ulaanbaatar.

*The Mongolian Dornod steppe region in the east of the country.* Groundwater in this zone has an average TDS of 950 mg/l, the average hardness is 5.6 mg-eqv/l and it is characterized by a high concentration of iron.

*Gobi region.* Groundwater typically has a TDS of 500–1,350 mg/l and the hardness is 3.4–7.0 mg-eqv/l. Salinity is the main water quality concern in the Gobi region due to high evaporation, low precipitation and flat surface morphology. Water quality in more than 100 soums does not meet the drinking water quality standard, since 60% of the soums have water with a too high TDS and 40% have high hardness.

The natural groundwater quality in Gobi and Dornod is generally unsatisfactory. Groundwater with a high concentration of fluorine is found in 25–35% of soums in the Dornod and Gobi aimags. Low concentrations of fluorine are found in most other aimags. Shallow groundwater constitutes only a small part of the total groundwater resources in the Gobi region. Most of the groundwater resources are located in deeper aquifers and are mainly fossil fuels. This water is not affected by current changes in climate variability, as it has been recharged in other climatic conditions and is now in deeper layers.

## References

Adiyabadam G (2013) Ecological and socio-economic baseline studies conducted in Ulz river basin. Ministry of Nature, Environment and Green Development, Ulaanbaatar

Baldandorj T (2017) Water resource and its' appropriate usage, vol II. Munkhiin Useg, Ulaanbaatar

Batima P, Dagvadorj D (2000) Climate change and its impacts in Mongolia. JEMR Publishing, Ulaanbaatar, Mongolia

Bichurin (1828) Zapiski o Mongolii (Notes on Mongolia), Samara, Agni Publ

Bogoyavlenskii B.A (1989) Atlas of Khuvsgul lake. In: MAR, AOS Institute of Geography (ed) General administration of Geodesy and Cartography, Russia, Moscow

Dagvadorj D, Batjargal Z, Natsagdorj L (2014) MARCC-2014: Mongolia second assessment report on climate change—2014. Ulaanbaatar

Dagvadorj D, Dorjpurev J, Namkhainyam B (2009) MARCC 2009: Mongolia assessment report on climate change 2009

Damdinsuren T (1957) Mongoliin nuuts tovchoo (The Secret History of the Mongols). State Printing Plant, Ulaanbaatar

Davaa G, Oyunbaatar D, Sugita M (2006) Surface water of Mongolia. Environmental Book of Mongolia. Tokyo, Japan, pp 55–82

Davaa G, Konya K, Purevdagva K, Davaadorj N, Baasandorj D, Batkhuu D, Yura, Khasherdene T, Sodnombaljir S, Bahitbol (2012) Variation in ice cup and ice rivers and trends of its mass balance and changes in Mongolia. Paper presented at the Workshop on climate change in high mountains, Khovd

Dulmaa A (1979) Hydrobiological outline of the Mongolian lakes. Internationale Revue der gesamten Hydrobiologie und Hydrographie 64(6):709–736

Egorov A (1993) Mongolian salt lakes: some features of their geography, thermal patterns, chemistry and biology. Hydrobiologia 267(1–3):13–21

Gendenjamts O (1984) Geochemical study of mongolian hot springs. Geochim Cosmochim Acta 48:1479–1492

Ivanov A (1958) Podzemnye vody Mongol'skoi Narodnoi Respubliki (Ground Waters of the MPR). Trudy Laboratorii Gidrogeologicheskikh Problem, 19

Jadambaa N (2002) Ground water of Mongolia. In: Жадамбаа Н (ed) White book of Mongolia, Ulaanbaatar, 199-214. Монгол орны газрын доорх ус, Монголын цагаан ном, Улаанбаатар, pp 199–214

Jadambaa N, Grimmelmann W, Kampe A (2003) Hydrogeological Map of Mongolia 1:1,000,000 Explanatory Notes. Geologisches Jahrbuch Reihe C 69:50

Jadambaa N, Grimmelmann W, Kampe A, Mongolia I (2004). Explanatory notes for the hydrogeological map of Mongolia-1: 1000000

Klimek K, Lomborinchen R, Starkel L, Sugar T (1976) Research problems of the first Mongolian-Polish physical geographical expedtion to the Khangai Mts. in 1974. Stanislaw Leszczycki (editor -in-chief) 107

Kondratiev S (1929) Telmen-Nur and West of Hangai plateau. Econ Mongolia:11–16

Krashnikov A (1973) Surface-groundwater relationships in Central Mongolia. Academy of Sciences USSR, Moscow

Kuznetsov N (1955) Main regularities rivers regime of Mongolian People's Republic. Academy of Sciences USSR, Moscow

Kuznetsov N.T (1951) To the question about water exchange between the lakes in hollow of great lakes in Mongolia. News of AS SSSR, *5*(Series Geogr)

Marinov N, Popov V (1963) Hydrogeology of the Mongolian Peoples Republic. Gosgeoltekhizdat

Marinov N, Tolstikhin N, Shkapskaya S (1974) A new scheme for regionalization of subsurface mineral waters of Mongolia. Int Geol Rev 16(6):697–703

MEGD (2012) Tuul River Basin Integrated Management Plan. Ulaanbaatar

MEGD (2013) Integrated water management plan of Mongolia. Ulaanbaatar

MEGD (2014) Ulz River Basin Integrated Management Plan. Ulaanbaatar

Murzaev E.M (1948) The Mongolian People's Republic. A physical geographical description. State Publication of Geographical Literature, Moscow (in Russian)

Myagmarjav B, Davaa G (1999) Surface water of Mongolia. Interpress, Ulaanbaatar, p 345

Myagmarjav B, Davaa G (2014) Surface water of Mongolia. Interpress, Ulaanbaatar, p 345

Namnandorj O, Nyamdorj Sh (1960) Springs of the Mongolian People's Republic. State press agency, Ulaanbaatar

Namnandorj O, Nyamdorj Sh (1966) Mineral waters of the Mongolia. Ulaanbaatar, Mongolia

Potanin G (1883) Ocherki severo-zapadnoj Mongolii (Essays of northwestern Mongolia). St, Petersburg

Pozdneev A.M (1892) Mongolia and the Mongols, vol. I, transl. from Russian

Przhevalsky N, Morgan ED (1876) Mongolia. Рипол Классик

Punsalmaa B, Nyamsuren B, Buyndalai B (2004) Trends in river and lake ice in Mongolia. Assessments of impacts and adaptations to climate change (AIACC). Working Paper (4)

Sevastyanov D.V, Dorofeyk N.I (2005) A short review: limnological and paleolimnological researches in Mongolia carried out by joint Russian-Mongolian expeditions. J Mt Sci 2(1):86–90

Smirnov V (1932) The report on the works of hydrochemical group of Mongolian Expedition in 1926. Proc Mongolian Comm 1:1–23

Smirnov V.A (1932) Account about hydro chemical works of Mongolian expedition of 1926. Leningrad

Tseesuren B (2001) Geothermal resources in Mongolia and potential uses (1670–7427). IS-108 Reykjavík, Iceland

Tsegmid Sh (1955) From the Past of Valley of Lakes in Mongolian Gobi. Natura 5:46–49

Tsegmid Sh (1969) Physical geography of Mongolia. Institute of Geography & Permafrost, Mongolain Academy of Sciences. State Press, Ulaanbaatar, p 405

Tserensodnom J (1971) Lakes of Mongolia. Institute of Geography & Permafrost, Mongolain Academy of Sciences. State Press, Ulaanbaatar, p 192

Tserensodnom J (2000) A catalog of lakes in Mongolia. Institute of Geography & Permafrost, Mongolain Academy of Sciences. Ulaanbaatar Shuvuun Saaral Publishing. p 141 (in Mongolian)

Tuinhof A, Buyanhisnig N (2010) Groundwater Assessment of the Southern Gobi Region. Mongolia Discussion Papers, East Asia and Pacific Sustainable Development Department. World Bank, Washington, DC

Tuvdendorj D (1985) The atlas of the climate and ground water resources in the Mongolian People's Republic explanetary notes
Vrba J (2015) Groundwater resources in shallow transboundary aquifers in the Baikal Basin: current knowledge, protection and management
Williams W (1991) Chinese and Mongolian saline lakes: a limnological overview. Hydrobiologia 210(1–2):39–66
Yembuu B, Tugjamba N, Gezel U, Gonchigjav Yu (2012) Geography III. Munkhiin Useg Publishing, Ulaanbaatar

# 6 Glaciers of Mongolia

Ser-Od Tsedevdorj

**Abstract**

In general, the mountains are elevated and the climate is dry due to the air barrier from the west and northwest in Mongolia. The cool climate is a condition for the preservation of Mongolian high mountain ranges in the glacial cover. The first part of this chapter deals with a brief history about the Mongolian glacial research, which considers factors such as climate, depressions, surface elevation, sedimentation, geographical distribution and latitude correlation. The following section shows the distribution of glaciers in the Mongol Altai, Khangai and Khuvsgul mountains, as evidenced by the changes in ancient and modern glacial traces and extent of the area as a result of the impact, causes and consequences of climate change. In the third part of the chapter, the impact of glaciers on the surface caused by surface forms and changes in the hydrological network due to global warming is considered as the example of some glacial mountains.

**Keywords**

Glacier · Glacial deposits · Glacial landforms · Glacial lake

## 6.1 Overview of Glacial Research in Mongolia

Although the glaciers of Mongolia are not well studied (Ohata et al. 2009), ancient glaciation has been investigated and described by various researchers who have focused more on glaciation and glacial landforms, such as Potanin G.N (1893), Pevtsov M.V (1951), Sapojnikov V.V (1911), Obruchev V.A (1940), Murzaev E.M (1952), Ivanov A.H (1949), Richter H, Nicholaev N.V, Marinov N.A (1967), Selivanov E.I (1972), Devyatkin E.V (1965), Grane I. G (1910), Byerki and Morris (1972). The glacial geomorphology of the high mountain ranges in Mongolia were studied (Tsegmid 1969; Jigj 1975; Munkhuu 1976), as well as the glacier landscape and its conservation and use by Ser-Od (Ser-Od 2019).

The first detailed description of the glaciers of the Tsagaan Gol (Akkol) river and Tsagaan-Us (Aksu) river basins in the Tavan Bogd mountains were edited by Sapojnikov (1911, 1949). Based the field data obtained during the 1905, 1906 and 1909 expeditions, he invented (Fig. 6.1) the scheme of glaciation of the Tavan Bogd

S.-O. Tsedevdorj (✉)
Department of Geography, Mongolian National University of Education, Ulaanbaatar, Mongolia
e-mail: serod@msue.edu.mn

© Springer Nature Switzerland AG 2021

B. Yembuu (ed.), *The Physical Geography of Mongolia*, Geography of the Physical Environment,
https://doi.org/10.1007/978-3-030-61434-8_6

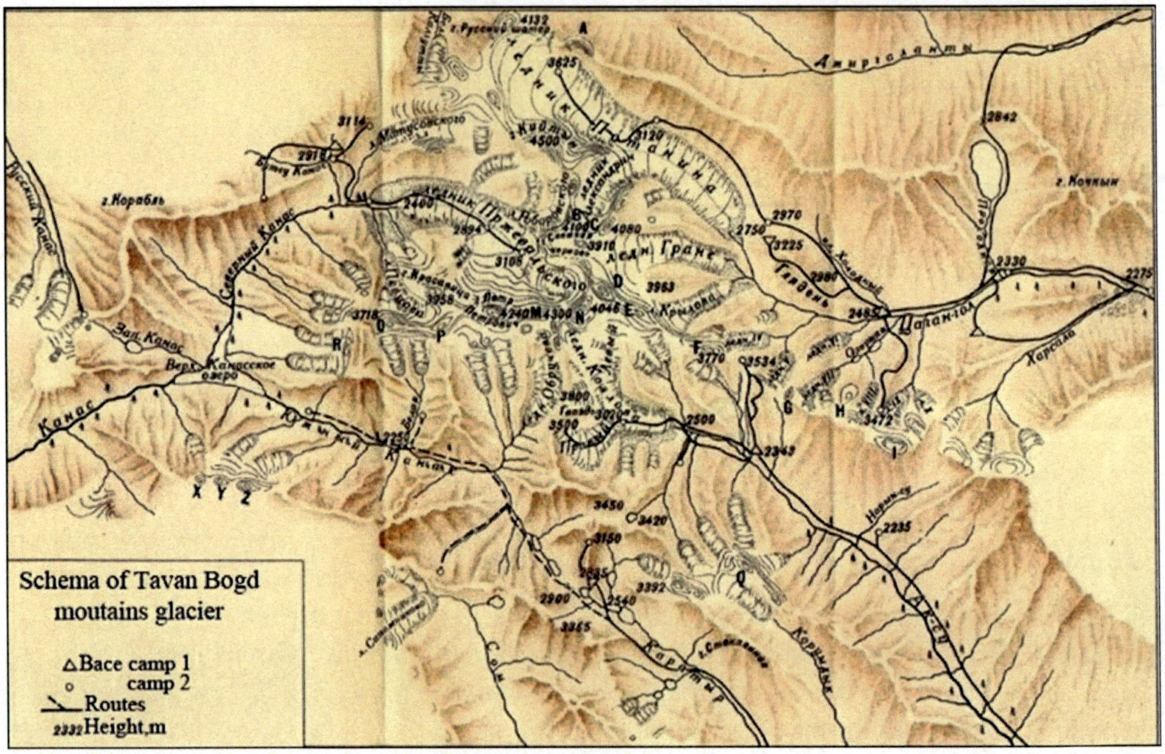


**Fig. 6.1** Glaciations near the Tavan Bogd mountain–Sapojnikov (1911). *(published "Across the Altai")*

mountains and detailed map of the Mongol Altai hydrography and described the current and former glaciation.

Tronov M.V and Tronov B.V (1916) had described glaciers of the Tavan Bogd mountains. A Catalogue of Altai glaciers and a map of the Khovd region on the basis of his research and materials collected by Sapojnikov were compiled (Tronov 1925). At that time, 25 glaciers were known in the Tsagaan-Us and Tsagaan-Gol river basins with a total area of 110 km$^2$. By the middle of the twentieth century, one of the largest glaciers in the Altai mountains, the Potanin glacier, was described in detail. Its length was estimated as 19 km and its area as 50 km$^2$ (Tronov 1949). The scheme of glaciation of the Tavan Bogd mountain (Sapojnikov 1949) is still used for research as the main source.

After a long break, the Altai State University (ASU) organized the Soviet-Mongolian expedition to the Tavan Bogd mountains including the phototheodolite survey of Potanin glacier in 1987. Its proportions were updated (length 11.5 km, area 38.5 km$^2$) as well as new glaciers were discovered and glacier dynamics since the middle of the nineteenth century was estimated. At that time, 60 glaciers in the massif with an area of about 223 km$^2$ were known.

The glacial areas and glaciers in Mongolia, as determined by various studies, such as Selivanov (1972) had 120 glaciers ranging in length from 0.5 to 20 km and measured a glacial area of 328 km$^2$; according to Dashdeleg (1983, 1990) 187 glaciers with a total area of 540 km$^2$ (Dashdeleg 1990). Baast (1998) estimated that the glacier area covers more than 262 glaciers with an area 300 km$^2$, and Klinge et al. (2013) identified 731 glaciers with an area of 655 km$^2$. In 1999–2002, Ohata et al. (2009) counted around 580 and 578 glaciers, respectively, with both surveys covering 423 km$^2$, which is roughly 24% of the Mongol Altai (Kamp et al. 2013). In 2001, ASU placed the marking stones near edges of Kozlov and Potanin glaciers and estimated the rates of large glacier retreat on the basis of field survey and literature data (Syromyatina et al. 2015).

In general, since 2000, studies of glaciers in Mongolia have developed gradually; recent glacier changes (Yabuki and Ohata 2009), its impacts on surface water resources (Orkhonselenge and Harbor 2018), mass balance (Konya et al. 2008) and detailed study of individual glaciers such as Mongol Altai (Kadota et al. 2011), Muknhairkhan, Tsambagarav (Davaa et al. 2008), Munkh Saridag and Ikh Turgen (Orkhonselenge 2014, 2017). In addition, paleoglacial researches are carried out by national and foreign researchers, by Lehmkuhl and Batkhishig (2012, 2016), Walter and Dashtseren (2017) and glacial geosystem by Plyushin (2013).

## 6.2 Distribution and Variation of Glaciers

One of the physical geographical features of Mongolia is the distribution of glaciers in the northern and western mountains. The geographical location in the northern hemisphere and severe cold climatic conditions, as well as Mongolia's high elevation, are the causes of glacier formation. The spatial distribution of glaciers in Mongolia is between 46°25′N and 50°50′N and 87°40′E and 100°50′E, and located at altitudes of 2750–4374 m above the sea level (Davaa et al. 2008).

The morphometric features and snow line of the mountains and the melting rate of the largest glaciers in Mongolia are presented in Table 6.1, which is based on glacial studies.

Mongolia's glacier masses are distributed within 42 mountain massifs (Fig. 6.2). The glacier massifs were classified into eight categories using glacier areas based on Landsat images by Davaa et al. (2014) (Table 6.2).

About 10% of fresh water is contained in the glaciers of Mongolia (Dashdeleg et al. 2015).

### 6.2.1 Glaciers of Mongol Altai Mountains

The Mongol Altai is located on the border between Mongolia and China, stretching from the northwest to the southeast for 800 km and ending in the Gobi Desert, which is the most extensive longitudinal mountain range in Central Asia and continues to the Sayan mountains of Russia. There are a number of isolated mountains greater than 4000 m altitude within the Mongol Altai with the highest elevations in the Tavan Bogd, where the highest Khuiten Peak reaches a maximum elevation of 4374 m. Glaciers occur at the highest peaks above the main leveling surface in the central part of various mountain systems within the Mongol Altai, and moreover, in plateau glaciers and cirque glaciers, many isolated ice patches have existed (Ser-Od 2019). According to air photos and satellite images, rock glaciers are common in the Mongol Altai at elevations of more than 2,200 m above the sea level (Lehmkuhl and Batkhishig 2003).

***Tavan Bogd mountain glaciers.*** The Tavan Bogd massif has an area of about 2600 km$^2$, and is situated in a remote intracontinental arid area. It is the largest glacial western of the Mongol Altai. The left headwater of Khovd river (Tsagaan-Us river) and its large tributary Tsagaan-Gol river begin from the glaciers of the Tavan Bogd in the eastern and southeastern slopes. In contrast, the western and eastern slopes of the massif are strongly dissected and have a pronounced alpine type of relief. These mountain peaks are elevated well above the current average firn line/equilibrium line altitude (ELA) (3285 m) persisting during the Little Ice Age (LIA) (3235 m), which makes it possible for glaciers to exist (Ganyushkin et al. 2015). There are eight morphological types of glaciers along the Tavan Bogd mountains (Ganyushkin et al. 2018).

Potanin glacier is currently one of the best-studied glaciers of the Tavan Bogd mountain. Dashdeleg (1983) presented that its length is 12 km and its area is 54 km$^2$; however, for which year is uncertain. During the last several years, Japanese–Mongolian group has been carrying out field investigations, including mass balance and meteorological observations on the main stream of Potanin glacier. Baast was measured in 1985 with an area of 47 km$^2$ (Baast 1998), and its length was determined by Kadota and Davaa in 2000 at only 11 km and an area of 43 km$^2$ (Kadota and Davaa 2004; Kadota 2007).

**Table 6.1** Morphometric indicators of Mongolian glaciers (based on National Atlas of Mongolia 1990, 2009)

| № | Glacier mountains | Mountain height, m | Height of snow line, m | Glacier area, km$^2$ | Glacier melting amount, mm/year |
|---|---|---|---|---|---|
| 1 | Khuiten | 4374 | 3628 | 105.71 | 527 |
| 2 | Samartai | 4082 | 3441 | 0.99 | 985 |
| 3 | Tsagaan khairkhan | 3628 | 3181 | 3.96 | 1287 |
| 4 | Rashaan | 3668 | 2900 | 0.99 | 2059 |
| 5 | Dungurukhiin barzgar | 3600 | 2600 | 1.76 | 1844 |
| 6 | Tsast mountain | 3669 | 3260 | 18.59 | 1230 |
| 7 | Tsengelkhairhan | 3943 | 3522 | 2.75 | 733 |
| 8 | Undurkhairkhan | 3914 | 3326 | 18.59 | 1130 |
| 9 | Khukh Sain | 3513 | 3410 | 11 | 509 |
| 10 | Tsagaan | 3732 | 3295 | 7.04 | 985 |
| 11 | Dush | | 3665 | 1.76 | 483 |
| 12 | Munkhkhairkhan | 4204 | 3781 | 48.51 | 440 |
| 13 | Bugat | 4041 | 3570 | 18.59 | 575 |
| 14 | Baatarkhairkhan | 3984 | 3743 | 5.39 | 401 |
| 15 | Sutai | 4090 | 3855 | 11 | 296 |
| 16 | Sair | 3684 | 3698 | 13.31 | 237 |
| 17 | Tsast | 4193 | 3904 | 92.51 | 237 |
| 18 | Munkh Tsast | | 3409 | 5.39 | 440 |
| 19 | Ik Turgen | 4029 | 3617 | 7.04 | 575 |
| 20 | Asgat | | 3426 | 7.04 | 853 |
| 21 | Turgen | 3965 | 3509 | 44 | 677 |
| 22 | Harkhiraa | 4037 | 3413 | 63.36 | 853 |
| 23 | Otgontenger | 4021 | 3752 | 1.32 | 446 |
| 24 | Munkhsaridag | 3491 | 3306 | 1.76 | 754 |

Kamp (2013) noted no significant change in glacier area between 1988 and 2000 (Konya et al.) (Kamp et al. 2013) (Fig. 6.3).

One of the most understudied glaciers of the Mongolian part of the Tavan Bogd mountains is Kozlov valley glacier. The glaciers and climate surveys were carried out from August 2013 to August. Using the HOBO Micro Station Data Logger tool, the temperature, moisture, and sediment of these glaciers are measured. At the same time, the glaciers on the north and south were sloped to the bottom of the glacier, and 22 poles of measurements were carried out in 2013–2015. According to data from the automatic station on the glacier, the intensity of the melting ice was increased by an average of 5.8 cm/day in connection with the increase in air temperature (Fig. 6.4).

Since the Altai State University (ASU) scientists measured the Kozlov glaciers in 2001, Landsat map had shown that the glacier melt was 130 m long and changed to 120 m from 2006 to 2013. Regarding the research results, the melting of Kozlov glacier was about 21 m/year in 2001–2013 (Syromyatina et al. 2015).

The first absolute optically-stimulated luminescence (OSL) dates for the southern slopes of Tsengel Khairkhan Uul were introduced by Lehmkuhl et al. in 2016. Walter noted that "he described three main moraine systems from:

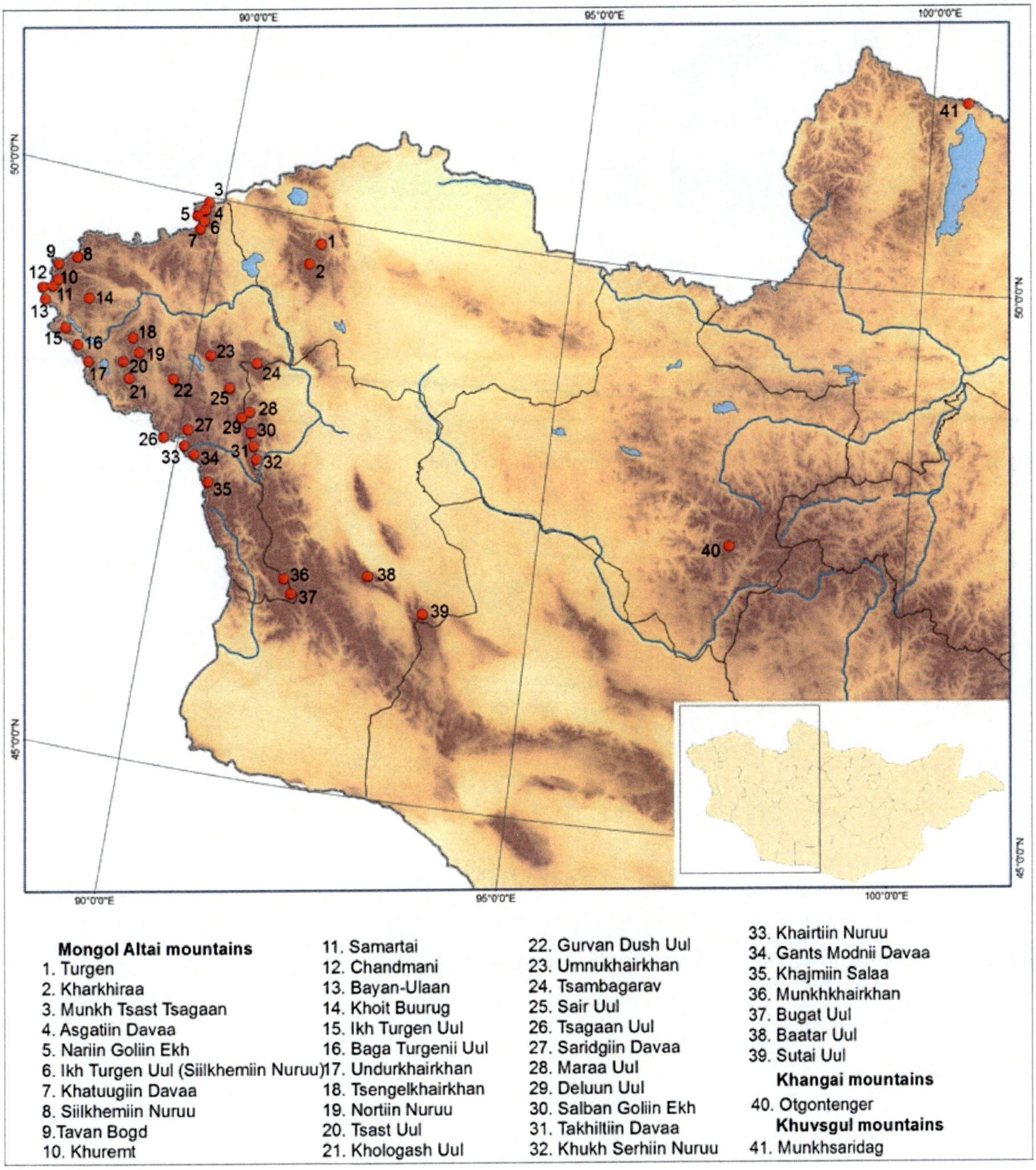


**Fig. 6.2** Glaciers distribution of Mongol Altai, Khangai and Khuvsgul mountains (modified by Ser-Od)

(i) the Holocene-and, in particular, the LIA-without any sediment and soil cover; (ii) the Local Last Glacial Maximum (LLGM; > 10.4 ± 0.6 ka) in three sub-stages of larger moraines ($M_{1a}$, $M_{1b}$, $M_{1c}$) further down-valley; and (iii) the penultimate glaciation ($M_2$; > 97.6 ± 30.1 ka and > 71.2 ± 4.8 ka) as relics of eroded moraines" (Walter et al. 2017).

***Munkhkhairkhan mountain glaciers.*** The morphostructure of the mountain range is homogeneous and consists of a block extending along a northwestward direction and comprising Paleozoic substrate bounded by sub-parallel faults rejuvenated at the neotectonic stage and are clearly expressed as the relief in the form of rectangular segments of valleys. Ice-firn field

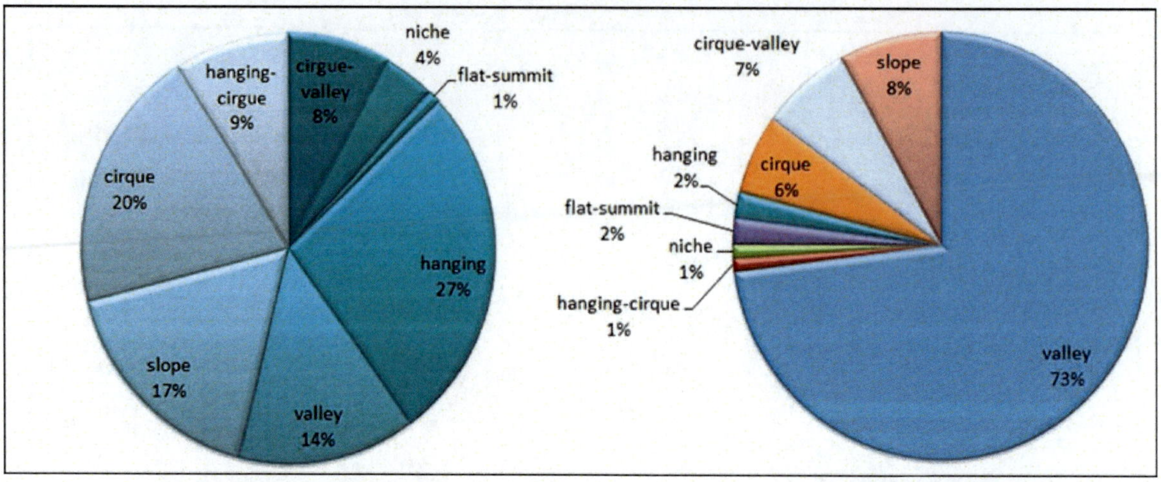


**Fig. 6.3** **a** Distribution of the number of glaciers morphological types. **b** Distribution of glacial area in morphological types (modified from Ganyushkin et al. 2018)

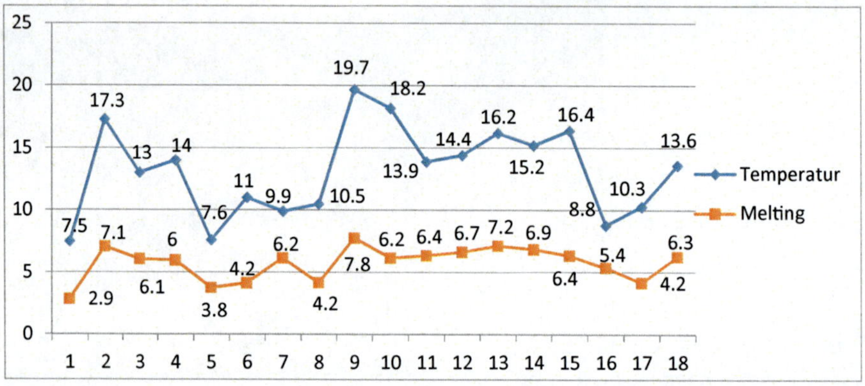


**Fig. 6.4** The melting of the Kozlov glacier and air temperature (modified from Syromyatina M.V)

**Table 6.2** The classification of glacier massive (Based on MARCC 2014)

| № | Massive area | Interval of area, km$^2$ | Mountains | Area of glaciers, km$^2$ |
|---|---|---|---|---|
| 1 | Large | 100< | Tavan Bogd | 106 |
| 2 | Big | 50–100 | Tsambagarav | 66 |
| 3 | Bigger | 20–50 | Kharkhiraa, Turgen, Munkhkhairkhan | 31.2, 29, 26.3 |
| 4 | Moderately big | 10–20 | Undurkharkhan, Sutai, Ikh Turgen (Asgat) | 11.61, 11.95, 10.08 |
| 5 | Medium | 5–10 | Khuremt, Nortiin nuruu, Sair, Deed Nariin Gol | 7.98, 6.14, 7.44, 5.1, 7.99 |
| 6 | Moderately small | 3–5 | Samartai, Tsengelkharkhan, Saridgiin Davaa, Takhilt, Khajmiin Salaa and others | – |
| 7 | Small | 1–3 | Chandmani, Bayan Ulaan, Khoit Bulag and others | – |
| 8 | Very small | <1 | Maraa, Bugat, Munkhtsast, Otgontenger, Munkhsaridag Kholagash | – |

*Source* MARCC 2014 (MEGD 2014)

occurs in its axial part on the flattened summits of Munkhkhairkhan (4362 m) (Fig. 6.5) and Bugat (4041 m), and on other summits (Plyushnin et al. 2013).

Klinge (2001) demonstrated that the glacier area of the Munkhkhairkhan mountains was 25.2 $km^2$ in the 1980s/1990s and reported that most individual plateau glaciers, like the one at the summit of Munkhkhairkhan (4364 m), and snowfields were common, while cirque glaciers were rare, and only three valley glaciers existed. During the last decades, glaciers in the Munkhkhairkhan mountains exhibited a general trend of recession. Krumwiede et al. (2014) show that from 1990 to 2006, the glacier area decreased by 28%, from 39 to 28 $km^2$, and almost all of the glaciers receded by 20–80 m at a rate of 1.3–5 m/a, and only the Khar Tsunkh glacier increased in area by 6% (Table 6.3).

In Munkhkhairkhan, 17 glaciers with 26.57 $km^2$ was distributed in the northwest (Doloon nuur glaciers), central (Shuurkhain glaciers) and in the southeast (Khukh nuur glaciers) side of the mountain (Otgonbayar 2011). The main types of these glaciers are icefields (6), hanging (4), cirque (3) and valley glaciers (4) (Otgonbayar 2012).

***Turgen-Kharkhiraa glaciers***. The first scientific descriptions of glaciers in the Turgen-Kharkhiraa mountains were investigated by Potanin (1914), Carruthers (1911, 1912, 1914), and Price (1968). During the expeditions of the Russian Geographical Society from 1876 to 1879, the eastern slopes of the mountains were mapped by Potanin. In addition, researches Carruthers and Price (1910) from the Royal Geographical Society expedition compiled a detailed topographic map of the Altai mountains and documented the extent of the glaciers with photographs. The result of this study was a collection of "*Unknown Mongol*" (Kamp et al. 2013).

The Quaternary glaciations and modern glaciers in the Turgen-Kharkhiraa mountains are relatively well studied (Fig. 6.6). One of the most detailed studies is the work of Khrutsky and

**Fig. 6.5** Munkhkhairkhan mountain glacier (photo by Nyamdavaa, G)

**Table 6.3** Change of the Munkhkhairkhan mountain glaciers, 1990–2006 (Krumwiede 2014)

| Year | Doloo Nuur glacier | Shuurkhai glacier | Khar Tsunkh Glacier | Arts glacieret | Arts glacier | Total Glacier area |
|---|---|---|---|---|---|---|
| 1990 | 2.94 | 0.84 | 1.49 | 0.26 | 1.24 | 6.77 |
| 2000 | 2.55 | 0.81 | 1.44 | 0.20 | 0.90 | 5.91 |
| 2002 | 2.25 | 0.77 | 1.37 | 0.15 | 0.80 | 5.34 |
| 2005 | 2.65 | 0.77 | 1.55 | 0.18 | 0.82 | 5.96 |
| 2006 | 2.53 | 0.81 | 1.58 | 0.17 | 0.80 | 5.88 |
| 1990–2006 ($km^2$) | −0.41 | −0.03 | +0.09 | −0.10 | −0.44 | −0.89 |
| 1990–2006 (%) | −13.86 | −4.08 | +6.19 | −36.25 | −35.74 | −13.13 |

**Fig. 6.6** Glacier of Turgen mountains (own photo, 2014)

Golubeva (2008), which stated that the mountains represent a significant center of glacierization in the Altai mountains, which is not exceeded by any other glacierized basin, such as Tsast-Uul, Munkhkhairkhan and Deluunii Nuruu. The maximum extent of past glaciation in the Ikh Turgen mountains has been identified and mapped in 2018 (Blomdin et al. 2018).

The southern slopes of the mountain ranges are precipitous, with relative altitudes of up to 500, whereas the northern slopes are gentle. The summit complex is represented by subdued watersheds with upland terraces covered in large-block deposits as well as by old-glacial forms: corries, cirques and through valleys with banks and ridges of morainal deposits (Plyushnin et al. 2013).

### 6.2.2 Glacier of Khangai Mountains

Khangai mountains are in the central part of Western Mongolia and extend for roughly 700 km from east to west. Otgontenger (4,021 m) is the highest peak of the range and is the only one covered by permanent snow and firn or (stagnant) ice. Baast (1998) measured a glacial area of 0.27 $km^2$. Richter et al. (1961) calculated a recent equilibrium line altitude (ELA) of 4,000 m in these mountains, while Lehmkuhl and Lang (2001), Rother H and Lehmkuhl F (2014) described two small ice fields reaching down to 3,200 m and estimated the recent ELA to be 3,700 m.

In the Khangai mountains, an ancient glacier in the form of morphosculpture and accumulation is clearly visible. From the base to the top of the mountain, it has a wall, sometimes a very large shroud. However, the lack of ice movement is typical for some river valleys. On the freezing of stationary dead glaciers in Khangai is written by Gerasimov and Lavrenko (1952) (Tsegmid 1969). Compared to the Mongol Altai and Khentii glaciers, the glaciers in the Khangai are poorly studied, and glacial lakes here are still not clearly estimated. The largest glacier in the Khangai is located in the valley of the Chuluut river at a height of 2,100 m (ELA), and its maximum length is about 60 km (Tsegmid 1969).

### 6.2.3 Glacier of Khuvsgul Mountains

The Khuvsgul mountain range is located in the north of Mongolia and is surrounded in the north by the Sayan mountains and the southeast end of the Eastern Sayan. Mount Munkhsaridag is located in the center of the Khuvsgul mountain range and rises to 3,491 m on the border of Mongolia and Russia (Fig. 6.7). The Khuvsgul mountains is a watershed that includes large lakes and rivers in the Arctic Ocean basin and is characterized by cool summers and extremely cold winters due to the high annual temperature amplitude (Kovalenko 2011; Orkhonselenge 2017).

Munkhsaridag is the center of Khuvsgul mountain range, had lost 42.6% of its total glacier area between 1970 (900 $m^2$) and 2007 (384 $m^2$) when the equilibrium line altitudes (ELAs) of the glaciers were retreated by 47 and 80 m on the north and south aspects, respectively.

The spatial analyses show the dynamic change of the modern glaciers in Mt. Munkhsaridag, and how this depends on topographical elements of elevation and aspects, durations of solar radiation and vulnerability to the solar insolation (Orkhonselenge 2017).

In the Khuvsgul mountain range there are numerous remnants of the paleoglacier, such as post-glacial gables, wide glacial depressions and large glacial lakes between the mountains of Khoridol-Saridag, Bayan and Ulaan-Taiga, the basin of the river Shishkhid and the upper part of the river Uur (Jigj 1976). Mount Munkhsaridag, which is on the northeastern edge of the Khuvsgul mountains, has steep slopes with precipices and paleoglacial remains, including circuses, U-shaped valleys and moraines (Tsegmid 1969).

The Khuvsgul mountain range has experienced glaciations several times in the Quaternary period (Tsegmid 1969; Jigj 1976; Grosswald and Rudoy 1996; Krivonogov et al. 2005; Gillespie et al. 2008, Kovalenko 2011, 2014; Orkhonselenge 2014; Blomdin 2016). Modern glaciers in Munkhsaridag mountain areas are estimated to have been 900 $m^2$ in 1970, 851 $m^2$ in 1986, 547 $m^2$ in 2000 and 384 $m^2$ in 2007, and have decreased by 42.6% between 1970 and 2007 (Orkhonselenge 2016).

### 6.2.4 Glacier of Khentii Mountains

The Khentii represents one of the typical land-locked high mountain systems, located within the

**Fig. 6.7** Munkhsaridag, Khuvsgul mountain range (photo by Batbold, D)

ecologically sensitive transition zone between Siberian forest area to the north and Mongol-Daurian steppe zone to the south and east. Khentii is a crucial region in reconstructing the Quaternary climate on both regional and global scales. Evidence of paleoglaciations in the Khentii mountains, such as glacial landforms (i.e., glacial troughs, moraine deposits, outwash plains, and glaciofluvial terraces), glacial sources, and their extent has been recorded and mapped by Mongol and Russian researchers (Tsegmid 1951; Jigj 1976, Devyatkin 1981). Recently, a study (Orkhonselege and Uuganzaya 2018) provided maps of glacial geomorphology and calculated extent of glaciation in Asraltkhairkhan and Baga Khentii Saridag, which are located about 30 km westward and 35 km eastward from the study area. While glacial landforms in northwestern Mongolia such as Altai, Khangai, and Khuvsgul were thought to have been formed multiple times during the Late Quaternary, no glacial chronological records have been reported for the Khentii (Kamp et al. 2013).

High-altitude glacial trimlines and abundant moraine deposits are evidence for the Late Pleistocene glacier advance in the Khentii mountains (Tsegmid 1951). Late Pleistocene and Local Last Glacial Maximum (LLGM) ELA of the past glacier were calculated from the five moraine sequences (Table 6.4).

## 6.3 The Trend of Changes in the Glaciers

The characteristics of the prevailing extreme continental climate condition locally differ with topography, and especially with respect to orography. The maximum air temperature reaches 11.6 °C (2009-7-12, at 1,700 m), while the minimum air temperature reaches −43.0 °C (2010-1-20, at 1,530 m), observed in Ulaan Tsunkhegiin Tsakhir and Munkhkhairkhan mountains. Air temperature inversion occurs in the Munkhkhairkhan area in winter, and continued for 93 days in 2008 and for 86 days in 2009,

**Table 6.4** Late Pleistocene ELA reconstruction in the Khentii mountains (modified from Purevmaa 2018)

| Moraine stratigraphy | Description | Aspect | Peak elevation (m)[a] | Headwall (m)[b] | Glacier terminus (m)[c] | ELA (m) | | |
|---|---|---|---|---|---|---|---|---|
| | | | | | | THAR (0.50) | THAR (0.58) | MELM |
| $M_{y2}$ | Latero-frontal | West | 2665 | 2545 | 1510[d] | 2028 | 2110 | – |
| $M_{y1}$ | Latero-frontal | West | 2665 | 2545 | 1430 | 1988 | 2077 | 1897 |
| $M_{k4}$ | Terminal | West | 2562 | 2380 | 2090 | 2235 | 2258 | – |
| $M_{k3}$ | Terminal | West | 2562 | 2380 | 2020 | 2200 | 2229 | – |
| $M_{k2}$ | Terminal | West | 2562 | 2380 | 1855 | 2118 | 2160 | – |
| $M_{k3}$ | Latero-frontal | | 2347[e] | 2240[e] | 1745 | 1993 | 2032 | – |

compared to meteorological station data at the Khovd. The coefficient of daily air temperature variation ($C_v$) is as high as 1.2 at the Tavan Bogd, measured with automated weather station (AWS) located at 3,084 m on Potanin glacier, 1.1 at 3,600 m on the Ulaan Tsunkhegiin Tsakhir mountain, Munkhkhairkhan mountains, and 1.8 at 3,567 m on the southern ice cap of the Ulaan valley, Tsambagarav mountains. This exceeds 5–10 times the coefficient of variation of the daily average temperature observed in meteorological Ulgii and Khovd stations which are located in the valley and foothill areas of these mountains. The coefficient of correlation between monthly air temperature observed at Tsambagarav and Ulgii is 0.94. The same dependency is revealed between air temperatures observed at Munkh-khairkhan and in the Khovd meteorological stations. These relationships enable observers to reconstruct air temperature time series in high mountains using long-term observation data at lower stations in Khovd and Ulgii. Thus, we derived the above-mentioned equations for the reconstruction of past climate in selected glacier mountains by regression of the meteorological data obtained in the last several years (MEGD 2014).

Detailed comparisons of areas of individual glacier massifs derived from different sources of data show that areas of glacier massifs tend to be overestimated by topographic maps compiled in 1940. The dynamics of glacier massif areas show that more intensive retreating occurs in flat top glaciers in the Tsambagarav, less retreating occurs in Corrie glacier dominated massif in the Munkhkhairkhan, an average retreating rate is observed in glacier complexes of the Tavan Bogd, Kharkhiraa, and Turgen mountains, and small glaciers are disappearing due to climate warming (Fig. 6.8).

There are 31 glaciers in the Kharkhiraa mountains revealed by LANDSAT ETM data with a total area of 34 km$^2$ in 2000 and 26.70 km$^2$ in 2011. The glacier retreat rate is higher in bigger glaciers, and small glaciers are disappearing in the Kharkhiraa mountains (Fig. 6.9).

There are 50 glaciers in Turgen mountains revealed by LANDSAT ETM data with a total area of 32.74 km$^2$ in 2002, and 27.51 km$^2$ in 2011 (Fig. 6.9).

Glacier area and thickness measurements allow estimating glacier volume. Glacier volumes are 1.91 km$^3$ at the Potanin glacier, 0.44 km$^3$ at the southern ice cap at upper Ulaan Am valley, 0.25 km$^3$ at northern icefield at upper Ulaan Am valley, and 0.22 km$^3$ at Baatar icefield. Davaa G et al. (2012) derived the equation for the estimation of the volume of glaciers in Mongol Altai, including some neighboring glaciers in Russia. Therefore, total glacier volume has been estimated with an area of 580 glaciers in 2000 as 21.48 km$^3$ of ice, and accordingly, 19.4 km$^3$ of glacier water resources are in

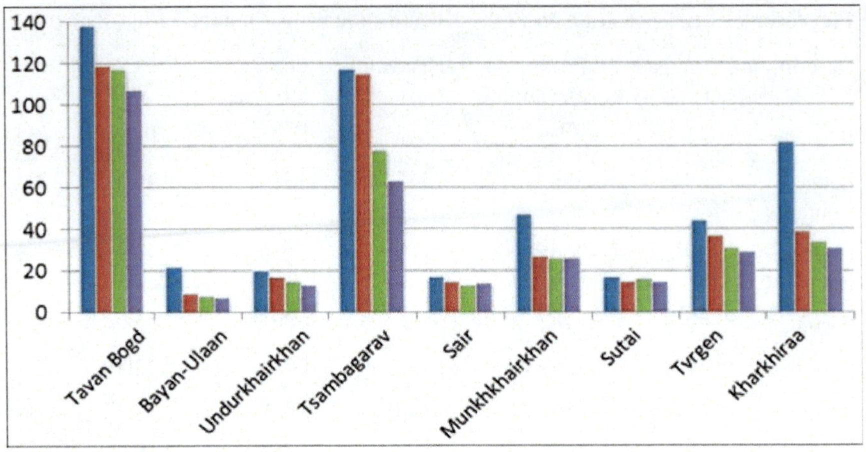


**Fig. 6.8** Dynamics of glacier massif areas (modified after Davaa 2014)

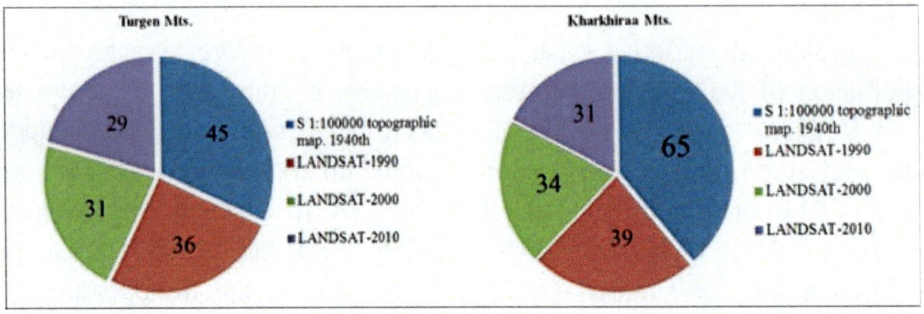


**Fig. 6.9** Glacier area change in the Kharkhiraa and Turgen mts. with area of glacier in 2000. (Based on Davaa, G 2014)

Mongolia. The average depth of glaciers in Mongolia is 31.3 m. The cumulative ablation rate at the altitude of 2,977–2,998, 3,033–3,057, 3,116–3,123, 3,234–3,247 and 3,339–3,366 m was 29.44–33.72, 25.34–28.06, 21.68–25.37, 19.54–23.00 and 13.05–19.24 m, respectively, on the Potanin glacier, in the period of 2004–2011 (Davaa et al. 2012) (Fig. 6.10).

The mass balance of the Ulaan-Am south ice cap in the Tsambagarav is reasonably simple due to the fact that ablation takes place at the top of the ice cap. The cumulative mass balances of the ice cap in the period of 2005–2011 were 10.94 m at the altitude of 3,607 m, 9.28 m at 3,621 m, 9.17 m at 3,700 m, 7.89 m at 3,732 m, 5.41 m at 3,771 m and 5.35 m at 3,814 m, respectively. This makes it possible to establish a regional equilibrium of mean annual ablation versus altitude for estimation of mean ablation in uncontaminated glaciers (MEGD 2014). The melting of the glacier of southern Tsambagarav mountain reaches an average 0.44 cm ice/°C day in warm season in 2010, 0.001 cm during in new snow and 0.76 cm/°C in the absence or no snow on the glacier surface. In 2011, this parameter ranges from 0.65 cm/°C or 0.40–1.22 cm/°C on average. Melting parameter of Potanin glacier will reach 12.06 mm/°C (Konya et al. 2008).

These equations, curves, observed hydrological data of the Kharkhiraa and Turgen streams, and the current climate data, simulated with a regional climate model, was used for snow runoff melt (SRM) model calibrations and verification of the current daily discharge of the Kharkhiraa river. Glaciers in the Kharkhiraa and Turgen mountains are distributed in several river basins. However, in 2011, 10.2% of the glacier area in the Kharkhiraa mountain and 53.4% of the glacier area in the Turgen mountains are located in the Kharkhiraa river basin. Topographic maps and Landsat ETM + data from various dates have been used for the estimation of change in

**Fig. 6.10** Khar nuur- Tsengel soum, Bayan-Ulgii aimag (own photo, 2013)

the glacier area in the Kharkhiraa and Turgen river basins. The glacier area in the Kharkhiraa river basin has decreased from 23.29 $km^2$ in 1992 to 17.06 $km^2$ in 2011. Consequently, the fraction of the glacier area in the Kharkhiraa river catchment of 738 $km^2$ above the Tarialan hydrostation has decreased from 3.16 % in 1992 to 2.31 % in 2011 (Table 6.5) (Davaa et al. 2015).

## 6.4 Impact of Glaciation

### 6.4.1 Impacts in Relief

The traces of ancient and modern glaciations are abundant in Mongol Altai, Khangai, Khentii and Khuvsgul mountain ranges, and even in the Gobi-Altai mountains. Common surfaces that are created by the glacier are the morphosculptural and accumulation forms. The landforms of the accumulation include the moraine hills, dams and large cobblestones, as well as a fluvioglacial terraces with the large proportion of gravels and stones (Tsegmid 1969).

The Pleistocene ELA increased continuously from 2,900 m in the central Mongol Altai to over 3,100 m in the southeastern Mongol Altai reaching over 3,590 m in the Gobi-Altai mountains. The late-Pleistocene ELA was 2,100–2,400 m basin and is continuously increased, therefore reached over –3,590 m in Gobi-Altai mountains. The Pleistocene ELA in Darkhad basin is 2,700–3000 m in western Khangai, and in eastern Khangai is 2,600 m (Jigj 1976; Purevmaa 2018).

The Mongol Altai mountains are largely covered by glacial traces. The reliefs of the ancient and modern glaciers are distinguished by the valley of rivers originating from the Tavan Bogd mountain. For example, Khovd, Mogoin Gol valley, Khoton and Khurgan lake depression are located along the Tavan Bogd mountain in the southeast direction to the Mongol Altai mountain range. The valley, the high and low Turgen mountains, and the valleys that extend to the Must mountain range include ancient moraine sediments. It is interesting that it is possible to create a variety of surface shapes, such hills of the moraine, glacial retreats, hills that originate from erosion, and lakes with different shapes (Ser-Od 2019).

The eternal snowy mountain and glaciers are abundant in the Tavan Bogd mountains, and extend

**Table 6.5** Total glacier areas in Kharkhiraa and Turgen mts. and Kharkhiraa river basin, km$^2$

| Data source | Total glacier area in Kharkhiraa Mts. | Kharkhiraa Mts. glacier area in Kharkhiraa river basin | Total glacier area at Turgen Mts | Turgen Mts. glacier area in Kharkhiraa river basin | Total glacier area in Kharkhiraa river basin |
|---|---|---|---|---|---|
| Topographic map, scaled as S1:100000 | 64.20 | 6.74 | 45.06 | 24.15 | 30.89 |
| Landsat ETM + 25/6/1992 | 39.18 | 4.31 | 36.10 | 18.98 | 23.29 |
| Landsat ETM + 10/9/2000 | 34.06 | 3.62 | 31.27 | 16.69 | 20.31 |
| Landsat ETM + 4/07/2002 | 33.15 | 3.49 | 37.75 | 19.70 | 23.19 |
| Landsat ETM + 29/8/2010 | 31.20 | 3.19 | 29.41 | 15.71 | 18.90 |
| Landsat ETM + 6/08/2011 | 26.73 | 2.43 | 27.49 | 14.63 | 17.06 |

*Source* MARCC 2014

to 3,500–3,800 m high of Tsagaanhairkhan, Ikh Rashaan Uul and Urt Khuiten Davaa, to the southeast of Khoton and Khurgan nuur, and the Khoton and Khurgan nuur front slope facing the depression is sub-divided into 2,500–2,700 m. In this area, the steep cliffs, the traces of the Quaternary glacial are very well preserved, and the tops of the modern glacier tongue are clearly observed. There is physical weathering, especially the cold weathering intensively.

### 6.4.2 Impacts in Hydrology

The glaciers of the Mongol Altai are the source of the most important water resources of the wider region in Western Mongolia. Glacial lakes are numerous in Mongolia, such as lake Khoton, lake Khurgan, lake Tolbo, lake Dayan and lake Tal in the Mongol Altai; Khukh, Otgonii Khukh, Tsagaan, Dood Tsagaan and lake Targan are located in the Khangai mountains. In the Darkhad depression there are more than 200 lakes of glacial origin, for example Dood Tagaan and Targan. Abundance of glacial lakes Khagiin Khar, Khukh, Kherkhluur, Mungunkhul and so on located in the mountains of Khentii are the center of the ancient glacial period (Dashtseren 2006). The lakes of glacial origin in Mongol Altai and Khangai were mostly closed by moraine sediments in the river valley. Depending on the frequency and intensity of glaciers, these lakes vary considerably across different ages and sizes. Some have runoffs of hundreds of kilometers, while most of them occupy only two hectares or less (Davaa et al. 2015).

Well-studied glacial lakes of the Mongol Altai are Khurgan and Khoton (48°40′N, 88°48′E), which are connected with each other by the Sirgal strait, whose width is 100–150 m and depth is 0.7–0.8 m (Tserensodnom 2000). The area of lake Khoton is 60 km$^2$ and is located at an altitude of 2,081 m above the sea level; the lake Khurgan is 77.5 km$^2$ and is located at an altitude of 2,073 m above the sea level. In the Khoton and Khurgan lake regions, glacial-originated morphosculpture and almost all types of depressions are distributed. These lakes were formed in a large glacial trough of the Pleistocene valley, extending from the Tavan Bogd to the southeast. The lakes are surrounded by more moraine dig hill, and there are many glacial-originated small lakes (Table 6.6). Many rivers and streams flow into the Khurgan and Khoton lakes from the Ikh Turgen and Baga Turgen mountains. Within these rivers, approximately 20 rivers are quite

**Table 6.6** Definition of some glacial lakes ( modified from Tserensodnom J)

| № | Lake name | Mountains | Height, m | Area, $km^2$ (catalogue) | Length, km | Width, km | | Depth, m | | Lake volume, $km^3$ | Lake area (LANDSAT), $km^2$ | |
|---|---|---|---|---|---|---|---|---|---|---|---|---|
| | | | | | | Average | Most | Average | Most | | 2010 | 2013 |
| 1 | Achit | Mongol Altai | 1435 | 279 | 28 | 16 | | | | | | |
| 2 | Khoton | Mongol Altai | 2083 | 50 | 22 | 2.3 | 4 | 27 | 58 | 1.341 | 56.5 | 53.4 |
| 3 | Khurgan | Mongol Altai | 2072 | 71 | 23.3 | 3.3 | 6 | 8 | 28 | 0.537 | 71.6 | 66.4 |
| 4 | Tolbo | Mongol Altai | 2079 | 84 | 21 | 4 | 7 | 7 | 12.7 | 0.571 | 82.7 | 79.3 |
| 5 | Dayan | Mongol Altai | 2232 | 67 | 18 | 4 | 9 | 2 | 4 | 0.157 | 71.0 | 63.1 |
| 6 | Tal | Mongol Altai | 2576 | 32 | 14.5 | 2.2 | 4.3 | 3.4 | 8.9 | 0.111 | 32.3 | 30.8 |
| 7 | Khukh | Khangai | 2649 | 17.2 | 12.8 | 1 | 1.7 | 15 | 43 | 0.206 | 13.0 | 10.8 |
| 8 | Tsagaan | Khangai | 1284 | 9.1 | 3.2 | 2.8 | 3.2 | 0.2 | 0.7 | 0.002 | 0.8 | 3,4 |
| 9 | Khukh | Khangai | 2668 | 6.7 | 4.7 | 1.4 | 2.4 | – | – | – | 6.7 | 5.8 |
| 10 | Dood Tsagaan | Khuvsgul | 1538 | 31.3 | 17.5 | 2.4 | 5.6 | 4.7 | 17 | 0.204 | 21.0 | 39.6 |
| 11 | Targan | Khuvsgul | 1538 | 19.5 | 7.5 | 2.6 | 4.6 | 2.8 | 5.5 | 0.054 | 20.3 | 21.4 |

large, such as Sumdairakh, Khar and Tsagaan Khovd, Zagastai and Chuluut (Tserensodnom 2000).

The only river originated from the lakes Khurgan and Khoton is the Khovd, which is the largest river in Western Mongolia. The river flows through a steep and narrow gorge across the moraine deposits, which filled the inter-mountain tectonic basin in the downstream part of the glacial trough. Within the hummocky terrain of the very extensive icemarginal zone, several systems of terminal moraines can be identified (Lehmkuhl 2012). An important feature is the incorporation of large amounts of glaciolacustrine clay and silt into the moraine, evidencing that a large proglacial lake filled the glacial trough before and during the late-Pleistocene ice advances and associated retreat stages. During glacial advances these sediments were pushed together, forming glaciotectonic moraine complexes (Lehmkuhl et al. 2016).

From the analysis of Holocene sediments dust the age of Davaanii lake (located at the top of Tarvagatai mountain in Khangai) is 11 180±120 years and the age of Dood Tsagaan lake (Darkhad depression) is 11 275±150 years, while the age of Achit lake (Altai mountains) is 11 500 ±150 years. Also, the age of clays of blue color, which are between the layers of glacial sediments at a depth of 3.2-3.5 m in lakes Khoton and Khurgan (Altai mountains) is almost the same period, 3340 ± 70 and 3270 ± 90 years (Tserensodnom 2000). These facts indicate that lakes of glacial origin in Mongolia are of the same age as the last glaciation, which ended in the country about 10-12 thousand years ago.

## References


Baast P (1998) Modern Glaciers and Hydrology of Mongolia. A Dissertation for Philosophical Doctor of Geography. Institute of Geoecology, Mongolian Academy of Sciences,Ulaanbaatar, 162 pp (in Russian)

Blomdin RHJ (2016) Glacial geomorphology of the Altai and Western Sayan Mountains, Central Asia. J Maps 12(1):122–136

Blomdin R et al (2018). Timing and dynamics of glaciation in the Ikh Turgen Mountains, Altai region, High Asia. Quat Geochronol (47):54–71

Dashdeleg N (1990) Modern Glaciers of Mongolia. In A. o. USSR, National Atlas of the Peoples Republic of Mongolia. Academy of Sciences of Mongolia and Academy of Sciences of USSR, Ulaanbaatar and Moscow (in Mongolian)

Dashdeleg N, Evilkhaan R, Khishigsuren P (1983) Recent Glaciations of Mongol Altai. In: Proceedings of IMH, vol 8. Institute of Meteorology and Hydrology, Ulaanbaatar, Mongolia, pp 121–126 (in Mongolian)

Dashtseren T (2006). Surface glacial appearance of Khoton and Khurgan lake, pp 85–91 (in Mongolian)

Davaa G et al (2012). Glacier mass balance and change of the Mongolia, climate change in the high mountain region, pp 22–35

Davaa G et al (2015) Surface water regime and resourses of Mongolia. (D. G, ed) Institute of Meteorology and Hydrology, Ulaanbaatar. Admon printing (in Mongolian)

Davaa G, Kadato T, Purevdagva K, Baasandorj D (2008) Preliminary results of study on flat-top glacier at Tsambagarav Mts. Mongolia. "Uncertainties in water resource management: causes, technologies and consequences"(Wrm-Mon2008), 123

Ganyushkin D, Christyakov K, Kunaeva E (2015) Fluctuations of glaciers in the southeast Russian Altai and northwest Mongolia mountains since the Little Ice Age maximum. Environ Earth Sci 74(74):1883–1904

Ganyushkin AD et al (2018) Present glaciers of Tavan Bogd Massif in the Altai Mountains, Central Asia, and their changes since the little ice age. Geosciences:3–15

Ivanov A (1949) About the glaciation of the north east Mongol Altai. Moscow (in Russian)

Jigj S (1975) Primary feature of Mongolian landforms. Institute of Geography & Cryology. Mongolian Academy of Sciences (MAS), Ulaanbaatar, 126 pp (In Mongolian with Russian abstract)

Jigj S (1976) A brief description of paleoglaciations and paleoglaciers. Institute of Geography & Permafrost. Mongolian Academy of Sciences. State Press, Ulaanbaatar (in Mongolian)

Kadota T, Davaa G (2004) A preliminary study on glaciers in Mongolia. In: Proceedings of the 2nd international workshop on terrestrial change in Mongolia. Institute of Meteorology and Hydrology, Ulaanbaatar, pp 100–102

Kadota T (2007) Recent glacier variations in Mongolia. Ann Glaciol:186

Kadota T et al (2011) Glaciological research in the Mongolian Altai, 2003-2009. Bull Glaciol Res 29:41–50

Kamp U, McManigal KG, Dashtseren A, Walther M (2013a) Documenting glacial changes between 1910, 1970, 1992 and 2010 in the Turgen Mountains, Mongolian Altai, using repeat photographs, topographic maps, and satellite imagery. Geogr J 179 (3):248–263

Kamp U, Krumwiede B, McManigal K, Pan C, Walther M, Dashtseren A (2013b) The Glaciers of Mongolia. INSTAAR Occas, Pap, p 61

Klinge M, Lehmkuhl F (2013) Geomorphology of the Tsetseg Nuur basin, Mongol Altai: lake development, fluvial sedimentation and aeolian transport in a semi-arid environment. J Maps (9):361–366

Konya K, Kadota T, Ohata T (2008) Mass balance of Potanin glacier. Mongolia Altai, AGU Fall Meeting Abstracts

Kovalenko SN (2014) O granicah i obiemah sovremennogo oledeneniy raiona g. Munku-Sardyk (Vostochniy Sayan). Vestnik Kafedry Geography. VSGAO, pp 19–31. http://kafgeo.igpu.ru/bulletin9/kovalenko14-1.pdf (in Russian)

Kovalenko SN (2011) Glacial Geomorphological region of Mount Munku-Sardyk. Forms of local glaciation of the rriver basin of Muguek and White Irkuta. Dept Geogr Newslett 2(1):44–69

Krumwiede BS, KU (2014) Recent glacier changes in the Mongolian Altai Mountains: Case studies from Munkhkhairkhan and Tavan Bogd. In: KJS (ed) Global land ice measurements from space. Springer Praxis Books, Ó Springer Verlag, Berlin Heidelberg, pp 481–508

Lehmkuhl F, Batkhishig O (2003) Rock glacier and periglacial processes in the Mongolian Altai. Permafrost, Phillips, Springman & Arenson, p 639

Lehmkuhl F et.al (2016) Distribution and timing of Holocene and late Pleistocene glacier fluctuations in western Mongolia. Ann Glaciol:175

Lehmkuhl F (2012) Holocene glaciers in the Mongolian Altai: an example from the Turgen-Kharkhiraa Mountains. J Asian Earth Sci 52:12–20

MEGD (2014) Mongolia second assessment report on climate change 2014 (MARCC 2014). Ministry of Environment and Green Development, Ulaanbaatar

Munkhuu Z (1992) General Geomorphology. State Press, Ulaanbaatar 37 pp (in Mongolian)

Ohata T, Kadota T, Konya K, Yabuki H, Gombo D (2009) Glacier condition and variation studies in the Mongolian part of the Altai Mountains. EOS (Transactions, American Geophysical Union) Fall Meeting Supplement, Abstract C23A-0484, 90

Orkhonselenge A (2017) Glacial geomorphology of Mt. Munkhsaridag in the Khuvsgul Mountain Range, Northern Mongolia. Geomorphologie: Reilef, Processus, Environnement:387–390

Orkhonselenge A (2016) Glacial Geomorphology of Mt. Munkh Saridag in the Khuvsgul Mountain Range, Northern Mongolia. Géomorphologie: relief, processus, Environment, 22(4):389–398

Orkhonselenge A, Harbor JM (2018) Impacts of modern glacier changes on surface water resources in Western and Northern Mongolia. J Water Resour Protect 10 (06):559

Otgonbayar D (2011) Contemporary glaciation of the Tsambagarav-uul Mountain Knot (Mongolian Altai). J Vestnic (Altai University):177–179 (in Russian)

Otgonbayar D (2012) Water-glacier resources of closed drainage basins of Western Mongolia: contemporary assessment and tendencies of change: extended abstracts of Candidate Science. (Geography) Degree Dissertation. Barnaul: Izd-vo AtTGU 24 p. (In Russian)

Plyushnin VM et al (2013) distinctive characteristics of formation and dynamics of nival-glacial geosystems in the South of East Siberia and on Mongolian Altai. Geogr Nat Resour 34:6

Purevmaa K (2018) Late Quanernary glacial history in Khentey Mountain Range, Central Mongolia. Seoul

Rother H, Lehmkuhl F, Fink D, Nottebaum V (2014) Surface exposure dating reveals MIS-3 glacial maximum in the Khangai Mountains of Mongolia. Quatern Res 82:297–308

Sapojnikov V (1911) Mongol Altai in the Irtysh and Khobdo River Heads. Tipolitogr, Sibirskogo T-va Pechatnogo, Tomsk (in Russian)

Ser-Od T (2019) Main characteristics of the landscape of the Mongol Altai Mountain Ridge and issues of its conservation and use. Doctor's dissertation of Geography Science, pp 28–33 (in Mongolian)

Syromyatina MV et al (2015) Current dynamics of glaciers in the Tavan Bogd Mountains (Northwest Mongolia). Environ Earth Sci 74:31–38

Tronov MB (1925) Catalogue of Altai glaciers. Moscow. Proc Russian Geogr Soc 57(2) (in Russian)

Tronov MV (1949) Sketches on the Altai glaciation. Geografgiz, Moscow (in Russian)

Tsegmid S (1951) The Physical geography and prehistorical glaciers of the Khentii mountain range. Moscow (in Russian)

Tsegmid S (1969) Physical geography of Mongolia. Institute of Geography & Permafrost. Mongolian Academy of Sciences. State Press, Ulaanbaatar 405 p. (in Mongolian)

Tserensodnom J (2000) Catalogue of Mongolian lake. Institute of Geography & Permafrost. Mongolian Academy of Sciences. Shuvuun saaral publishing, Ulaanbaatar (in Mongolian)

Walter M et al (2017) Glaciers, permafrost and lake levels at the Tsengel Khairkhan Massif, Mongolian Altai, During the Late Pleistocene and Holocene. Geosciences 7(3):121–132

Yabuki H, Ohata T (2009). The recent glacier changes in Mongolian Altai Mountains. In: AGU Fall Meeting Abstracts

# 7 Permafrost in Mongolia

Avirmed Dashtseren

**Abstract**

In this chapter, the general characteristics of permafrost and cryogenic features are described, and the degradation of permafrost is simply analyzed too. The permafrost in Mongolia lies at the southern edge of Siberian permafrost with different permafrost zones. The distributions of continuous and discontinuous permafrost zones are usually controlled by the climate in Mongolia while the local environmental factors are mean force that persist the existences of sporadic and isolated permafrost zones. Due to the location and climate condition, permafrost temperature in Mongolia is mostly close to 0 °C, and making it vulnerable to climate warming and anthropogenic impacts. The results from permafrost monitoring indicate that permafrost in the country has degraded more intensively during the last decades and it has a negative effect on ecosystem services, because the permafrost in Mongolia overlaps considerably with forest and the source area of river water. In Mongolia, several features of cryogenic develop in the permafrost region, such as patterned ground forms, frost cracking, ice wedge, karst, thermokarst lake, aufeis, pingo, seasonal pingo, hummock, frost heave, frost sorting, dog hole.

**Keywords**

Permafrost · Active layer · Cryogenic features · Permafrost degradation

## 7.1 Overview of Permafrost Research in Mongolia

The study of permafrost worldwide is usually started belatedly compared with other earth sciences, and it hampers several difficult reasons. For instance, the permafrost table is situated underground that makes a problem to know, and permafrost occurs in the cold area with complex conditions of environmental and geological. An intensive scientific study on permafrost issues in Mongolia did not begin until the mid-twentieth centuries, although permafrost is widespread in the country. In 1962, the Institute of Geography and Permafrost was established within the structure of the Mongolian Academy of Sciences. From this time, permafrost study was systematically started to conduct over Mongolian territory. The first scientist who had conducted research on the permafrost and seasonally frozen ground in Mongolia was Lonjid (1963), and he described the general characteristics of permafrost and

A. Dashtseren (✉)
Institute of Geography and Geoecology, Mongolian Academy of Sciences, Ulaanbaatar, Mongolia
e-mail: dashka.ig@gmail.com

© Springer Nature Switzerland AG 2021



B. Yembuu (ed.), *The Physical Geography of Mongolia*, Geography of the Physical Environment,
https://doi.org/10.1007/978-3-030-61434-8_7

seasonally frozen ground. Furthermore, he compiled the first map of the distribution permafrost and features of permafrost in Mongolia. From the beginning of the 1970s, a joint team of the Mongolian–Soviet geological expedition had intensively worked in the field of permafrost studies and conducted many field works. As a result of the joint study, the general features of permafrost, distribution of permafrost and seasonally frozen ground, the dynamic of permafrost in Mongolia have been studied. Additionally, they found that permafrost covered 63% of the total area of the country with different distribution types (Gravis et al. 1971).

The scientist Tumurbaatar (1971) created the map freezing and thawing of ground within Ulaanbaatar city, and he also explored the normative depths on freezing and thawing of grounds in Mongolia. The features of cryogenic in Mongolia are more clearly described by Luvsandagva (1978). The scientist Sharkhuu (2001) published several permafrost maps that cover from regional to countrywide scales, and he estimated the temperatures of permafrost in many boreholes. In addition, they together described the general characteristics of Mongolian permafrost, the features of permafrost, the distributions of permafrost and seasonally frozen ground, and temperatures of permafrost, which was very useful for the development of permafrost study in Mongolia. However, a full understanding of permafrost situation and its nexus with other environmental fractions were still lacking.

In recent years or after the 2000s, permafrost researchers who work in Mongolian territory have carried out more detailed studies. For instance, the heat insulation survey on the different vegetation covers was done by Sharkhuu et al. (2007) and Sharkhuu and Anarmaa (2012), and their result shows that the vegetation cover is one of the natural factors that protect the permafrost from overheating during summer. Furthermore, Dashtseren et al. (2014) and Ishikawa et al. (2018) found that the forest at the marginal permafrost in the Khentii mountain allows permafrost to persist, because the forest cover effectively blocks incoming shortwave radiation. The snow cover does not directly affect the seasonal thawing regime, while the summer air temperature has a major impact on seasonal thawing regime, which is directly related to the depth of thawing procedures (Dashtseren et al. 2014).

During the last decade, 180 permafrost monitoring sites (boreholes) were established in the different permafrost zones over Mongolia. These boreholes were considered to be representative of the most landscape types in the permafrost region. Monitoring of the ground temperatures over the past decades has indicated that the permafrost has been degrading in the Mongolian territory, but the ratio of permafrost degradation depended on permafrost distribution types (Ishikawa et al. 2018; Dashtseren et al. 2017a; Jamvaljav 2017; A. Sharkhuu et al. 2007). This different permafrost degradation in Mongolia also partly affects the development of thermokarst lakes, and it is confirmed by Saruulzaya's (2016) literature. According to her study, thermokarst lakes are shrinking more at the marginal permafrost or isolated and sporadic permafrost zones, where permafrost also has degraded in the last decades while the areas of thermokarst lakes are increasing in continuous permafrost zone due to melting water from permafrost (Saruulzaya et al. 2016). Based on the temperature information at the monitoring sites and modeling method, permafrost map of Mongolia was produced by (Jambaljav et al. 2016).

## 7.2 Geographical Distribution and Zones of Permafrost in Mongolia

Permafrost is ground (soil or rock and included ice and organic material) that remains at or below 0 °C for at least two consecutive years, and around 22% of the exposed land surface in the Northern hemisphere is underlain by permafrost (Obu et al. 2019; French 2007). The existence of permafrost is controlled by many environmental factors, but it largely depends on climate conditions globally (Harris et al. 2017).

The permafrost in Mongolia is located at the southern boundary of Siberian continuous permafrost, therefore, being the marginal permafrost. As the permafrost in Mongolia is usually marginal and underlies mountains, its distribution is characterized by arid and mountainous terrain underlain by permafrost with a continuous, discontinuous, sporadic and isolated. The distribution of permafrost in Mongolia is shown in Fig. 7.1, which is based on the TTOP model and MODIS LST dataset. The permafrost in Mongolia encompasses about 29% of the total land area of the country (Jambaljav et al. 2016). Based on the ground temperatures at the top of permafrost, they classified the permafrost zones as follows (Jambaljav et al. 2016):

- Continuous permafrost: < −2.0 °C
- Discontinuous permafrost: −2.0 to −1.0 °C
- Sporadic permafrost: −1.0 to 0.0 °C
- Isolated permafrost: 0.0 to +1.0 °C.

As shown in Fig. 7.1, almost all areas in the most northern part of Mongolia and around the top of high mountains are underlain by continuous permafrost zone, especially at the Darkhad depression, the top areas of the Khentii, Khuvsgul, Khangai and Altai mountains with the taiga and mountain steppe. Other permafrost zones are located outside the continuous permafrost zone, with discontinuous, sporadic and isolated.

The high mountain steppe and mountain taiga in the western and northern parts of the country have high annual precipitation (> 300 mm) and long cold season, and the average annual air temperature in these areas is usually lower than −4 °C (MARCC 2014). Therefore, permafrost is distributing with its continuous zone. The average annual air temperature in the forest steppe and steppe regions ranges from −4 to 0 °C (MARCC 2014), and these regions roughly overlap with discontinuous, sporadic and isolated permafrost zones. This result indicates that permafrost distribution is controlled by the topography, vegetation cover and soil type in the latter regions.

Most of the major mountains in Mongolia glaciated at least two times, as a result of it, the current permafrost and glacier in Mongolia are formed (Tsegmid 1969). Japan–Mongolian joint expedition revealed the age of pingos in the permafrost area of Mongolia using isotopic techniques (Yoshikawa et al. 2013; Ishikawa and Yamkhin 2016). Yoshikawa et al. (2013) drilled 30 m deep borehole on the top of a pingo at the toe

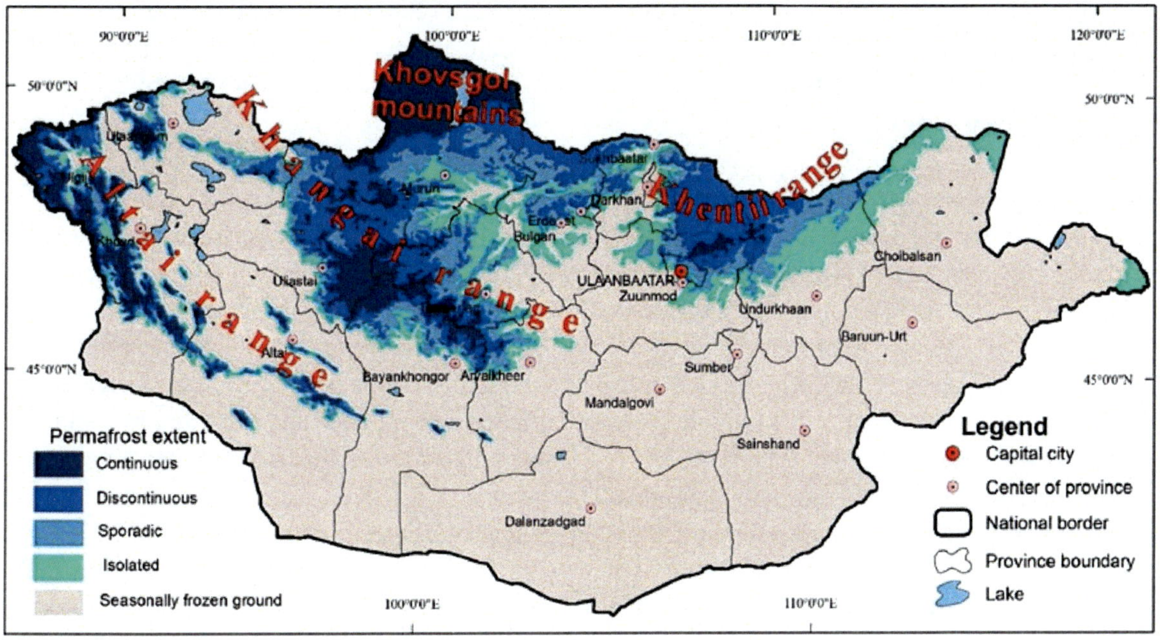


**Fig. 7.1** Permafrost distribution and zones in Mongolia (Jambaljav et al. 2016)

of alluvial fans in the Tes valley, northwestern Mongolia. They pointed out that the pingo began to develop about 8790 years before the present (BP). The Arsian pingo in the Darkhad Basin, northern Mongolia was drilled by scientists Ishikawa and Jambaljav with 35 m deep borehole, and their study shows that the growth of pingo began after 4500 years BP, when Paleo Lake was completely drained (Ishikawa and Yamkhin 2016). Pingos must develop after the permafrost formation; therefore, permafrost in Mongolia probably formed approximately 11,000 years BP or during the last glacial period in the Northern hemisphere (Dashtseren 2018).

Permafrost acts as a water-resistant buffer against water penetration to sediments and plays an essential role in arid ecohydrological system and herders in Mongolia. For instance, the permafrost in Mongolia directly sustains the livelihoods of inhabitants because it produces locally wet soil conditions, even under a low annual rainfall (Dashtseren et al. 2014). Furthermore, the forests are distributed in a mosaic pattern and overlap considerably with permafrost regions, and river discharges originate entirely from the high mountains and northern territory where permafrost occurs extensively (Dashtseren et al. 2014; Ishikawa et al. 2018).

## 7.3 Cryogenic Processes and Features

Several types of cryogenic landforms develop in permafrost region of Mongolia, namely patterned ground forms, frost cracking, ice wedge, karst, thermokarst lake, aufeis, pingo, seasonal pingo, hummock, frost heave, frost sorting, dog hole and mass movements. However, detailed information on cryogenic features in Mongolia is not relatively adequate.

The patterned ground includes non-sorted and sorted circles, nets and polygons. These features are formed under the process of frost sorting of the coarse and fine materials. Based on their origin, they are divided into sorted and non-sorted patterned ground (Harris et al. 2017). Patterned ground is a widely distributed cryogenic landform in permafrost regions, Mongolia; however, a number of each patterned ground phenomena is varied with the different landscape conditions. The sorted patterned ground is usually common in the upper side of Altai, Khangai, Khuvsgul and Khentii mountains. The soil of these areas is characterized by fine sediment, stones and boulders with abundant soil moisture, and this condition can lead to sorted patterned. Figure 7.2 shows an example of patterned ground that is surrounded by stones with an inner core of finer sediment ground in the Altai mountain. This patterned ground is usually found in groups, and its average diameter is approximately 3 m. William and Erdenebat (2005) found two types of patterned ground in the Kharkhiraa Mountains. According to their result, both sorted stone-bordered polygons and non-sorted stone circles were found on a gentle slope at 3000 m a.s.l. adjacent to a glaciated valley in the Kharkhiraa Mountains (William and Erdenbat 2005).

Hummocks, polygons and circles are often formed in the flats and valleys landscape with fine sediment and high soil moisture. The dominant height of hummock in Mongolia ranges from 10 to 50 cm, while its diameter ranges from 30 to 100 cm. Vegetation develops well in the hummock area compared with its adjacent area (Fig. 7.3). Polygons usually distribute in the Darkhad depression, but it could be noticed in other permafrost regions of Mongolia.

According to common knowledge, thermokarst lake is an important indicator whether permafrost is degrading or not, and thermokarst lakes only forms within the permafrost area. One of the well-studied cryogenic phenomena is thermokarst lake by Saruulzaya et al. (2016). As shown in her study, thermokarst lakes in the continuous permafrost zone have increased significantly in number and area due to permafrost melting, while in the isolated permafrost zone she observed a decrease in both number and area owing to disappearing permafrost and deepening of the active layer over the last 45 years. The biggest thermokarst lake with an area of 10500 ha is located in the continuous permafrost zone at the Darkhad depression, while small thermokarst lakes are mainly lied at the other

**Fig. 7.2** Sorted circles on the north slope of the Tsengelkhairkhan uul in Altai mountain (photo by Dashtseren)

**Fig. 7.3** Hummock area in the valley of Zuunsalaa, Ulaanbaatar city (photo by Dashtseren)

permafrost zones (Saruulzaya et al. 2016). The general scenery of thermokarst lake in Mongolia is shown by Fig. 7.4.

Pingos are a prominent cryogenic feature in permafrost regions and form only in the area with permafrost. A pingo has an ice core with a conical in shape that is developed by injection of water due to hydrostatic or hydraulic pressure from below. There are two types of pingos: closed-system (hydrostatic) pingos and open-

**Fig. 7.4** Thermokarst lake in the Darkhad depression, with an approximately diameter of 200 m (photo by A. Dashsteren)

system (hydraulic) pingos (French 2007). In Mongolia, both types occur, and the former is mainly continuous permafrost zone, while the latter is more common in other permafrost zones. Pingos in Mongolia vary from a few meters to over 20 m in height and up to approximately 150 m diameter (Lomborinchen 2000). Groups of pingos are mostly found in the Chuluut valley and Olgoi lake basin in the Khangai mountain as well as at the Darkhad depression. These areas have mainly fine sediments, wet ground, and many small lakes. As mentioned before, pingos at some sites in Mongolia began to develop from 4500 years BP to 8790 years BP, but the formations depend on the historic to present site-specific conditions (Yoshikawa et al. 2013; Ishikawa and Yamkhin 2016). Numerous pingos exist and are degrading at the south edge of permafrost in Mongolia. For example, roughly 5 m high pingo is located in the Nalaikh, Khentii mountain. According to leveling measurement data, the southeastern part of the pingo top was subsided by 50 cm in 2005, 40 cm in 2006 and 45 cm between 2007 and 2009. This subsidence was caused by melting of the pingo ice core from its table (Sharkhuu and Anarmaa 2012). In the continuous permafrost zone or northern Mongolia, the top area of the pingos is mainly smooth, indicating no active degradation is noticeable (Fig. 7.5).

Ice wedges are observed in steep exposures along the thermos erosional banks of the Sharga-Ganga River, and the ice wedge polygons on the land surface are widespread here. The melting of ice wedges along polygonal hollow channels resulted in 2–3 m deep subsidence cavities on the permafrost table, and incidents of large herds (yaks and horses) falling into the cavities have occurred recently (Sharkhuu et al. 2007). The number and spatial distribution of cryogenic features are much higher in the Darkhad depression in Khuvsgul mountain with continuous permafrost zone than in the Altai, Khangai and Khentii mountains. This difference between the former areas is mostly attributed to the local climate, soil characteristics and permafrost conditions.

## 7.4 Active Layer

The thickness of the active layer and seasonally frozen ground show large spatial variations in Mongolia. Such variations depend on the microclimate associated with the topography, thermal properties of the soils and the vegetation cover on a local scale (Dashtseren et al. 2014). Real active layer thickness (ALT) can be estimated from several methods such as mechanical probing, reading from thaw tubes with liquid,

**Fig. 7.5** Closed-system pingo at the Darkhad depression, with approximately 12 m height and 150 m diameter (photo by A. Dashsteren)

and linear interpolation using the 0 °C isotherm penetration based on soil temperature data. However, ALT can be mainly determined by the latter two methods in Mongolia, because ALT is quite deep. Monitoring of the ground temperatures over the past decade has indicated that ALT over Mongolian permafrost ranges from 1.9 to 8.0 m, with predominantly average of 2.5 to 4.0 m. The maximum seasonally thawed active layer is mainly observed at the end of December or January, while full freezing of the active layer continues until April. Differences in ALT among the monitoring sites may be explained by site-specific conditions such as local climate, permafrost distribution type and ice content. ALTs in the continuous and discontinuous permafrost zone are generally shallower than that in the other permafrost zones (Fig. 7.6). Furthermore, temporal variability in the active layer is less in ice-rich sites (e.g. SH-river, Ts-nuur, Arsai pingo) than that in the sites with bare ground, poorly vegetated and non-vegetated (e.g. Tsengel 1, Tsengel 4, Bayanbulag).

Few studies have demonstrated the clear linkages between local climate, vegetation and development of active layer in Mongolia. Dashtseren et al. (2014) concluded that ALT at the forest steppe mosaic in the Khentii mountain was strongly related to air temperature in the summer. In addition, he pointed out that the active layer usually begins to thaw at the end of April, with an average thawing ratio of 0.022 m/day while refreezing of the active layer usually begins in mid-October, with a rapid freezing ratio of 0.043 m/day. That high ratio on freezing is due to the thick organic layer beneath the snow cover that enhances the freezing rate (Dashtseren et al. 2014).

The long-term trends in ALT in some permafrost sites are shown by Sharkhuu and Amarnaa (2012). Their results confirmed that ALT over Mongolia increased at different rates depending on changes in the mean annual air temperature, the environmental and ground conditions of regions, and human activities. For instance, no substantial change of ALT was observed in the Chuluut, Terkh and Sharga boreholes, which have ice-rich and fine-grained sediments and a shallow thawing depth. In contrast, the active layer depth in the Burenkhan and Argalant increased at the rate of 25–40 cm per decade, and this rate appeared to be characteristic of areas with a deep active layer. Regional trends exist regardless of these differences in ground characteristics; the estimated average increase in ALT varied over the range of 0.2–1.5 cm/year in

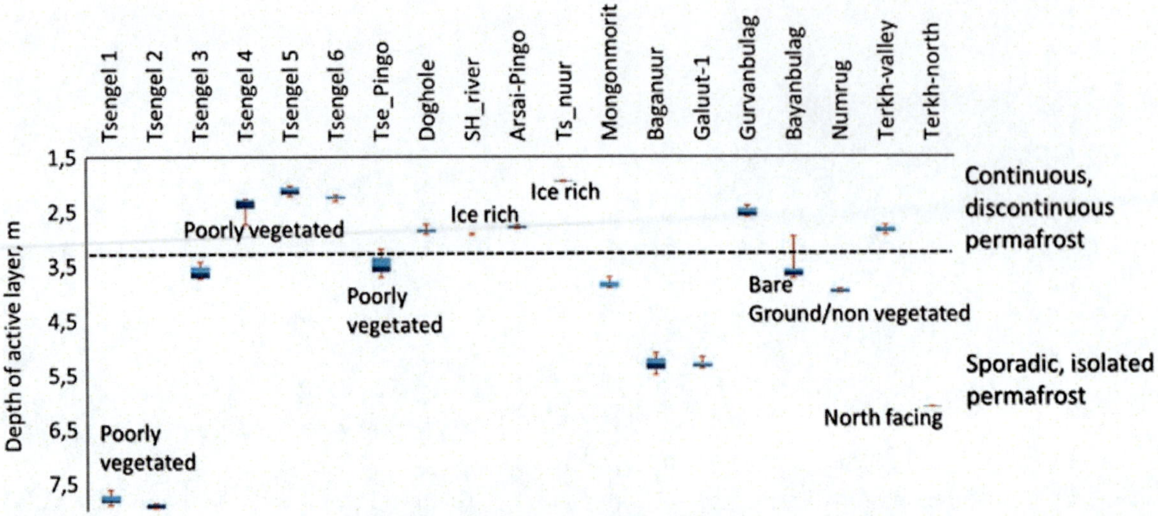


**Fig. 7.6** Temporal variation in active layer thickness at the selected monitoring sites in Mongolia. Ts-1, Ts-2, Ts-3, Ts-4, Ts-5, Ts-6 and Tse-Pingo sites are located in the Altai mountain; Dog-hole, Darkhad-khur, Arsai-Pingo and TS-nuur sites are located in the Khuvsgul mountain; SH-river, Galuut-1, Gurvanbulag, Bayanbulag, Numrug, Terkh-valley and Terkh north sites are located in the Khangai mountain; Kh-50, Mongonmorit, Baganuur and Nalaikh sites are located in the Khentii mountain, respectively

the Khangai and Khentii regions and 0.3–2.4 cm/year in the Khuvsgul region (Sharkhuu and Anarmaa 2012). Of course, this thinning of ALT is caused by an increase in air temperature over Mongolia.

If the freezing front of active layer in winters does not completely reach permafrost table, then talik (unfrozen ground) is formed between permafrost and active layer. This phenomenon usually occurs in the marginal permafrost due to the effect of global warming and groundwater state. The ground temperatures monitoring at the southern territory of the Khangai mountain indicate the occurrence of talik. For example, a thin talik was formed at the Bayanovoo site, in the southern territory of Khangai. It was evident that permafrost was partly disconnected from year surface temperature fluctuations due to talik effect, and permafrost in this area is degrading from both permafrost table and permafrost base because of groundwater movement and increase in air temperature.

## 7.5 The Temperature State of Permafrost

As permafrost is usually the product of climate, the temperature of permafrost strongly depends on environmental factors like air temperature, incoming solar radiation, snow and vegetation cover, moisture and composition of ground, type of soil, as well as the thermal conductivity of materials (Harris et al. 2017). During the last decade, studies on the temperature state of permafrost have received much attention in the world scientific community because of changes in permafrost that led to dramatic variations in heat and water cycles, which in turn exert a profound influence on climate.

As mentioned before, permafrost in Mongolia is located at Siberian permafrost zone. On the other hand, Mongolia has a harsh continental climate due to its geographical location in the central Eurasian continent, and the shift of mean

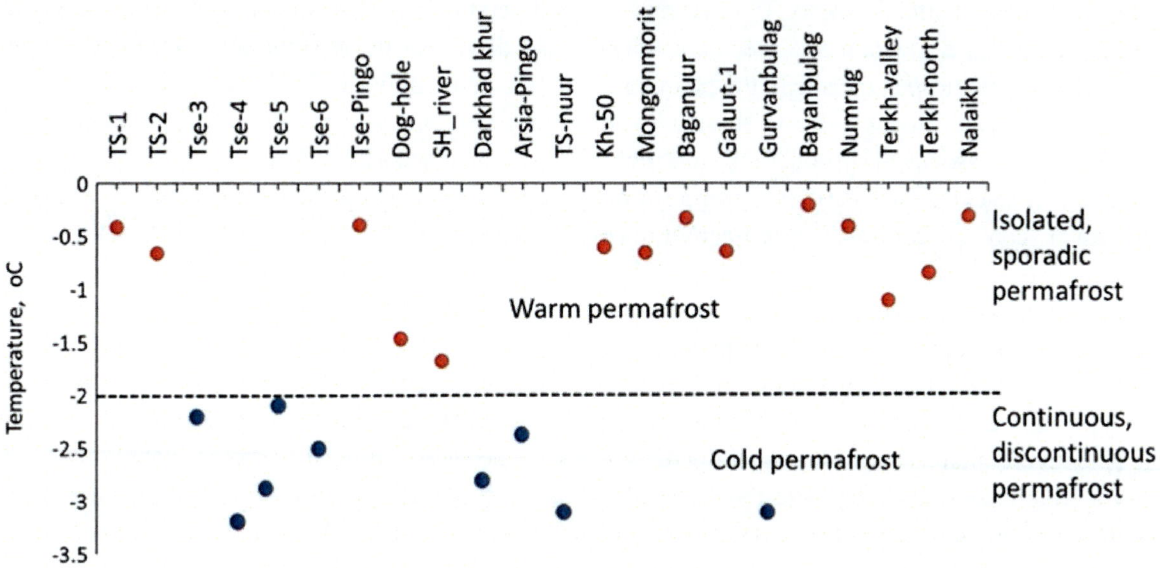


**Fig. 7.7** Permafrost temperatures from the monitoring sites in the Altai, Khangai, Khuvsgul and Khentii mountains with different permafrost zones. Ts-1, Ts-2, Ts-3, Ts-4, Ts-5, Ts-6 and Tse-Pingo sites are located in the Altai mountain; Dog-hole, Darkhad-khur, Arsai-Pingo and TS-nuur sites are located in the Khuvsgul mountain; SH-river, Galuut-1, Gurvanbulag, Bayanbulag, Numrug, Terkh-valley and Terkh-north sites are located in the Khangai mountain, Kh-50, Mongonmorit, Baganuur and Nalaikh sites are located in the Khentii mountain, respectively

annual air temperature from positive to negative lies in Mongolian territory (MARCC 2014). Due to the above-mentioned reasons, permafrost temperature in Mongolia is mainly close to 0 °C, and it also makes that permafrost in Mongolia can be vulnerable to climate and environmental changes.

In Mongolia, the measurement of permafrost temperature has been conducted since the end of the 1960s. However, this measurement was made only single or few times within a year. During the last several years, 180 permafrost monitoring sites (boreholes) were established within the different permafrost zones over Mongolia, and about half of the boreholes have been equipped with HOBO U-12-008 temperature data logger and TMC-HD temperature sensors. These boreholes were considered to be representative of the most landscape types in Altai, Khangai, Khuvsgul and Khentii mountains. The depth of boreholes ranged from 1.5 to 50 m, with the majority below 15 m. Also, some of these boreholes have a single temperature measurement from the late 1960s and early 1980s. Therefore, it is possible to determine the current temperature state of permafrost from those monitoring sites in Mongolia. Also, numerous studies have reported the current temperatures permafrost conditions in Mongolia (Ishikawa et al. 2018; Dashtseren et al. 2017a; Jambaljav 2017; Sharkhuu and Anarmaa 2012). Based on average ground temperature data from some permafrost monitoring sites, mean annual ground temperature (MAGT) varies from −0.1 to −3.4 °C, which is almost similar characteristic (Sharkhuu and Anarmaa 2012). In general, MAGT in continuous and discontinuous permafrost zones was considerably lower than that in sporadic and isolated permafrost zones. MAGT within the continuous zone was ranging from −2.1 to −3.4 °C, reaming the cold permafrost, while MAGT in the other permafrost zone was ranging from −0.1 to −2.1 °C, satisfying the warm permafrost (Fig. 7.7). This result indicates that both the warm permafrost and cold permafrost are co-exciting in Mongolian territory. The depth of yearly zero amplitude is more than 10 m over permafrost in Mongolia.

Ishikawa et al (2018) have well described the specific differences in permafrost thermal state in local areas of Mongolia. For example, in the middle Khangai mountain, MAGT variations in permafrost at the floodplain site were considerably smaller than those at the site in the pediment. This difference was probably due to abundant soil moisture and associated production and consumption of latent heat. Also, soil composition and water content control the temperature of grounds at the Galuut site in the southern Khangai. At the Galuut 1 site, the temperature fluctuated greatly as underlying the dry permafrost, while the fluctuation of temperature at Galuut 2 site was small due to the high latent heat effect. In the Khentii mountain, the small annual fluctuation in ground temperatures at the forest site was also due to forest shading effect and low conductivity of organic-rich soil. In the Altai mountain, ground temperature mainly decreased with elevation. This phenomenon was clearly seen at Tsengel (1–6) sites. The thermal offset was −1.5 °C at Ts–1, −0.2 °C at Ts–2, −0.9 °C at Ts–3 and 0.8 °C at Ts–6, respectively. Also, the yearly zero amplitude increases from Ts–1 to Ts–6 with together elevation (Ishikawa et al. 2018).

## 7.6 Permafrost Degradation

According to the definition of permafrost (French 2007), it can be thought that permafrost is one of the products of climate. Therefore, aggradation and degradation of permafrost can be generally related to climate change from local to regional. It is evident that the climate has changed in the world with a warming trend as well as in Mongolia. Due to geographical location, Mongolian climate is the clearly distinctive four seasons which leads to high fluctuation of air temperature, and low precipitation. The average annual air temperature in Mongolia has increased by 2.14 °C during the last 70 years, and it is almost three times higher than the rise of average global temperatures. The warming trend indicated in all natural zones during the last four decades (MARCC 2014). However, climate warming is less pronounced in the Gobi-desert, while it is high in the high mountain areas and their valleys, where permafrost widely occupies (Batima et al. 2005). Due to climate change, permafrost temperature in Mongolia is not only rising but also disappearing in some areas (Ishikawa et al. 2018; Jambaljav 2017; Dashtseren et al. 2017a; Sharkhuu and Anarmaa 2012; Sharkhuu et al. 2007) as discussed below.

### 7.6.1 Temperature Changes in Permafrost

Several studies have described the temperature changes in the permafrost of Mongolia over the last decades, and all studies agree with an increasing trend in permafrost temperature (Ishikawa et al. 2018; Jambaljav 2017; Dashtseren et al. 2017a; Sharkhuu and Anarmaa 2012; Sharkhuu et al. 2007). According to temperature monitoring data in the Altai mountain, MAGT at a depth of 10 m at the Tsagaannuur-1 on the foot of south-facing slope was −0.8 °C in 1985, −0.42 °C in 2010 and −0.39 °C in 2016, while MAGT at the north-facing slope was −1.2 °C in 1985, −0.62 °C in 2010 and −0.67 °C in 2016 at the Tsagaannuur-2, respectively. In addition, this warming trend is observed at all monitoring sites in the Altai such as the Khashaat-2, Tsengel-1, Tsengel-3 and Tsengel-4 (Fig. 7.8). The southernmost site (Erdene-2) is located in a valley, where MAGT at the depth of 8 m was −0.3 °C in 1982, −0.17 °C in 2010 and 0 °C in 2017, likely indicating that permafrost has disappeared recently. The same phenomenon was observed in the Khangai mountain. MAGT at Gurvanbulag was −3.7 °C in 1982, −3.0 °C in 2010 and −2.9 °C in 2017, and at Therk khundii was −2.0 °C in 1982, −1.4 °C in 2010 and −1.1 °C in 2016, while in the Sharga khundii it was −2.35, −1.78 and −1.19 °C for 1963, 2010 and 2016, respectively. Overall, the average rate of increase in MAGT in the Altai and Khangai mountains ranged from 0.1 to 0.2 °C/year (Dashtseren et al. 2017a; Jambaljav 2017). In contrast, the decreasing temperature in permafrost has not been observed in both mountains.

Similar temperature changes pattern in permafrost have been observed from the monitoring sites in the Khentii and Khuvsgul mountains. For instance, MAGT at Tsagaannuur site was −3.9, −3.0 and −2.88 °C for 1989, 2010 and 2016,

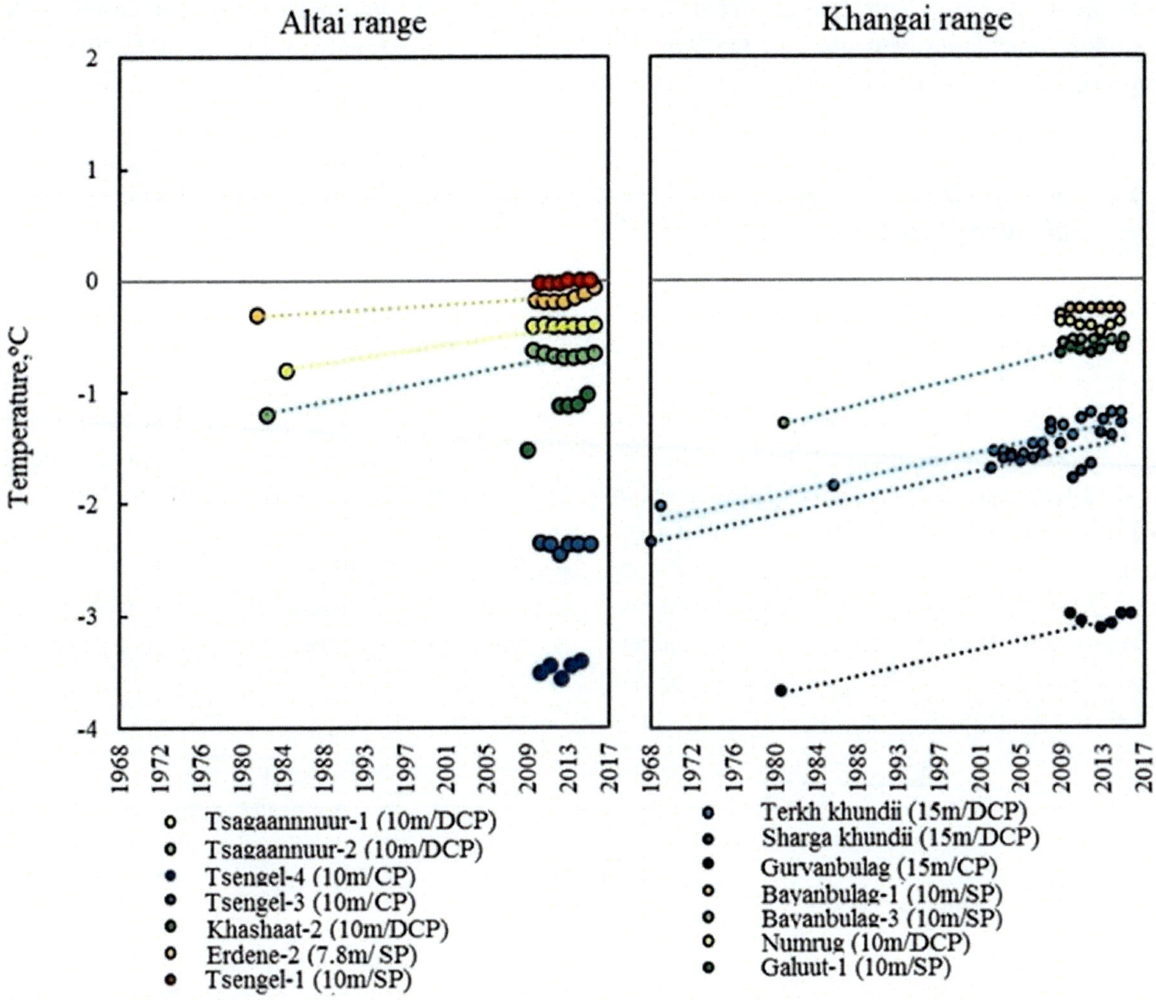


**Fig. 7.8** Mean annual ground temperatures at the monitoring sites in the Altai and Khangai mountains

respectively. At the Munguush, MAGT has increased from −1.87 °C in 1987 to −0.25 °C in 2017. Small permafrost warming rates have been reported in the southern part of Khentii mountain. MAGT at Baganuur site was −0.31 °C in 2012, −0.31 °C in 2013, −0.31 °C in 2014 and −0.28 in 2016, respectively. A similar trend was observed at the Mungunmorit site with MAGT of −0.73, −0.76, −0.76 and 0.53 °C for 2011, 2012, 2013 and 2014 (Fig. 7.9). In general, MAGT increased from 0.2 to 0.3 $°Cyr^{-1}$ in the Khuvsgul mountain and from 0.1 to 0.2 $°Cyr^{-1}$ in the Khentii mountain (Dashtseren et al. 2017a; Jambaljav 2017). These rates in the permafrost temperature in the mountains are similar to the previous study findings from Sharkhuu et al, (2007).

As shown in Figs. 7.8 and 7.9, increases in MAGT are greater in continuous permafrost zone with cold state than they are in the other permafrost zones with warm state. This higher increase in MAGT cold permafrost is due to the domination of sensible heat rather than latent heat effect. While the warm permafrost is mainly under the state shift from ice to water, it consumes a high amount of heat, dominantly leading by latent heat effect. It can be assumed that the increase in permafrost is mainly in response to increased air temperature over Mongolia.

A comparative study shows that permafrost in high latitudinal continuous permafrost zones (in Siberia and Alaska) has degraded more rapidly than at the southern fringe of continuous permafrost zone (in Mongolia). Meanwhile, the permafrost in the Khuvsgul mountain has degraded more intensively than in the Khangai and Khentii mountains (Sharkhuu et al. 2007).

### 7.6.2 Spatial Degradation of Permafrost

As mentioned before, several evidences of permafrost degradation have been reported in Mongolia such as the increase in permafrost temperatures, deepening of active layer and the areal changes in thermokarst lakes (Ishikawa et al. 2018; Jambaljav 2017; Dashtseren et al. 2017a; Saruulzaya et al. 2016; Sharkhuu and Anarmaa 2012; Sharkhuu et al. 2007; Batima et al. 2005). Furthermore, significant permafrost thawing has been reported at the southern boundary of permafrost in Mongolia. For instance, thin permafrost disappeared completely in the period 1984–2008 in the Khentii-36 sites at the southern part of Khentii mountain (Fig. 7.9). Similarly, in the Erdene-2, there was permafrost in 1985 with a temperature of −0.8 °C at the depth of 7.8 m, while the temperature at the site was 0 °C in 2016, also indicating no permafrost existence. Furthermore, the areal degradation and shift of the southern boundary of permafrost can be estimated by comparison between permafrost map in 1971 (Gravis et al. 1971) and permafrost map 2016 (Jambaljav et al. 2016). About 63% of the territory was underlying by permafrost in 1971 (Gravis et al. 1971), while it is now approximately 29% according to the result from Jambaljav et al. (2016). However, it should be noted that we could not conclude

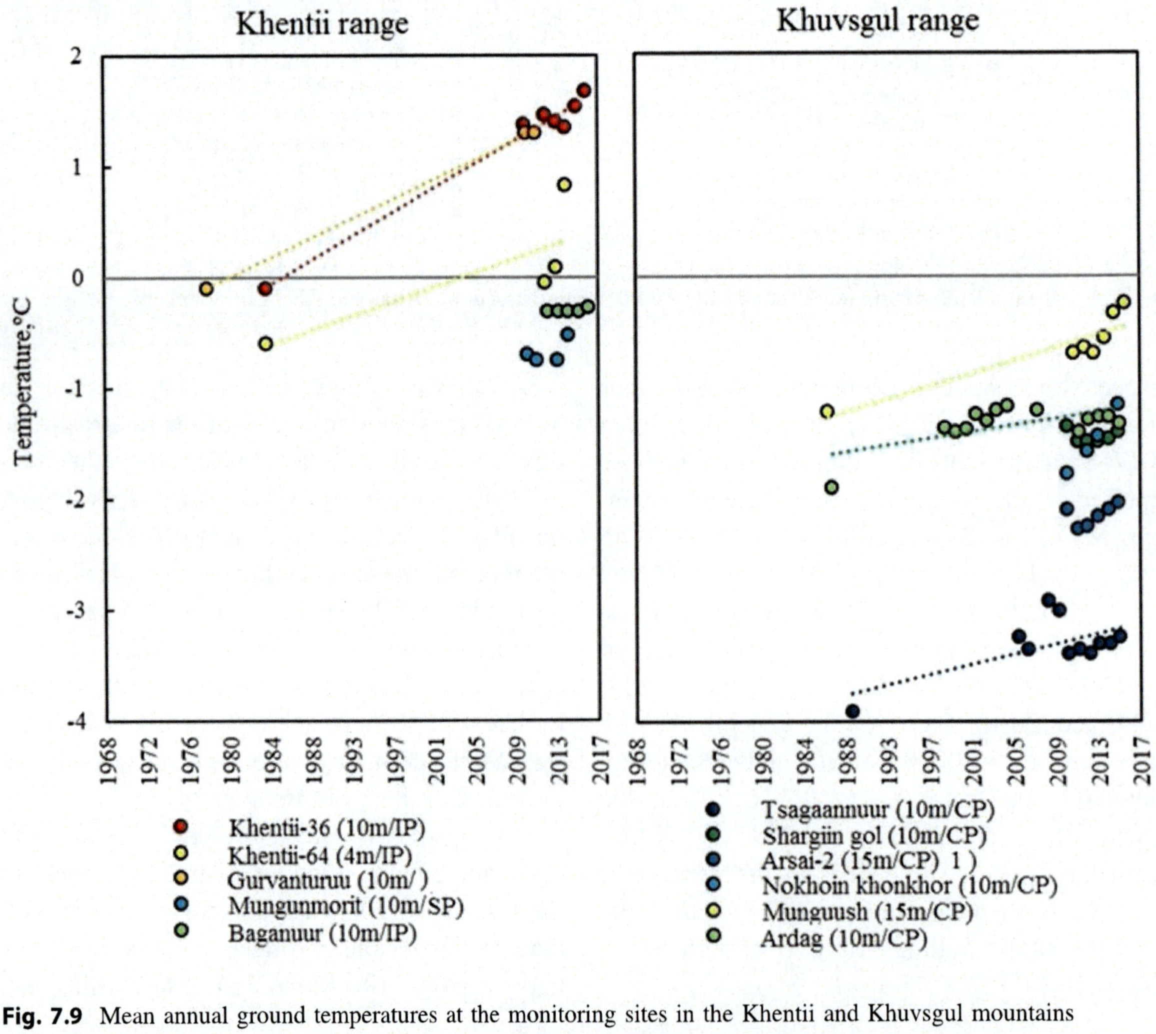


**Fig. 7.9** Mean annual ground temperatures at the monitoring sites in the Khentii and Khuvsgul mountains

that the area of permafrost in Mongolia has decreased by 50% during the last 45 years using the former two results. They used different methods and classifications for permafrost zonation because each permafrost zone for the maps had different percentages of real permafrost existence. Comparison of the maps in 1971 and 2016 roughly shows that the southern permafrost boundary has moved to the north by 96 km in the Khangai mountain, 184 km in the Khentii mountain and 60 km in the Altai mountain, respectively.

## 7.7 Permafrost Problems on Infrastructure

It is well known that most of the population of Mongolia live in the northern part of Mongolia where the ecosystem services are more abundant with usually underlying permafrost than that in the southern part of Mongolia. Furthermore, the productivity of vegetation cover and groundwater is higher in permafrost areas of Mongolia. This indicates that permafrost has a positive effect on the ecosystem in the arid Mongolian environment. However, a number of engineering problems resulting from the occurrence of permafrost have been reported in Mongolia. For the most cases, they first relate to the water and/or ice content of permafrost. These can be summarized as being either frost heave, thaw-subsidence or hydrologic in nature (French 2007).

Due to the negative effect from permafrost, some sum centers have been moved from permafrost area to non-permafrost area. For example, Galuut sum in the south part of Khangai mountain, Umnudelger sum in the south part of Khentii mountain, Erdene and Altai sums in the southern Altai mountain, and Khyargas sum in the northern Mongolia moved into non-permafrost area, respectively (Dashtheren et al. 2017b; Figs. 7.10 and 7.11). Of course, these movements had negative impacts on the local economic conditions. Furthermore, many broken buildings in the sum centers have been registered in all permafrost zones in Mongolia within the last few decades. This indicates that buildings in permafrost regions are under a high risk. As Mongolia has a harsh-cold winter, local people

**Fig. 7.10** The current situation of the old sum center of Khyargas sum in the northern Mongolia (photo by G. Tsogt-Erdene)

**Fig. 7.11** Result of permafrost thawing under the wooden house in Chandmani-Undur sum in north part of Mongolia (photo by A. Dashsteren)

use much heat for warming the buildings in winter. Moreover, they often utilize firepot for making meals during other seasons. These actives can lead to an increase in ground temperature under buildings, leading to the melting of permafrost and damage to the structure of the building. Of course, several types of constructional techniques in Mongolia are used to reduce the thermal changes under buildings, but there are many broken buildings in relation to permafrost effect.

Another permafrost impact on infrastructure in Mongolia is the paved road construction. During the construction of a road, the ground surface condition completely changes (e.g. removal of vegetation, the color changes in surface), which in turn affects the ground thermal equilibrium. In Mongolia, the paved roads cross both the regions of permafrost and seasonally frozen ground. Some parts of these roads are heavily damaged by permafrost melting, mainly resulting in subsidence of the road. Such phenomenon generally forms many valleys and depressions with ice-rich permafrost. For example, in the Chuluut valley where ice-rich permafrost is abundant, the subsidence of the road has been seen with approximately the maximum depth of 60 cm. In addition, a similar phenomenon has been observed in the Erkhel nuur depressions in the Khuvsgul mountain.

## References


Batima P, Natsagdorj L, Gombluudev P, Erdenetsetseg B (2005) Observed climate change in Mongolia. Assessments of Impacts and Adaptations to Climate Change (AIACC) working paper, isssue no 12, p 25. https://doi.org/10.1890/08-0144.1

Dashtseren A, Jambaljav Y, Ishikawa M, Iijima Y, Gansukh Y, Temuujin K, Tsogterdene G (2017a) Monitoring and thermal state of permafrost in mongolia. In The 2nd Asian conference on permafrost (ACOP2017), Sapporo, pp 2–3

Dashtheren A, Jambaljav Y, Gansukh Y (2017b) Permafrost impact on the development of social and economic condition in Mongolia. In: The 3rd international conference proceeding on natural condition and territorial location aspects influencing in Socio-Economic development. Ulaanbaatar

Dashtseren A (2018) Cryosphere, Dorjgotov D (ed). Tepe Publishing House LLC. Ulaanbaatar

Dashtseren A, Ishikawa M, Iijima Y, Jambaljav Y (2014) Temperature regimes of the active layer and seasonally frozen ground under a forest-steppe mosaic, Mongolia. Permafrost Periglacial Process 25(4):295–306. https://doi.org/10.1002/ppp.1824

French HM (2007) The periglacial environment. Wiley, The Atrium, Southern Gate, Chichester, West Sussex PO19 8SQ, England

Gallin W, Erdenbat M (2003) Formation and organization of patterned ground in the Kharkhiraa Uul, pp 140–144. https://keckgeology.org/files/pdf/symvol/18th/mongolia/gallin.pdf

Gravis G, Jamsran S, Zabolotnik S, Sukhodrovskiy V, Sharkhuu N, Lisun A, … Tumurbaatar D (1971) Geocryological map of the Mongolian People's Republic

Harris SA, Brouchkav A, Geocryology C (2017) Geocryology: characteristics and use of frozen ground and permafrost landforms. CRC Press/Balkema

Ishikawa M, Jamvaljav Y, Dashtseren A, Sharkhuu N, Davaa G, Iijima Y, Yoshikawa K (2018) Thermal states, responsiveness and degradation of marginal permafrost in Mongolia. Permafrost Periglac Process 29(4):271–282. https://doi.org/10.1002/ppp.1990

Ishikawa M, Yamkhin J (2016) Formation chronology of Arsain Pingo, Darhad Basin, Northern Mongolia. Permafrost Periglac Process 27(3):297–306. https://doi.org/10.1002/ppp.1877

Jambaljav Y (2017) Distribution and changes in permafrost of Mongolia. Publishing House of Colorful, Ulaanbaatar

Jambaljav Y, Gansukh Y, Temuujin K, Tsogt-Erdene G, Undrakhtsetseg T, Saruulzaya A, … Narangerel S (2016) Permafrost map of Mongolia. (1:1 000 000, in Mongolian). Ulaanbaatar

Lomborinchen R (2000) Short communication frost heaving and related landforms, Mongolia, 90(July 1998):85–90

Lonjid N (1963) Seasonally frozen ground and permafrost of Mongolia. Publishing House of Mongolian Academy of Sciences, Ulaanbaatar

Luvsandagva D (1978) Permafrost in Khangai and Khuvsgul mountains. Publishing House of Mongolian Academy of Sciences, Ulaanbaatar

MARCC (2014) Mongolia second assessment report on climate change 2014. Ulaanbaatar

Obu J, Westermann S, Bartsch A, Berdnikov N, Christiansen HH, Dashtseren A, Delaloye R, Elberling B, Etzelmüller B, Kholodov A, Khomutov A, Zou D (2019) Northern hemisphere permafrost map based on TTOP modelling for 2000–2016 at 1 km$^2$ scale. Earth-Sci Rev (193):299–316. https://doi.org/10.1016/j.earscirev.2019.04.023

Saruulzaya A, Ishikawa M, Jambaljav Y (2016) Thermokarst Lake Changes in the Southern Fringe of Siberian Permafrost Region in Mongolia Using Corona, Landsat, and ALOS Satellite Imagery from 1962 to 2007. Advances in Remote Sensing 05(04):215–231. https://doi.org/10.4236/ars.2016.54018

Sharkhuu A, Sharkhuu N, Etzelmüller B, Heggem ESF, Nelson FE, Shiklomanov NI, Brown J (2007) Permafrost monitoring in the Hovsgol mountain region, Mongolia. J Geophys Res: Earth Surf 112(2):1–11. https://doi.org/10.1029/2006JF000543

Sharkhuu N, Anarmaa S (2012) Eurasian Steppes. Ecological problems and livelihoods in a changing world. In: Werger M, van Staalduinen MJA (eds) https://doi.org/10.1007/978-94-007-3886-7_17

Sharkuu N (2001) Permafrost mapping in Mongolia. In: International symposium on mountain and Arid land Permafrost, Ulaanbaatar

Tsegmid S (1969) Physical geography of Mongolia. State Publishing House, Ulaanbaatar

Tumurbaatar D (1971) The map of seasonal freezing and thawing in Ulaanbaatar city. Publishing House of Mongolian Academy of Sciences, Ulaanbaatar

William G, Erdendebat M (2005) Formation and organization of patterened ground in the Kharkhiraa uul. https://keckgeology.org/files/pdf/symvol/18th

Yoshikawa K, Sharkhuu N, Anarmaa S (2013) Groundwater hydrology and stable isotope analysis of an open-system pingo in Northwestern Mongolia. Permafrost Periglac Process 24(3):175–183. https://doi.org/10.1002/ppp.1773

# 8 Soils of Mongolia

Sandag Khadbaatar

**Abstract**

Soil science is one of the integral parts of physical geography, and has evolved into an independent science due to its role and importance in nature and society. Soil science is the study of soil as a natural resource on the surface of the earth, including soil formation, classification, and mapping; physical, chemical, biological, and fertility properties of soils; and these properties are in relation to the use and management of soils. Mongolian soil scientific concepts such as nomenclature, classification units, and soil-geographical regions have been developed through Dokuchaev's theory, and the terminology has become a tradition through Russian soil science. According to the history of soil research, the work of Russian researchers is the main, and there are many works of Mongolian scientists. Among them, the works of academician D. Dorjgotov on soil geography are considered the largest, and there are many fundamental research materials based on the theory of physical geography and landscape. Therefore, the works of Sh. Tsegmid, D. Dorjgotov, O. Batkhishig, and N.D. Bespalov are mainly used in this chapter. Due to its different physical geographical conditions many soil types are distributed throughout Mongolia. Based on the latitudinal characteristics of these soils, the definition of each zone is given as an example of predominantly distributed soil. In the content of the chapter, the soil is considered in terms of the components of the landscape geosphere; the distribution patterns, types, classifications, and characteristics are briefly described, and the principles of soil-geographical zoning, soil use, and protection are compiled.

**Keywords**

Soil types · Soil classification · Soil-geographic region · Soil erosion and degradation

## 8.1 Overview of the Soil Research

The works of the Russian Geographical Society occupy an important place for the research of Central Asia and Mongolia, which began to be published in the second half of the nineteenth century. Since that time, a great deal of research material has been accumulated on the geography, history, ethnology, economy, climate, vegetation, geology and hydrology of Mongolia. There were no notes about the soil of Mongolia but in their literatures of the physical geography, such as N.

S. Khadbaatar (✉)
Department of Geography, Mongolian National University of Education, Ulaanbaatar, Mongolia
e-mail: khadbaatar@msue.edu.mn

© Springer Nature Switzerland AG 2021

B. Yembuu (ed.), *The Physical Geography of Mongolia*, Geography of the Physical Environment,
https://doi.org/10.1007/978-3-030-61434-8_8

M. Przewalski, GN Potanin, P.K. Kozlov, VA Obruchev, brief notes on stony, sandy, and gravelly soils are included (Bespalov 1951). The first scientific paper on the soil of Mongolia was published in 1912 by L.I. Prasolov, and it was a research report aimed at revealing the geographical patterns of soils in the southern part of Lake Baikal. This research was to map soil characteristics from Khiagt to Ulaanbaatar, and the study combined soil-forming bedrock, permafrost, and vegetation to identify local landscapes in that area.

After that, B.B. Polynov, I.M. Krasheninnikov, V.I. Lisovskii, N.N. Lebedev, and O.N Mikhailovskaya's 1925 studies of the soil in the Ar and Uvur Jargalant river basins are considered to be a massive contribution. In the course of this study, a detailed description of the physical geography and soil vegetation of the area was given and thick, fully developed black earth and brown soils were found to be distributed in the area. In 1926, B.B. Polynov, V.I. Lisovsky, and others conducted two surveys in the northern Gobi.

The first route is from Ulaanbaatar to Ikh Tukhum Lake, Sangiin Dalai, and Ongi River Khoshuu monastery. In terms of height, soils are divided into three main types: mountain-steppe chernozem and dark brown, carbonate steppe brown, steppe salt marshes, and salt marshes (Bespalov 1951).

The second route runs from Ulaanbaatar to the northern Gobi, including Choir, Sain-Us (Sainshand), and Baga-Uud (northeast of Dornogovi). The study determined that the upper part of the Gobi brown soil has two layers, humus and compacted, and that there is no gypsum in the steppe and desert steppe soils but it had high contents of manganese and silicon. So, it was noted that the Gobi region should be the object of detailed research in the future.

The work of a soil-agronomic research expedition led by Baranov in 1930 is important for the study of soil in Mongolia's agricultural lands. In this study, well-known Russian soil scientists G.V. Dobrovolsky, S.I Andreev, and others conducted research on food and fodder plants, crop development, and fertility of hay and pasture soils in Western Mongolia in Khovd, Buyant, Uyench, Bodonch, Bulgan river basins, Mongol Altai, Tsenkher, and Durgun lakes. This research remains an important fact in the study of arable soils. As a result, the distribution of saline and alluvial meadow soils in the Khovd, Buyant, and Khar Us Lake basins was mapped as a basis for subsequent research.

At the same time, Andreev surveyed 5,000 hectares of arable land in the Selenge Basin and identified the first terrace 40–50 cm above the water level of sandy-gravelly sediments in connection with the distribution of soil. The second terrace with saline and meadow soils, the third terrace with brown soils, and the fourth terrace with dark brown soils were mapped at a scale of 1:25,000. In addition, V.N. Ivanov identified the zoning patterns of fertile, alkaline dark brown and brown soils on 4,000 hectares of the Baru-unturuun hay-pasture zone. In 1931, the research team returned to Mongolia to conduct an independent soil study of the Great Lakes depression, and the first explanation of the zonal patterns of altitude and latitude contributed to the development of soil geography in Mongolia.

In 1940–1943, a research team led by Tsatsenkin conducted a soil survey in southern Mongolia, collected hay and pasture material in the eastern part of the Gobi, and they collected a lot of material about plants and designed a 1:1,000,000 soil map, which became a major work used in many fields. Although these surveys were conducted at different levels depending on the purpose and content of the work at the time, they have not lost their relevance as valuable evidence for soil surveys (Bespalov 1951).

In 1940–1942, a soil research team led by soil scientist N.D. Bespalov surveyed more than 20,000 km of land, covering more than 70% of Mongolia, collected more than 3,000 soil samples, and analyzed 13 groups of soils. This study was published in Moscow in 1951 in Russian under the title "Soils of the Mongolian People's Republic". This book is divided into seven regions of geomorphology based on physical geography regions, and for the first time, the characteristics of vertical soil zoning, soil properties, resources, and the area of soil types are determined. The importance of this book was emphasized by academician D. Dorjgotov noting that "until the 1970s, it was the only significant work on the soil of our country". In 1962, the Institute of Geography was staffed by national soil specialists. In 1970, a comprehensive joint Russian–Mongolian biological expedition began and independent soil research was carried out. In 1974, the Institute of Geography was expanded to include an independent soil laboratory, which intensified soil research. Since this period, the work of national researchers grew and new branches of research were created. Of these, D. Dorjgotov (soil formation and taxonomy), G. Undral (features of the geographical distribution of mountain taiga soils), U. Bekhtur (characteristics, distribution and origin of brown soil), D. Batbayar (chemical properties of steppe soil), B. Batjargal (soil characteristics of Khuvsgul mountainous area), J. Garidkhuu (Orkhon-Selenge basin distribution patterns and characteristics), Ch. Gonchigsumlaa (dry steppe soil geochemistry), R. Baatar (characteristics and use of floodplain soils), Z. Sanjmyatav (soil washing), O. Batkhishig (Tuul river basin soil geochemistry), O. Battulga (Gobi brown soil moisture and thermal regime), J. Mandakhbayr (characteristics of saline soils), D. Avaadorj (soil characteristics and evaluation), Ch. Lkhagvasuren (Mountain soils geochemistry), N. Nyamsambuu (soil erosion), there are many works in the field of crop yields, fertilizers, irrigation regimes, and soil agro-chemistry. Thus, soil science has become a major independent branch of physical geography in Mongolia. This chapter does not discuss the history of soil research in detail, but about the patterns, characteristics, and classifications of soil geographical distribution.

## 8.2 Soil Zones, Belts and Characteristics

The natural condition of Mongolia is very different and there are intercontinental location, geological history of territory, high position and specifics of mountain, plain, intermountain depression features. On the other hand, marked substantial contrasts of natural factors in different parts of the country define specific soil formation. V.V. Dokuchaev explains that the soil depends on physical and geographical conditions of the place of origin, and defines the soil as a "mirror" of the landscape. Soil is one of the parts of the ecosystem that is formed as a result of the interaction of other natural spheres. Soils develop over time under the influence of chemical, physical, and biological processes. They develop where rocks and sediments (lithosphere) are influenced by flora and fauna (biosphere), water (hydrosphere), and climate (atmosphere). Based on these characteristics, soils can be identified, named, explained, and widely used in the classification of natural zones and landscapes.

Mongolia's topography, climate, and latitude are the main factors in soil distribution. They also differ in the nature and land use, which is formed in different relief conditions. For example, mountain tundra soils have a long snow cover, low thermal activity due to low biological activity, incompletely decomposed organic residues, acidic and rocky properties, while brown soils have high weathering products, well-drained, carbonated and relatively good biological activity. The effects of climate such as heat transfer to great depths are relatively large. Steppe and mountain soils differ not only in geomorphological features but also in soil source

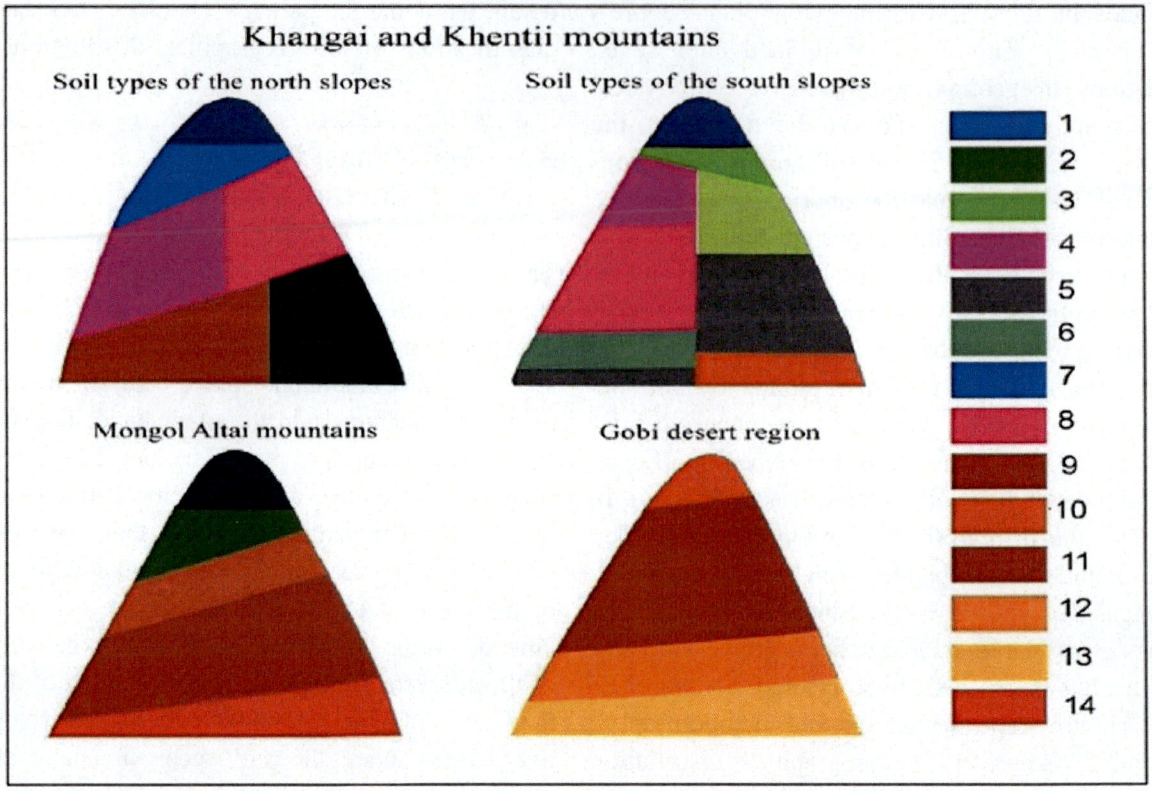


**Fig. 8.1** Soil vertical zones (National Atlas of Mongolian People's Republic 1990). 1. Mountain tundra, 2. High mountain steppe raw humic, 3. Mountain meadow, 4. Mountain meadow steppe, 5. Mountain taiga cryomorphic surface ferrimorphic, 6. Mountain taiga cryomorphic, 7. Mountain derno-taiga depth permafrost, 8. Mountain meadow-forest freezing depth, 9. Mountain chernozem meal carbonated, 10. Mountain dark kastanozem, 11. Mountain kastanozem, 12. Mountain light kastanozem, 13. Mountain semi-desert brown soil, 14. Mountain desert gray brown soil

processes. Due to the fact that mountain soils are stable in conditions of surface fragmentation, slope, and extreme climatic conditions, slope erosion and temperature weathering are intensive, organic matter decomposes and mineralizes slowly, and the moisture-thermal regime depends on the mountain (Dorjgotov 2003) (Fig. 8.1).

Despite these differences, mountain and steppe soils have similar characteristics. Thin, gravelly soils on hilly plains are difficult to distinguish from mountainous soils, and thick-walled powdered soils formed on mountain plateaus or hollows and terraces are difficult to distinguish from steppe. This is due to the fact that latitude has a significant effect on the formation of steppe soils, while this feature is also observed in mountainous areas. In the mountainous part of Mongolia, the vertical belt is clearly visible and the latitude is sometimes lost. Although such differences do occur, geomorphological features are taken into account in distinguishing soil zones according to the latitudinal patterns of soil distribution in Mongolia. In the northern part of Mongolia, most of the mountains are located along the latitude, which is a key factor in differentiating the latitude of the soil. Therefore, there are two belts and three zones of soil from north to south: (1) mountain taiga belt with cryomorphic-taiga and derno taiga soils, (2) mountain forest steppe belt with chernozem, dark kastanozem, forest dark colored, and derno taiga soils, (3) dry steppe zone with

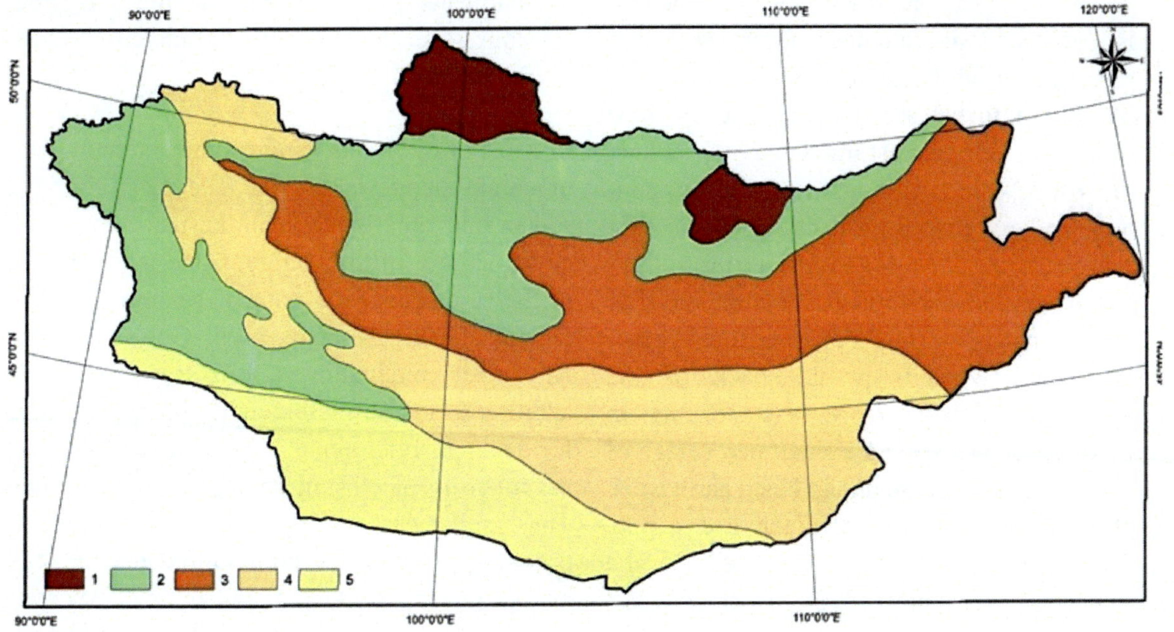


**Fig. 8.2** Soil belt and zones (modified from Dorjgotov 2009). 1. mountain taiga belt with cryomorphic-taiga and derno taiga soils, 2. mountain forest steppe belt with chernozem, dark kastanozem, forest dark colored, and derno taiga soils, 3. dry steppe zone with kastanozem soils zone, 4. semidesert brown soils zone, 5. gray-brown desert soils zone

kastanozem soils zone, (4) brown semi-desert soils zone, (5) gray-brown desert soils zone. The wide distribution of taiga and forest steppe soils in Northern Mongolia cannot be possibly explained by just vertical zone; there is also influence of the horizontal zone of the territory (Fig. 8.2) (Dorjgotov 2003).

A brief introduction at the characteristics of these zones, belts, and some key soil characteristics is given below.

1. ***Mountain taiga belt with cryomorphic-taiga and derno taiga soils***. Belongs to the taiga of Khuvsgul and Khentii mountains. Because this belt is entirely in the permafrost region, the effect of permafrost on soil moisture and soil formation is significant. It is rare for the taiga turf to cover all sides of the mountain evenly, but the meadow steppe black earth soil is quite widespread on the mountain slopes. Permafrost soils in the mountain taiga are generally thin, with a total soil thickness of no more than 60–80 cm, and the reaction environment being acidic in all layers. The humus content usually fluctuates between 11 and 20% and decreases sharply. The high carbon/nitrogen ratio in the top soil indicates that the humus of the organic residue is weak. Researchers have noted that the upper layers are mostly small, granular, loamy-light loamy texture, and that the high content of sand and dust is a characteristic feature of permafrost taiga soils (Dorjgotov 1976). In the middle and upper part of the soil layer, gravel accumulates under the influence of permafrost and the formation of the layers is lost. Larch-birch-cedar mixed forests are dominated by taiga turf and turf-ash soils, and permafrost boggy are abundant in the foothills. In the turfy taiga soils of mountain turf, a thin layer of light gray or light spots are formed under

the humus layer. The southern part of the belt is dominated by mountain forest gray and mountain meadow steppe black earth soils.

2. **Mountain forest steppe, forest gray, and mountain steppe kastanozem belts** include the Khangai mountainous area, the Orkhon-Selenge basin, and the Khentii lowlands. The transfer of black earth and forest gray soils is found at the back of the mountain. The characteristics of this belt are forest gray, brownish-gray soils at the back of the mountain, mostly steppe kastanozems in mountainous areas, and a small percentage of forest gray soils in mountain black earth soils. Due to the high impact of the dry steppe, forest soils are spread only behind the mountains (Dorjgotov 1992). Mountain forest steppe, forest gray, and mountain steppe kastanozem belts include the Khangai mountainous area, the Orkhon-Selenge basin, and the Khentii lowlands. The transfer of black earth and forest gray soils is found at the back of the mountain. The characteristics of this belt are forest gray, brownish-gray soils at the back of the mountain, steppe kastanozem soils in mountainous areas, and, in most parts, a small percentage of forest gray soils in mountain black earth soils. Due to the high impact of the dry steppe, forest soils are spread only behind the mountains. Permafrost and temperature weathering have a significant impact on the origin of mountain forest soils. The content of texture varies greatly, depending on the intensity of weathering, gravel is up to 60%, and sand and dust are predominant above the foot of the mountain, while the amount of clay increases in the middle and lower part of the foot and is about 32–45%. The clay content of schist and sandstone, which is common in the soil, is very low. The concentration of exchangeable cation in sediments formed from acidic (5.1) alkaline (8.6), weathering granite and sandstone soils reaches 28 mg/eq, and in sediments formed from dolomite and limestone reaches 28.0–32.0 mg/eq (Krasnoshchekov 2013) (Table 8.1).

Most of Mongolia's mountainous terrain has a mountain steppe landscape. The most common soil in this zone is mountain kastanozem soil. It covers 29.6% of the total area and is distributed in the mountains of Khangai, Khentii, Orkhon, Selenge basins, Mongol Altai, Gobi-Altai, and Khyangan mountains (Dorjgotov 2003). In addition to the absolute height, the main criteria for distinguishing mountain kastanozem soils are the intensity of erosion on the mountain slopes. These soil layers are replenished by weathering products, so there is no accumulation of carbonate due to the large number of debris, thin underdeveloped strata, poor carbonate composition, and moisture regime that is washed away by mountain runoff, and the bedrock is close to or exposed to the surface. Depending on the slope of the mountains, the thickness of the humus layer varies between 10 and 40 cm and soils are thin, less stability, and low humus than slopes of the back side and covered stones. As the mountain slope increases, the fine earth layer becomes thinner, the gravel layer becomes more abundant, and the soil layer is no clear. The humus content in the upper layer of mountain kastanozem ranges from 1.5 to 4.5%. Effervescence of hydrochloric acid varies, in some cases from the surface, and sometimes even to a depth of 10–40 cm. Mountain kastanozem is not suitable for cultivation due to its agriculture and can be used for rangeland.

3. **Dry steppe zone with kastanozem soils zone**. This is widely distributed in the eastern part of Mongolia and extends westwards through the Khangai mountain range to the Great Lakes depression. Mongolian kastanozems are characterized by carbonate accumulation, which is associated with extreme climates and precipitation distribution. Steppe kastanozems are distributed in flat and hilly

**Table 8.1** Characteristics of forest soil (adopted from Krasnoshchekov 2013)

| Profile number | Depth, m | Sum of exchangeable cation, mg-eq/100 g | pH | The content of elements, % | | | | | Texture, % | | |
|---|---|---|---|---|---|---|---|---|---|---|---|
| | | | | $SiO_2$ | $Fe_2O_3$ | $Al_2O_3$ | CaO | MgO | Sand | Silt | Clay |
| *Soil of upper part of mountain slope* | | | | | | | | | | | |
| 865 | 80–90 | 5.6 | 7 | 62.74 | 6.3 | 16.59 | 2.03 | 1.24 | 77 | 17 | 24 |
| 866 | 65–75 | 12 | 8.6 | 47.5 | 10.23 | 19.92 | 7.22 | 4.4 | 74 | 21 | 17 |
| 867 | 75–85 | 17.6 | 6.7 | 60.95 | 6.23 | 16.02 | 2.88 | 1.98 | 41 | 47 | 32 |
| 869 | 50–60 | 9.5 | 5.9 | 60.07 | 7.22 | 15.7 | 2.3 | 1.4 | 57 | 32 | 39 |
| *Soil of middle, lower part of mountain slope* | | | | | | | | | | | |
| 862 | 40–50 | 20 | 6.8 | 59.14 | 6.52 | 17.83 | 1.37 | 1.15 | 46 | 22 | 77 |
| 868 | 30–40 | 8 | 5.8 | 63.31 | 6.54 | 16.82 | 2.88 | 1.98 | 68 | 25 | 26 |
| 782 | 80–90 | 20.3 | 6.5 | 56.36 | 6.04 | 14.62 | 2.33 | 1.61 | 50 | 38 | 43 |
| 873 | 60–70 | 10 | 6.6 | 68.37 | 6.08 | 16.43 | 1.09 | 0.93 | 55 | 34 | 36 |
| 895 | 20–30 | 17.3 | 5.5 | 65.09 | 3.98 | 18.1 | 6.2 | 4.44 | 27 | 59 | 51 |

steppes, dry steppes, and dry valleys of the Khangai and Khentii mountains, and are divided into three types according to humus content and layer thickness: dark kastanozem, typic kastanozem, and light kastanozem.

*Dark kastanozems* are distributed in the northern part of Mongolia in the valleys between Khangai and Khentii mountains, mountain slopes, foothills, valleys of large rivers, and in the steppes of Dornod and do not form a zone alone. The thickness of the humus layer is usually 30–40 cm, sometimes more than 50 cm, and the humus content is 3.5–4.5%. Carbonate accumulation occurs at 45–50 cm and is not gypsum. The predominant textures are loam, light loam, and sand. Dark kastanozems are important for agriculture due to their high nutrient content (Tables 8.2 and 8.3).

*Typic kastanozems* spread throughout the east and the central part of Mongolia, up to the west southern part of Khangai. The humus layer of this soil is light kastanozem in brightly color, with an average thickness of 20–30 cm, the amount of humus is 2.0–3.5%, and it has a dry dusty texture. In some cases, the humus content of the AB layer of kastanozems are more than the top layer. This has been explained by the nature of the kastanozem residual which is moister than now. Kastanozem usually boils in hydrochloric acid from under humus layer, and in sandy and loamy sandy soils from a depth of 60–80 cm. The carbonate accumulation layer is about 35–40 cm, and 70–90 cm in sandy and loamy soils. The soil contains less soluble salts and no gypsum. These soils are loamy, light loamy and sandy loam texture, and mixed gravels and stones. Kastanozems are used for hay and rangeland due to their lack of nutrients, especially water and moisture (Fig. 8.3).

*Light kastanozems* are mainly distributed in the southern part of the steppe zone and occupy a large area in the Great Lakes depression. The thickness of the humus layer in the soil is 12–18 cm, and the amount of humus in the upper layer is about 1.0–1.5%. The morphological features of these soils are brownish, thin, dry, with a lot of gravel, unclear granular structure, low density, and frequent cracks. It is dominated by gravelly loam, light loam, sandstone, and sand texture. All layers of soil are rich in gravel, 8–12 cm in hydrochloric acid effervescence, and 80–100 cm in sandy and loamy soils. Light kastanozems are poor in nutrients, rich in gravel, and have a thin humus layer, so they are widely used in rangeland that are not suitable for agriculture (Fig. 8.4).

**Table 8.2** Genetic horizon thickness and humus content in loamy kastanozem (soil cover and soil of Mongolia 1984)

| Criteria | Loamy | | | | | | |
|---|---|---|---|---|---|---|---|
| | Horizon A | | | | Horizon AB | | |
| | Number of samples | Average | Minimum | Maximum | Average | Minimum | Maximum |
| *Dark kastanozem meal-carbonated* | | | | | | | |
| Thickness, cm | 26 | 24 | 9 | 40 | 13 | 10 | 40 |
| Humus, % | 16 | 3.2 | 2.3 | 6.1 | 1.3 | 0.5 | 2.9 |
| *Kastanozem meal-carbonated* | | | | | | | |
| Thickness, cm | 25 | 13 | 8 | 25 | 18 | 10 | 25 |
| Humus, % | 8 | 2.7 | 2.2 | 3.0 | 2.3 | 1.4 | 4.2 |
| *Light kastanozem meal-carbonated* | | | | | | | |
| Thickness, cm | 9 | 7 | 4 | 12 | 14 | 10 | 23 |
| Humus, % | 9 | 1.4 | 0.5 | 1.8 | 1.1 | 0.4 | 1.7 |

**Table 8.3** Genetic horizon thickness and humus content in sandy loam, sandy kastanozem (Avaadorj 2018)

| Criteria | Sandy loam and sandy | | | | | | |
|---|---|---|---|---|---|---|---|
| | Horizon A | | | | Horizon AB | | |
| | Number of samples | Average | Minimum | Maximum | Average | Minimum | Maximum |
| *Dark kastanozem meal-carbonated* | | | | | | | |
| Thickness, cm | 4 | 21 | 15 | 30 | 15 | 15 | 25 |
| Humus, % | 4 | 2.9 | 1.2 | 4.0 | 1.0 | 1.8 | 2.3 |
| *Kastanozem meal-carbonated* | | | | | | | |
| Thickness, cm | 15 | 14 | 3 | 25 | 19 | 5 | 35 |
| Humus, % | 15 | 1.6 | 0.6 | 2.5 | 1.4 | 0.2 | 2.2 |
| *Light kastanozem meal-carbonated* | | | | | | | |
| Thickness, cm | 25 | 10 | 6 | 12 | 9 | 7 | 12 |
| Humus, % | 5 | 1.0 | 0.3 | 1.7 | 0.9 | 0.1 | 1.9 |

4. **The desert steppe (Gobi) brown soil zone** includes the Great Lakes depression, the Lakes Valley, the Gobi-Altai Mountains, and the eastern part of the Gobi. Due to low rainfall, high winds, and dry heats in this area, soil formation takes place in dry conditions. The general characteristics of the soil are well-defined layers, loamy, low humus, effervescence in hydrochloric acid from the surface or topsoil, and not solonchak, lack of moisture, no gypsum, covered with gravel. The Gobi brown soil humus content reaches

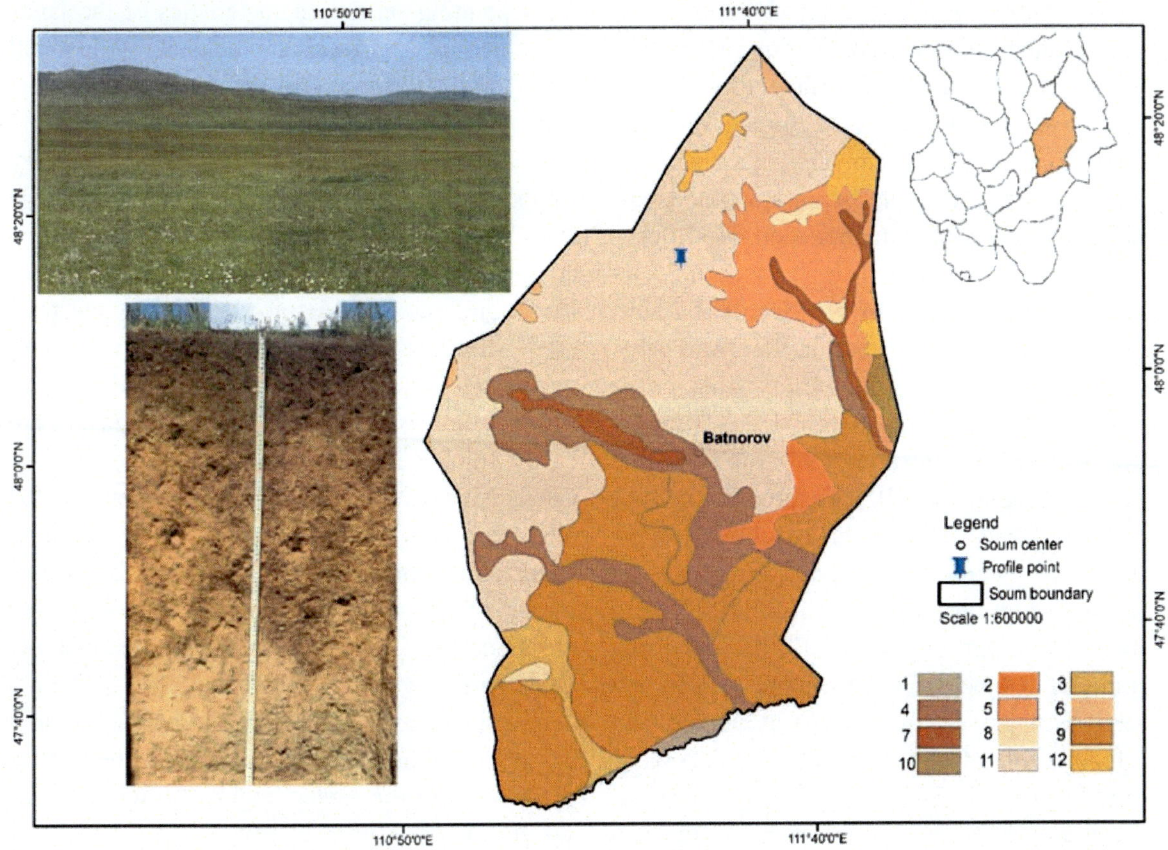


**Fig. 8.3** Light kastanozem layers (Batnorov soum, Khentii): 1. Alluvial meadow, 2. Saline boggy, 3. Dark kastanozem weakly developed, 4. Dark kastanozem solonetzic non-carbonated, 5. Meadow dark kastanozem, 6. Chernozem non-carbonated, 7. Meadow cryoturabated, 8. Meadow solonchak, 9. Stony shallow dark kastanozem, 10. Mountain dark kastanozem non-carbonated, 11. Mountain chernozem noncarbonated, 12. Mountain forest derno dark colored

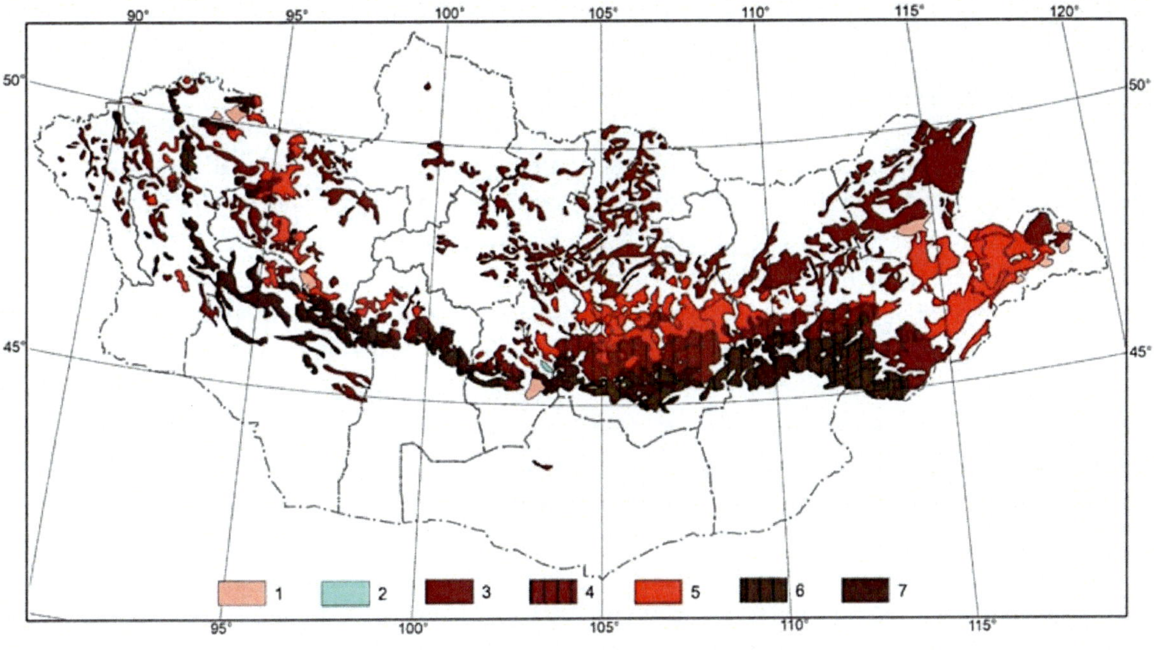


**Fig. 8.4** Distribution of Kastanozem in Mongolia (Batkhishig et al. 2018). 1. Kastanozem 2. Crypto-gleyed, 3. Dark kastanozem, 4. Hill kastanozem meal carbonated, 5. Kastanozem residual meadowish, 6. Hill light kastanozem meal carbonated, 7. light kastanozem

up to 0.3–0.8% and the lower layer has a more humus content. This is due to the high temperatures of the soil surface, the mineralization of humus, and the spread of plant roots to about 10–30 cm. This is a feature of Gobi Desert soils. The depressions and hollows are saline and loamy, and sand accumulation is widespread.

5. **The northern edge of the Central Asian deserts** forms a desert belt in the Trans-Altai region. Semi-desert and desert soils cover 28.5% of the total area, of which 21.9% are in the steppes and 6.6% in the mountains and hills (Dorjgotov 1992). The three main types of desert and desert soils distributed in the highlands are desert steppe brown, desert brown gray, and borzon soils. Soil-forming bedrock in this area is mostly proluvial and diluvial sediments. The soils are clearly layered, with gravel, light loam, sandy loam, and alkaline soils. The humus content is 0.2–0.5% and is easily soluble in salts and gypsum, but is abundant in the lower layers of the soil formed on Cretaceous and Paleogene saline sediments.

Desert soils are predominantly gray-brown and reddish colors and have a granular, subungular structure, with almost no humus layer, and only toneless in the upper part of the illuvial layer of secondary gypsum accumulation (Avaadorj 2018). The depth of soil carbonate varies even in the different side of the profile, and sometimes occurs as a spot that does not effervesce in hydrochloric acid. There is a light-colored porous solid layer under the gravel cover of the soil surface. The main distinguishing feature is the formation of a light gray thin layer that is beneath this layer. The classification is also complicated by the fact that light brown soils are distributed along the boundaries of the desert brown-gray soils.

The mountain foothills of the region, intermountain depressions, gorges, and cracked hills are rich in dry gravel, and there are many sediments transported by temporary runoff during the rainy season. The surface of the soil is covered with gravel from the desert hamada, which is a common feature of desert soils (Fig. 8.5).

Depending on the characteristics of the landscape in these soil zones and belts, individual soil types are distributed. This includes *halomorphic, hydromorphic, and floodplain* soils.

***Halomorphic soils*** are also found in dry steppe, Gobi, desert and forest steppe zones. It is most common in intermountain depressions, around salt lakes, in river floodplains and lowlands, and accounts for about 1.7% of the total area. All saline soils in Mongolia are divided into two main groups: solonetz and solonchack. A saline soil is one, or all of the soil layers are easily soluble in salts (NaHCO, NaCl, Na, SO, CaCl, MgSO, Ca ($HCO_3$)). Depressions in the arid steppe, Gobi, and desert areas are prone to salinization, and easily soluble salts are transported by precipitation from the highlands and are associated with groundwater mineralization. Due to its high sodium content, soil is highly alkaline, dense, columnar, and impermeable, so it has a low moisture content that is beneficial to plants, and in the lower stages, it often contains salt that is harmful to plants.

Wind has a significant effect on the formation and accumulation of salt in arid areas. Saline soils are usually found in the central part of the depression, so they are carried by the wind and affect the surrounding soil. The process of salt accumulation in meadow soils is regulated by normal leaching, such as overflow of snow and spring floods (Pankova 1992) (Fig. 8.6).

***Hydromorphic soils*** are distributed in small areas through mountain valleys, ravines, wet meadows, wetlands, and in more humid conditions. Extra moisture is due to the proximity perched groundwater to the surface, fluctuations of groundwater levels, more infiltration of precipitation into the soil, and permafrost thawing. In these cases, some chemical compounds can accumulate in the soil. Organic matter, iron, and manganese compounds accumulate in the humid mountainous areas of Khuvsgul, Khentii, and Khangai, while carbonates, soluble salts, and gypsum easily accumulate in the Gobi and steppe regions. The geochemical processes of moist soils take place in an anaerobic environment, so gleying is common. Hydromorphic soils may be

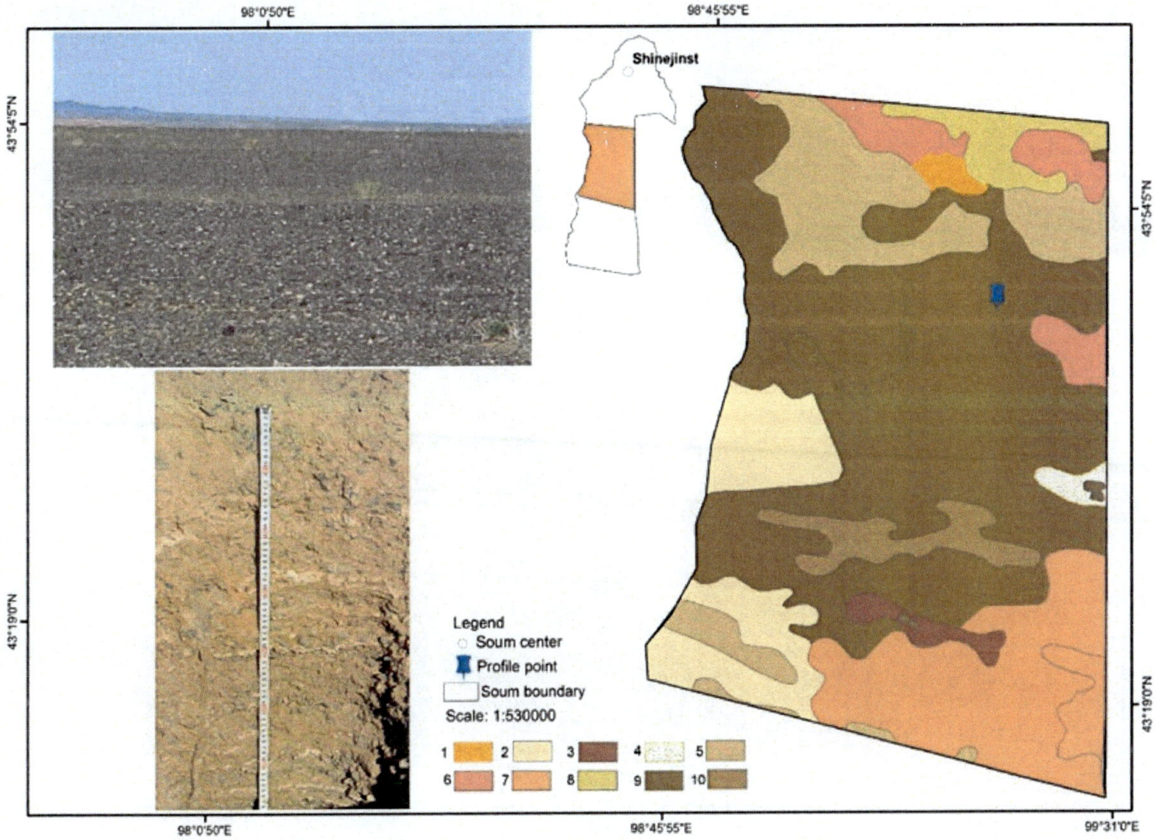


**Fig. 8.5** Desert soils and profile (Shinejinst soum, Bayankhongor): 1. Weakly fixed sands, 2. Extra arid desert borzon, 3. Extra arid desert solonetzic borzon, 4. Solonchak weakly fixed sands, 5. Hills shallow gray brown, 6. Desert gray brown covered sand, 7. Desert gypsum gray brown, 8. Desert solonetzic gray brown, 9. Desert sairic gray brown, 10. Mountain steppe desert shallow gray brown

classified into following three types referring to their conditions for additional moisture formed in soil, accumulation of organic substances, and shapes and forms of gley: (1) meadow humic gleyed; (2) meadow boggy raw humic gleyed; and (3) boggy peaty gleyed (Tsegmid 1969). In soils saturated with moisture, the reduction reaction is activated and the bluish, light gray color is weakened, the oxidation/reduction capacity is weakened, and ochre-brown and yellowish colors appear. Hydromorphic soils are spread mosaic in various natural zones and are formed in meadow and wetland landscapes.

***Floodplain soils*** vary in origin depending on the geological structure, and origin of the river valley, the size of the river, and the flood and feeding regimes. Depending on the development of the riverbed, the floodplain soil consists of a thin layer of sand and clay. Depending on the distance from the main riverbed of the flood river there will be soil differences. Derno soils is formed in the high areas along the river banks, meadow soils in the central part, and boggy soils in the edge of the floodplain. Alluvial soils are formed by the inundation of rivers, that is, they are composed of rock particles, which rivers carry from the upper to the lower course and leave on their banks during the flood. This material is called alluvium. It is very fertile, as rivers lay down not only minerals, but also biological remains of plants and animals. Groundwater is a major factor in the formation of alluvial soils. As the central floodplain usually has 1–1.5 m of water, it is constantly moistened at the bottom of

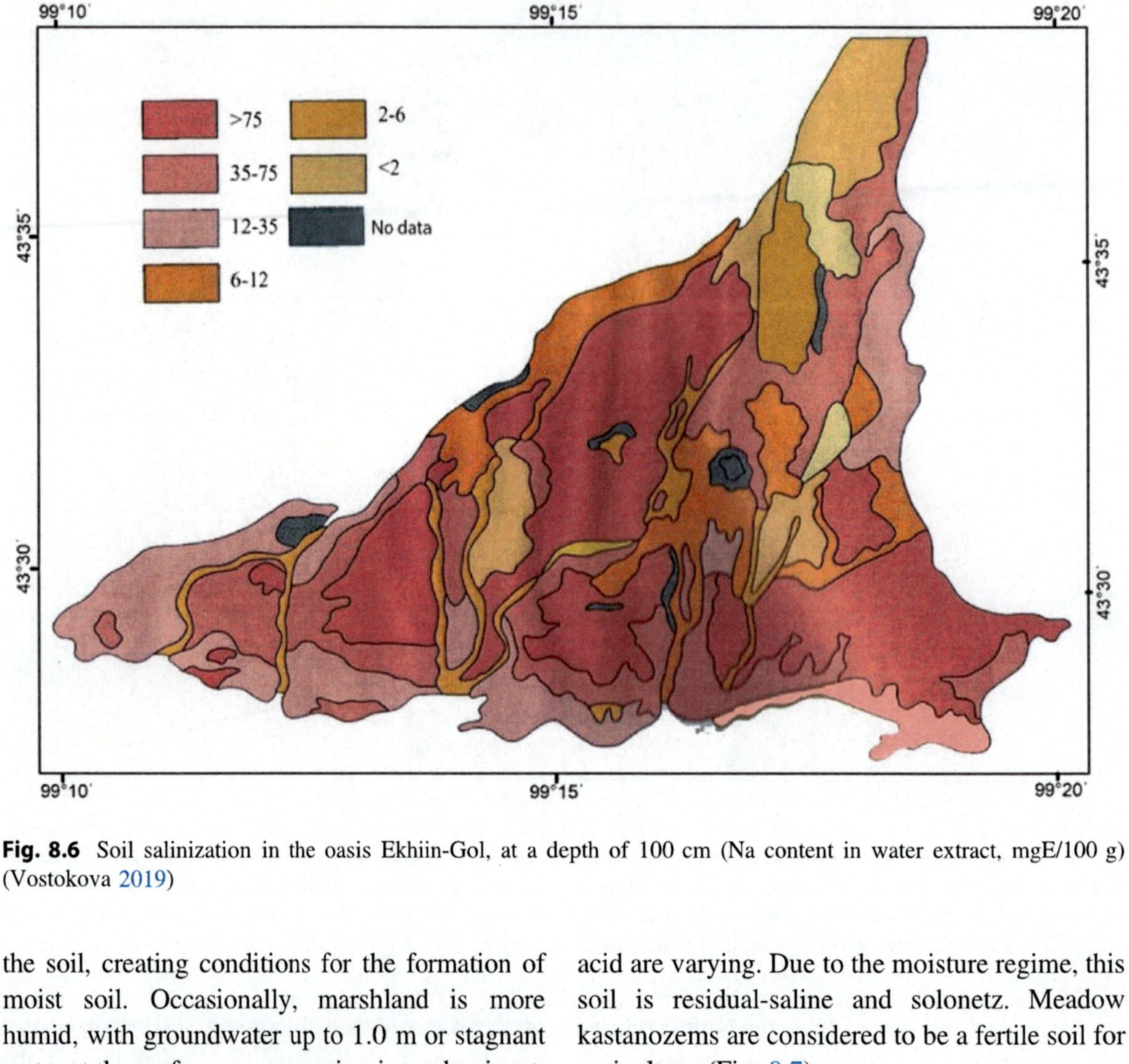


**Fig. 8.6** Soil salinization in the oasis Ekhiin-Gol, at a depth of 100 cm (Na content in water extract, mgE/100 g) (Vostokova 2019)

the soil, creating conditions for the formation of moist soil. Occasionally, marshland is more humid, with groundwater up to 1.0 m or stagnant water at the surface, so swamping is predominant. At the same time, permafrost soils are common. However, it has little effect on the peripheral soils along the river banks, where groundwater levels are relatively low. In river floodplains, various minerals are transported from the plateau, deep and surface runoff, so the accumulation process in the floodplain soils is clearly visible.

Floodplain kastanozems are distributed in the lowlands between the mountains. It is morphologically different from other kastanozems due to the influence of groundwater. The humus layer is gray brown, 45–80 cm thick; in some soils it is more than 1.0 m, and the humus content is 3.0–4.5%. It gradually decreases to the depth, and there are several layers of compacted humus in the lower layers. Granular structure, dusty, light loam, loam, effervescence depths in hydrochloric acid are varying. Due to the moisture regime, this soil is residual-saline and solonetz. Meadow kastanozems are considered to be a fertile soil for agriculture (Fig. 8.7).

## 8.3 Soil Classifications of Mongolia

Depending on the geographical location of Mongolia, its topography, and climatic conditions, many types of soils have been formed, and researchers have followed different criteria to classify them. The first soil classification in Mongolia was made in the early 1950 by N.D. Bespalov, who classified 13 groups of soils (Soils of the Mongolian People's Republic 1951). Since then, scientists and researchers have classified Mongolian soils into different levels, such as groups, types, subtypes, and so on, and are compared in Table 8.4 with some other classifications (Table 8.4).

**Fig. 8.7** Soil types (photos by Khadbaatar, S). 1. Mountain tundra soil, 2. Mountain meadow steppe soil, 3. Mountain forest soil, 4. Mountain meadow steppe soil, 5. Mountain desert steppe, 6. Steppe soil, 7. Semidesert and desert soil, 8. Hydromorphic soil, 9. Halomorphic soil, 10. Floodplain soil

**Table 8.4** Checklist of the soil classification

| Year | Soil classification | | | |
|---|---|---|---|---|
| | Group | Type | Subtype | Genus |
| Bespalov (1951) | 13 | 17 | – | 5 |
| Dorjgotov et al. (1974, 1981, 1986) | 13 | 38 | – | – |
| Nogina et al. (1980, 1984) | 10 | 37 | – | 100 |
| Dorjgotov (2003) | 10 | 34 | 66 | 100 |
| Krasnoschekov (classification of forest soil) 2013 | 6 | 18 | 50 | – |
| Batkhishig (2016) | 12 | 34 | – | 150 |

Since 1970, the Mongolian Academy of Sciences has been conducting soil research in Mongolia in several stages, focusing on the development of a unified soil classification. As a result of these studies, Russian and Mongolian researchers have developed an updated version of soil classification and identified 38 types and more than 100 subtypes of soil (Nogina and Dorjgotov 1982). A 1:1,000,000 scale map of the soil of Mongolia (edited by Mikhailov, Dorjgotov) is published in 1981. This map shows the distribution of more than 220 soils in 13 groups. Until the 1990, large, medium, and small-scale thematic maps of aimag, soum, hay, rangeland, and agricultural soils were published in large numbers, and soil classifications varied depending on the purpose and scope of the study. Subsequent versions of the soil classification were developed by national researchers, who elaborated on large- and small-scale thematic drawings, detailing soil types, subtypes, and types (Dorjgotov and Batbayar 1986). Following the theory of Dokuchaev, the relief plays a decisive role in the formation of mountain soils. Soils were divided into two main groups: mountain and steppe. Due to their natural state, mountain soils are divided into independent taxonomic groups, but due to insufficient study of high mountain soils, there is not enough material for their detailed classification (Dorjgotov and Batbayar 1986) (Fig. 8.8).

The International Soil Classification considers only the characteristics, composition, and bedrock characteristics of soil layers, regardless of landscape differences. For example, thin, gravelly soils formed on bedrock are classified as Leptosol, while rocky, permafrost, crumbly, limestone as mollic, and coarse humus soils are classified as umbric (Jahn et al. 2006). In the case of Mongolia, it is located in the center of the continent and is very different from similar soils of the same latitude due to its mountainous, geological structure, and climatic characteristics. Mongolia has an arid and cold climate due to its geographical location of inland and mid-latitude highlands. Due to this, it makes different from other soils in the temperate zone of the world. Most of Mongolia's mountains are latitudinal, and natural zones transfer in relatively short distances in mountainous areas. As a result, in the northern part of the country, in addition to the latitudinal features, there is a mixed landscape that retains the characteristics of a vertical belt. In such landscapes, mountain latitudes, taiga, and mountain black soils are affected by both latitude and altitude differences. Therefore, the soils of Mongolia in the group of mountain soils are classified, in addition to the vertical belt soils, which included latitudinal features. This classification of soils is based on the principles of geomorphological features and soil origin. The classification of soils in this category retains the terms used in the previous version, and uses some traditional local names, such as takyr and borzon. Therefore, this is a classic classification of soil science that is based on characteristics of the landscape.

There are 18 types of mountain soils that are grouped into five groups: tundra, meadow and meadow steppe, forest, steppe, desert steppe, and

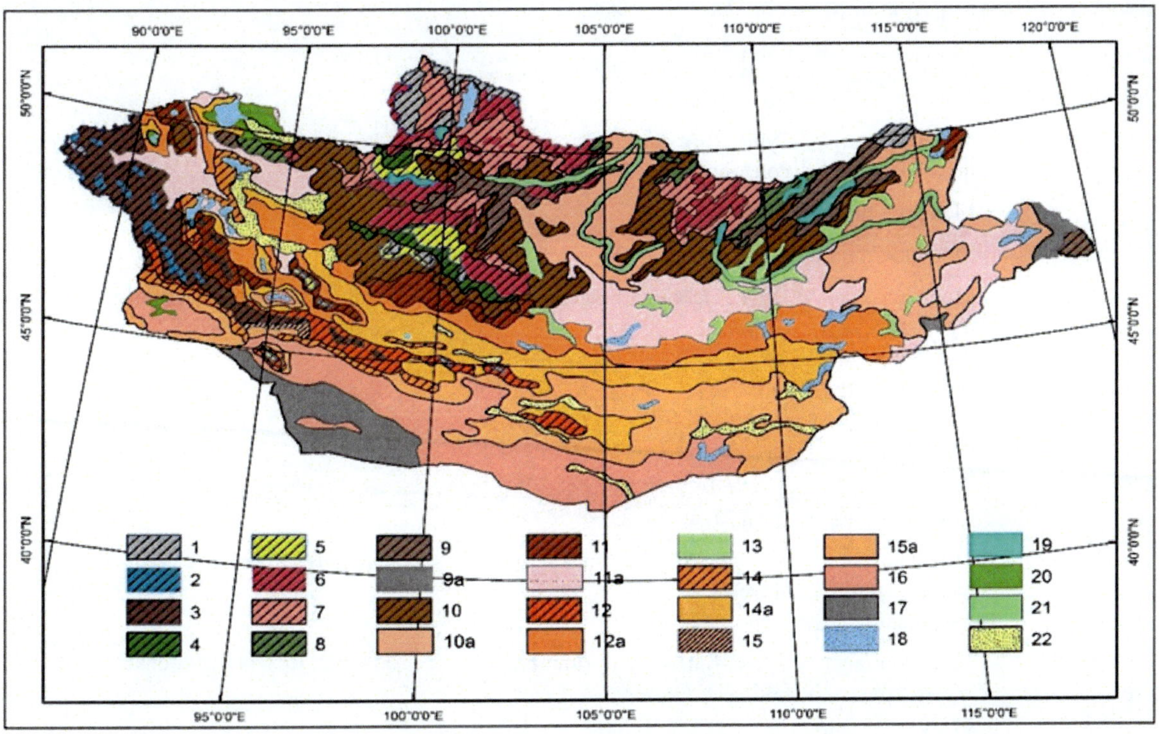


**Fig. 8.8** Geographic distribution of soil (modified from Nogina 1982). 1.Tundra, 2.Tundra and high mountain steppe, 3. High mountain steppe, 4. Mountain dry steppe now humic, 5. Mountain meadow, 6. Mountain meadow steppe podzolic, 7. Mountain cryomorphic-taiga and podzolic, 8. Mountain derno-taiga, 9–9a.Chernozem meal carbonated and noncarbonated, 10–10a.Dark kastanozems meal carbonated, noncarbonated, 11–11a. Carbonated meal carbonated, 12–12a. Light kastanozems meal carbonated, 13. Meadow-kastanozems, 14–14a. Semidesert brown, 15–15a. Dessert stepped brown, 16. Desert gray-brown, 17. Extra arid desert gray-brown, 18. Soloneztic and solonchak, 19. Meadow-boggy, boggy-cryomorphic, 20. Meadow-boggy, halomorphic, 21. Alluvial, 22. Sand

desert; Steppe soils divide 16 types into five groups: steppe, some desert and desert, wet origin, saline and river floodplain, which is 10 groups in total divided into 32 types (Table 8.5).

The last version of soil classification in 2016 by Batkhishig is characterized by organic content, stony, thickness, and thickness in terms of the international classification principle (Batkhishig 2016). It introduced new classification of soils that were developed and clarified by authors based on field investigations and other published materials (Fig. 8.9).

Soils of Mongolia are characterized by severe frost in the upper part of winter with deep freezing up to 3–4.5 m, a long period with seasonal permafrost (6–9 months a year). The coincidence of warm months with rainiest seasons of the year promotes biological activity. During this period, the amount of carbon dioxide in soils increased comparatively, which influenced the shift of air carbonates to soluble forms; therefore, a more migratory form, and at the same time, the regime of moisture penetration prevailed in soil profiles. All these situations provided geochemical migration of carbonates in steppe soils of Mongolia; therefore, the top part of soils was usually leached with carbonates. In some cases it occurred without a carbonate variant. Such very clear migration of carbonates was not usually observed in steppe soils of other regions.

In this classification, he focuses on two main criteria: soil source process and soil properties, and uses numerical values. For example, each soil type is classified according to detailed criteria, such as humus layer thickness of at least 40 cm, permafrost within 1 m, and rocks more than 40% within 1.0 m (Table 8.6). In order to make the soil classification criteria simple and clear, the organic content, color, humus layer thickness, and rocky properties have been quantified, and some traditional

**Table 8.5** Soil classification of Mongolia (Dorjgotov 2003)

| Group | Type | Sub-type | Genus |
|---|---|---|---|
| *Mountain soils* | | | |
| 1. Mountain tundra | Mountain tundra ochre | Typic; Row humic Podzolic; Gleyed | Ordinary; Residual carbonated; Stony |
| | Mountain tundra peaty-gleyed soil | | |
| | Mountain tundra cryoturabated | | |
| 2. Mountain meadow, mountain meadow-steppe soil | Mountain meadow peaty raw-humic Mountain meadow dark-colored raw-humic | Typic; Ochric; Chernozem-like; Gleyed | Ordinary; Residual carbonate; Non-complete developed |
| | Mountain meadow peaty raw-humic | | |
| | Mountain meadow steppe mull-humic | Typic; Meal carbonated Crypto-gleyed | Ordinary; Residual carbonatic |
| 3. Mountain forest soil | Mountain taiga cryomorphic ochro soil (podbur) | Typic; Surfacely ferrimorphic; Podzolic; Gleyed | Gleyed; Residual carbonatic |
| | Mountain taiga cryomorphic peat-muck humic soil | Typic; Gleyed | Ordinary; Residual carbonatic |
| | Mountain taiga podzolic soil | | |
| | Mountain derno-taiga (ferrimorphic) soil | Typic; Raw humic; Podzolized; Crypto-gleyed | Ordinary; Residual carbonatic |
| | Mountain forest, dark colored derno soil | Typic; Meal carbonated; Crypto-gleyed; Gleyed; Raw humic; Buried layers | |
| | Forest slightly podzolic sandy soil | | |
| 4. Mountain steppe soil | High mountain steppe raw humic | Typic; Ochric; Pillow like | |
| | Mountain chernozem | Non-carbonated; Meal carbonated; Crypto-gleyed | Low humic; moderate humic, High carbonated; Deep carbonated; Warped; Non-complete developed; |
| | Mountain kastanozem | Non-carbonated dark kastanozem; Meal carbonated dark kastanozem; Meal carbonated kastanozem; Meal carbonated light kastanozem; Crypto-gleyed | Ordinary; High carbonated; Deep carbonated; Warped; Non-complete developed; Aeolian sandy covered |
| 5. Mountain desert-steppe and desert soils | Mountain semi-desert brown soil | | |
| | Mountain desert gray brown soil | | |
| *Plain and valley soils* | | | |

(continued)

**Table 8.5** (continued)

| Group | Type | Sub-type | Genus |
|---|---|---|---|
| 6. Steppe soil | Chernozem | High humic; low humic; Crypto-gleyed; Anthropogenic | Ordinary; High carbonated; Deep carbonated; Low carbonated (or leached); Warped; Residual solonetzic; Buried gravel layer in humus; Non-complete developed; Sandy |
| | Kastanozem | Dark kastanozem; Kastanozem; light kastanozem; Crypto-gleyed; Anthropogenic | Ordinary; Warped; Residual meadowish; Paleocryogenic; Non-carbonated (low carbonated); High carbonated; Deep carbonated; Buried gravel layer in humus; Solonetzic; Solonchak; Non-complete developed; Seolian sandy covered; Sandy |
| 7. Semi-desert and desert soils | Semi-desert brown soil | Brown; light brown; Crypto-gleyed; Anthropogenic | Ordinary; Solonetzic; Solonchak; Gypsic; Aeolian sandy covered; Sandy; Sairic |
| | Desert gray-brown soil | Typic; Crypto-gleyed | Ordinary; Solonetzic; Solonchak; Gypsic; Aeolian sandy covered; Sandy; Sairic |
| | Extra-arid desert 'Borzon' soil | Ordinary; Gypsic; Sairic | |
| | Takyr soil | Takyr; Takyr like | Ordinary; Solonetzic; Crypto-gleyed (residual meadowish) |
| 8. Hydromorphic soil | Meadow dark humic gleyed soil | Crypto-gleyed; Gleyed | Ordinary; Solonchak; Stepped |
| | Meadow-boggy raw humic gleyed soil | Raw humic gleyed; Gleyed | Ordinary; Residual carbonatic; Solonchak |
| | Boggy peaty gleyed soil | | |
| 9. Halomorphic soil | Solonchak automorphic | – | Chloric-sulphatic; Sulphatic-chloridic |
| | Solonchak hydromorphic | Gleyed; Crypto-gleyed | Sodic; Chloridic-sodic; Chloridic-sulphatic; Sulphatic-chloridic; Sodic-chloridic; Sodic-sulphatic; Sulphatic; Residual humic (sulphatic-chloridic) |
| | Solonetz automorphic | – | Ordinary; Solonchak |
| | Solonetz hydromorphic | – | Ordinary; Solonchak |
| 10. Floodplain soil | Alluvial boggy gleyed soil | Alluvial boggy gleyed soil; Alluvial meadow boggy derno gleyed; Alluvial boggy clayed, gleyed | Ordinary; Carbonatic; Solonchak; Cryoturbated |
| | Alluvial meadow gleyed soil | Alluvial meadow derno gleyed; Alluvial meadow dark-colored crypto-gleyed | Ordinary; Layered; Solonetzic; Solonchak; Stepped |
| | Alluvial derno soil | – | Ordinary; Layered; Stepped |

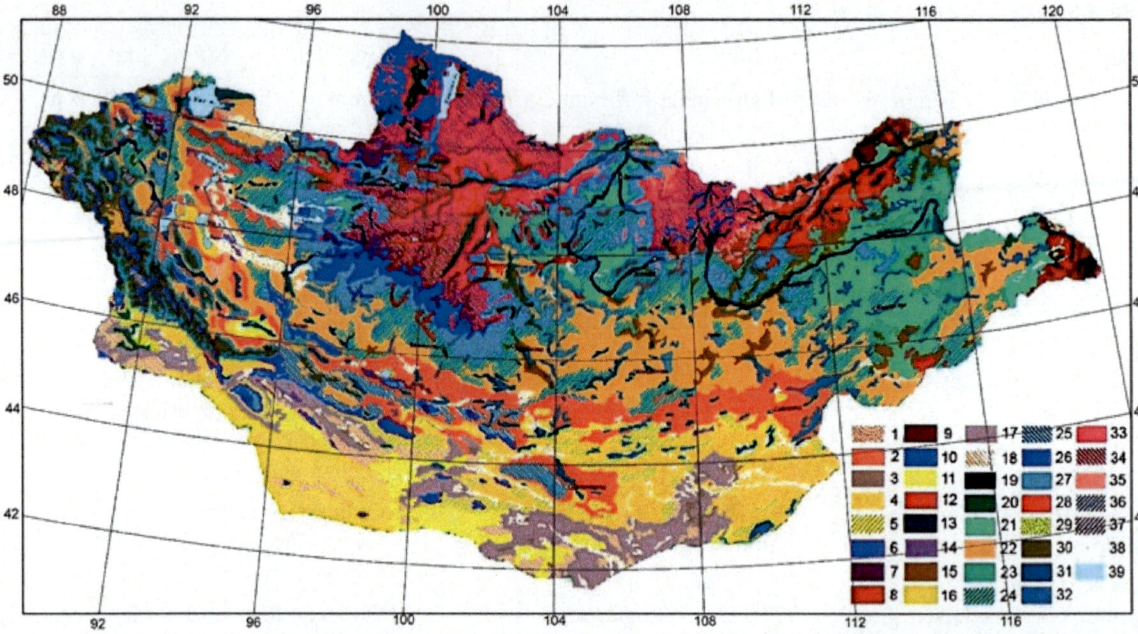

**Fig. 8.9** New classification of soil types in Mongolia based on the FAO (by Batkhishig 2016). 1. Arenosols, 2. Calcisols Aridic, 3. Calcisols Takyric, 4. Calcisols Yermic, 5. Leptic calcisols, 6. Cambisols, 7. Mollic cambisols, 8. Chernozems, 9. Haplic chernozems, 10. Cryosols, 11. Gypsisols, 12. Leptic Chernozems, 13. Fluvisols, 14. Aridic Gleysols, 15. Mollic Gleysols, 16. Gypsisols Yermic, 17. Calcic Gypsisols, 18. Leptic Gypsisols, 19. Gelic Histosols, 20. Salic Histosols, 21. Kastanozems, 22. Aridic Kastanozems, 24. Leptic Kastanozems, 25. Leptosols, 26. Calcic Leptosols, 27. Mollic Leptosols, 28. Phaeozems, 29. Arenic podzols, 30. Regosols, 31. Solonchaks, 33. Umbrisols, 34. Gelic umbrisols, 35. Mollic umbrisols, 36. Glacier, 37. Rock cliff, 38. Sand, 39. Lake

**Table 8.6** Soil classification criteria

| Characteristics | Criteria |
|---|---|
| Peaty horizon (H) | Organic content more than 20% or carbon content more than 12% |
| Layer of organic accumulation (O) | Organic content 10–20%, or carbon content 6–12%, plant residue |
| Humus horizon (A) | Organic content 1–10%, carbon content 0.6–6% |
| Carbonate soil (k) | First 5 cm effervesce with 10% HCl, or in 0.5 m carbonate layer more than 15 cm, $CaCO_3$ content more than 2% |
| Gypsum soil (y) | Throughout 0.5 m soil the gypsum content is more than 5%, horizon thickness more than 15 cm |
| Stoneless soil | Stone size bigger than 2 mm makes up less than 5% |
| Sairic | Stone size with 2–20 mm makes up 5–40% |
| Stony | Stone size with 2–20 mm makes up more than 40%, or bigger than 20 mm is more than 20% |

Mountain peat: Organic peat layer, organic content less than 20%, 1.0 m permafrost, rock less than 40%, rocky,
Mountain dark: Organic layer 30–50 cm, no carbonate accumulation, less than 40% per 1 m of rock

nomenclature has been retained. This classification will be useful in updating the Guidelines for Soil (2006) to suit the soil characteristics of Mongolia. For example, high mountain coarse humus soils are regosol humic, mountain taiga soils are umbrisols, dark brown soils are kastanozem, brown soils are calcic kastanozem, light chestnut soils are aridic kastanozem, and international classifications (gypsic, mollic, calcic) are used (IUSS Working Group WRB 2015). Based on these criteria, a total of 12 groups, 34 types and 150 genus of soils were identified and mapped in Mongolia. In the field, quantitative data are important for soil identification, diagnosis, and distillation using laboratory test results (Table 8.6). International criteria show that the organic content of the humus layer of brown soils is more than 1.0% (Mollic) and less than 1.0% for the Gobi soils (Table 8.7).

**Table 8.7** New classification of soil types in Mongolia (adopted from Batkhishig 2016)

| Group | Type | Genus |
|---|---|---|
| High mountain, tundra | Mountain peaty | *Mountain; Mountain stony; Mountain cryomorphic* |
| | Mountain dark colored | *Mountain; Mountain stony; Mountain peaty; Mountain cryomorphic* |
| | Raw humic | *Raw humic; Stony raw humic* |
| Forest, taiga | Mountain taiga cryomorphic | *Taiga cryomorphic; Taiga cryomorphic ferrimorphic* |
| | Mountain derno-taiga | *Taiga derno; Stony; Carbonated* |
| | Podzolic | *Podzolic; Sandy weakly podzolic* |
| | Forest dark colored derno | *Forest dark colored; Carbonated; Gleyed* |
| Mountain steppe | Stony humus | *Stony humic; Shallow stony humic* |
| | Stony dark kastanozem | *Stony dark kastanozem; Shallow stony dark kastanozem* |
| | Stony kastanozem | *Stony kastanozem; Shallow stony kastanozem* |
| | Stony light kastanozemara> | *Stony light kastanozem; Shallow stony light kastanozem* |
| Gobi mountain, hills | Stony brown | *Stony brown; Shallow stony brown* |
| | Stony gray brown | *Stony gray brown; Shallow stony gray brown* |
| Steppe | Chernozem | *Chernozem; Sairic; Stony; noncarbonated* |
| | Dark kastanozem | *Dark kastanozem; Sairic; Noncarbonated; Carbonated; Stoneless; Thickness sandy; Covered sand; Sandy; Crypto gleyed; Solonchak; Solonetzic; Bazaltic* |
| Dry steppe | Kastanozem | *Kastanozem; Sairic; Carbonated; Stoneless; Thickness sandy; Covered sand; Sandy; Crypto gleyed; Solonchak; Solonetzic; Solonetzic gleyed; Noncarbonated* |
| | Light kastanozem | *Light kastanozem; Sairic; Shallow sairic;Covered stony; Carbonated; Thickness sandy; Covered sand; Sandy; Gleyed; Solonchak; Solonetzic gleyed* |
| Gobi-desert steppe | Brown | *Brown; Sairic; Shallow sairic; Solonchak; Carbonated; Thickness sandy; Covered sand; Sandy; Gleyed; Reddish; Solonchak; Solonetzic; Solonchak gleyed* |
| | Light brown | *Light brown; Sairic; Shallow sairic; Covered stony; Carbonated; Thickness sandy; Covered sand; Sandy; Solonetzic; Solonchak; Gypsic* |
| | Gobi reddish | *Gobi reddish; Sairic covered stony; Covered sand; Solonchak* |

(continued)

**Table 8.7** (continued)

| Group | Type | Genus |
|---|---|---|
| Desert | Gray-brown | *Gray brown; Sairic; Shallow sairic; Covered stony; Carbonated; Covered sand; Sandy; Reddish; Solonetzic; Solonchak; Gypsic* |
| | Borzon soil | *Borzon; Covered stony; Sairic; Covered sand; Solonchak; Gypsic* |
| Meadow-boggy | Peaty | *Peaty; Gleyed; cryomorphic; Carbonated* |
| | Dark colored | *Dark colored; Peaty; Peaty dark colored cryomorphic; Solonchak; cryomorphic* |
| Floodplain, sairic | Alluvial | *Alluvial; Derno; Loamy; Saline; Gleyed; cryomorphic* |
| | Alluvial peaty | *Alluvial Peat; Alluvial Peaty* |
| | Alluvial coarse | *Alluvial coarse; Alluvial sand* |
| | Sairic Soil | *Sairic; Coarse, Sandy; Stony* |
| Halomorphic soil | Solonetzic | *Solonetzic; Solonchak* |
| | Solonchak | *Solonchak; Humic solonchak* |
| | Takyr like | *Takyr like* |
| Anthropogenic | Agrozem (Антросол) | *Loamy; Clayey; Brown* |
| | Urban soil (Technosol) | *Technosol, degraded* |

## 8.4 Soil-Geographic Zoning of Mongolia

The division of territories into regions and provinces with different natural conditions is based on the scientific planning, location and efficiency of urban, economic and industrial development. These divisions are divided into relief, climates, surface- and groundwater, soils, and vegetation. From the physical geographic component, criteria such as soil zoning, its characteristics, landscape appearance, bioclimatic regime, and geomorphological features are considered. Geographical zoning of soils is important, not only for land management, ecology, and conservation but also for the construction and transportation sectors.

Bioclimatic, lithological, and historical and genetic factors in the spatial differentiation of the soil cover of Mongolia are of first-rate importance for orographic factors. Large mountain structures, intermountain depression, foothills, and other orographically well-distributed areas always differ clearly peculiar soil cover and the natural-landscape conditions, and they occupy a position as independent structural units at appropriate levels of the hierarchical zoning system. In such areas, the soil cover structure is determined by the influence of a complex combination of different geographical patterns: latitudinal, depression-vertical, humid-piedmont, and arid zones. Materials obtained from the map of soil in Mongolia are used at a scale of 1:1,000,000 (1981) and are compared with soil-geographical zoning schemes with those developed for Russia and the World soil zones. (Dorjgotov1992)

It is divided into 2 major provinces, 7 provinces, 5 latitude sub-regions, and 53 districts, taking into account the soil distribution characteristics and bio-climatic factors of Mongolia (Dorjgotov 2009). A brief description of the soil regions is given.

***Khangai soil region***. Extreme climatic conditions in the northern part of the country significantly soften the soil cover. The steppe kastanozem are mainly distributed here, and depending on the internal differences of the provinces, they can be dark kastanozem, kastanozem, light kastanozem, mountain kastanozem, meadow kastanozem, and forest kastanozem. There are also turf-ash and mountain forest gray, mountain black earth, meadow-swamp and saline soils. Soil-forming bedrock consists mainly of granite, schist, and bedrock, as well as thin alluvial sediments, diluvium, and thick proluvial sediments formed as a result of their weathering. Loess-type loam of wind and proluvial origin occurs in the mountainous areas of the Khangai region. In this region, most of the annual rainfall falls in a short period of time in late summer, so the soil is washed away intensively. According to the geomorphological characteristics, the Khangai soil region is divided into four provinces of Khuvsgul, Khentii, Khangai and Dornod, and one sub-region of latitude, and each province is divided into a sub-province. The general features of soil formation in this region are: (1) significant accumulation of humus as a result of sod process; (2) continuous distribution of long-seasonally frozen and deep frozen soils; (3) powdery forms of carbonate excreta; (4) non-salinity of automorphous soils and absence of easily soluble salts, gypsum and salinity in them; (5) peculiarity of geochemical migration of carbonates, connected with peculiarities of water-heat river and frequent occurrence of carbonate-free steppe soils; (6) short biologically active periods; (7) relatively low thickness of humus horizon in the whole soil profile; (8) wide spread of relic features in modern soils (paleohydromorphism, paleocryomorphism, residual salinity, residual carbonated, etc.); (9) light granulometric composition and crushed stone, causing high water permeability. In this area, there are territories with all three types of structural zonality. All mountains belong to the sub-humid type of vertical soil zonality. The East Mongolian plain is characterized by a wide spread of dark-kastanozem soils and the existence among them of extensive depressions with more arid kastanozem soils as a result of the manifestation of depression vertical zonality. At the outlying areas of the Eastern plain of Mongolia, the humid-mountainous zonality is clearly defined. A significant part of dark kastanozem soils has signs of contact meadowiness. In the south of the Khangai region there is a vast area with a latitudinal type of structural zoning, covering subzones of kastanozem and dark kastanozem.

***Gobi soil region***. In the northern part of the Gobi, light brown and brown soils are spread to the south by light desert brown and gypsum soils, and the area of solonetz and solonchak soils increases. Due to the influence of wind on the soil formation process, the surface of the Gobi is characterized by a thin or coarse layer of gravel. It has three provinces, Mongol Altai, Gobi-Altai and Baruun Khuurai, and four latitudinal sub-provinces (Gunin and Vostokova 2005). Soils of the Gobi region are characterized by (1) development of arid soil formation processes without modern salt accumulation; (2) formation of rounded and obtuse-rounded gravel on the soil surface; (3) low humus content; (4) lack of readily soluble salts and gypsum in the profile; (5) residual salinity, carbonated, salinity; (6) retardation of the ecological activity of soils due to acute lack of moisture; (7) strong development of deflation processes to the erosion of sairic; (8) profile shortening. Within the limits, the region is distinguished by territories with all three types of structural zoning. Mountainous areas are divided into two types of vertical zonality: sub-arid and arid. The territory with a mixed type of zoning is represented by two provinces: the Great Lakes depression and the Baruun-Khuurai Basin. The province of the Great Lakes depression occupies a vast depression located between the mountain systems: the Mongol Altai and Khangai, and is divided into several independent lake depressions. Accumulative and denudation inclined plains prevail within the limits of the basin, and there are island mountains and hills and low-mountain ranges along its periphery. A huge area is occupied by semi-fixed and wavy sands. This province sharply differs from other territories of the region by its climatic conditions. It is characterized by a more extra continental dry cold climate (Fig. 8.10).

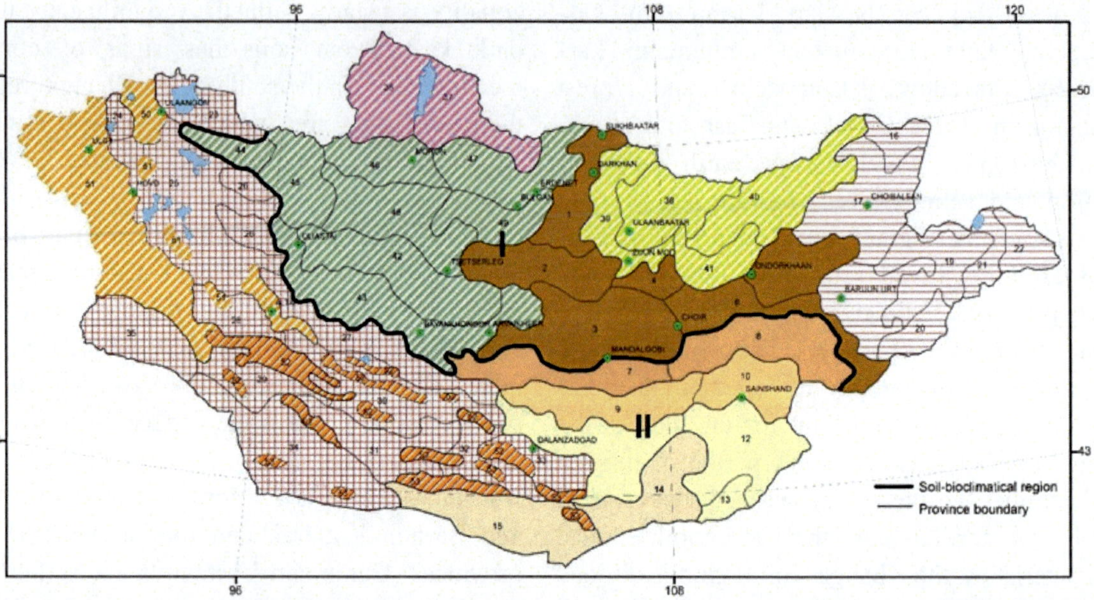


| Soil-bioclimatic region | Soil-geographical regions | | |
|---|---|---|---|
| | Latitudinal | Depressional | Altitudinal |
| I Khangai | **Chernozem, kastanozems sub region**<br>*Provinces:*<br>1. Orkhon-Shaamar<br>2. Tuul-Dashinchilen<br>3. Burd-Bayantsagaan<br>4. Maant<br>5. Undurkhaan-Tumentsogt<br>6. Borundur - Uulbayan | **Eastern region**<br>*Provinces:*<br>16. Ereentsav<br>17. Bayantumen<br>18. Baruun-urt<br>19. Menen-Matad<br>20. Dariganga<br>21. Tamsagbulag<br>22. Khyagan | **Khuvsgul region**<br>*Provinces*<br>36. West Khuvsgul<br>37. East Khuvsgul<br>Khentii region<br>Provinces:<br>38. Central Khentii<br>39. Khentii out<br>40. Eastern Khentii<br>41. Kherlenbayan-Ulaan<br>Khangai region<br>Provinces:<br>42. Central part Khangai<br>43. Southern Khangai<br>44. Khankhukhii<br>45. Telmen<br>46. Delgermurun<br>47. Selenge<br>48. Northern Khangai<br>49. Khanui-Orkhon |
| II Gobi | **Light kastanozems sub region**<br>*Provinces:*<br>7. Dundgovi<br>8. Bayandelger<br>Semidesert brown soil sub region<br>Provinces:<br>9. Khuld-Undurshil<br>10. Altanshiree<br>**Semidesert light brown soil sub region**<br>*Provinces:*<br>11. Manlain<br>12. Dornogvi<br>13. Sulinkheer<br>**Desert gray brown soil subregion**<br>*Provinces:*<br>14. Galba gobi<br>15.Zag shuuj-Borzon gobi | **Southern region**<br>*Provinces:*<br>23. Uvs lake<br>24. Achit lake<br>25. Khyargas-Khar lake<br>26. West Khangai<br>27. Valley of lakes<br>28. Sharga gobi<br>29. Zakhui gobi<br>30.Biger-Bayangol<br>31. Ingen khuuvur-Lig<br>32. Bayandalai<br>33. Dalanzadgad<br>34. Trans Altai<br>35. Baruun khuurai | **Mongol Altai region**<br>*Provinces:*<br>50. Kharkhiraa-Turgen<br>51. Mongol Altai<br>**Gobi Altai region**<br>Provinces:<br>52. Gobi Altai<br>53. Trans Altai Gobi |

**Fig. 8.10** Soil-geographic zoning of Mongolia (based on Dorjgotov 2009)

## 8.5 Soil Degradation

There are two main factors, natural and social, and both influence each other. Natural factors include wind, water, climate, precipitation, and from a social point of view, urbanization, population concentration, transportation, livestock numbers, grazing capacity, and agriculture. The most important diagnostic degradation of soils, the transformation of the soil profile, includes: from morphological characteristic to the thickness of the humus horizon, from chemical to the content of humus, and from physical to the structural composition. In these indicators the peculiarities of soil formation processes and genetic characteristics of soils are shown with completeness, and under the influence of anthropogenic factors in them shows the most significant changes (Undarmaa et al. 2018). Due to Mongolia's generally dry climate and high wind speeds in the spring, soil erosion is quite common. Soil erosion is relatively high in the Orkhon-Selenge and Onon-Kherlen basins, which are Mongolia's main agricultural areas. Due to the fact that the driest and windiest periods of the year coincide, soil erosion intensifies in the spring.

Water degradation moves the top layer of fertile soil and gradually causes an adverse impact on circulation of fertile substance of soil, vegetation shrouds development, and providing ecosystems balance. The process of soil is degraded by water and is determined by climate conditions, land surface inclines, and physical properties of soils. As of the territorial space, the soil degradation grade has relatively high value, stable soil is distributed in hilly steppe, typical steppe, southern part of dry steppe, deserts where takyr and takyr-like are dominantly spread. Any soils susceptible to degradation processes include dry steppe kastanozem, floodplain soil, hill hydromorphic soil, alluvial soil, and sand assimilation. Landscape features should be taken into account in the scientifically based development of the proper location of agriculture and its efficiency, which is the basis for regulating various environmental and social processes such as determining agricultural areas, selecting tillage technologies, soil protection, and improvement will occur. In the Orkhon-Selenge basin, the agricultural region of Mongolia, soil erosion is relatively high due to the high density of arable land and because of the inadequacy of soil improvement technologies. For example, compared to the dry steppe zone, the compost content of pasture and arable brown soil decreased by 50–60% in the plowed area, the amount of fine grains decreased by 40–50%, while the forest steppe zone reduced the humus content by 30–50%, and the sand and dust by 20–40% (Bazha et al. 2018).

### 8.5.1 Water Degradation

On estimating the soil, degradation grade is high by the edges and boundaries of mountainous areas. Steppe zones and basins of major rivers are in the moderate grade. Great Lakes depression, central part of the Valley of Lakes, and lowlands of eastern Gobi are included in territories which are not incurred in soil degradation or improved in comparison with the previous years. Areas with dramatic changes on soil degradation are not noticed after comparison, and a slight reduction of territories with high degradation grade is noticed. In extremely arid regions, soil salinization is increasing due to the evaporation of groundwater, where its water table is near the surface. One of the most remarkable examples of soil salinization is the Ekhiin-Gol oasis. The slight changes are also noticed in the southern part of Sukhbaatar aimag, at the neighboring area of Uvurkhangai and Dundgovi aimags and along the Khovd river (Khudulmur et al. 2013) (Fig. 8.11).

### 8.5.2 Wind Erosion

Wind erosion is a common sight in dry and arid areas where vegetation doesn't hold soils in place. Soil erosion can be a slow process that continues relatively invisibly or can occur at an alarming rate, causing serious loss of topsoil. There are number of reasons for the plant shrouds to get sparse, the most influential of which are droughts, pastureland pressure, and mining operations. Soil degradation by wind effects is determined by such factors as the climate, plantation and physical properties of soil. The wind causes deterioration which drives the coarse granule soil to move which is a very common natural phenomenon that takes place in dry and droughty regions (Garidkhuu 1975; Nyamsambuu 2004; Khadbaatar 2010). Due to the impact of climate change, desertification and soil degradation are the most serious problems in Mongolia. Thus, soil quality issues are studied in

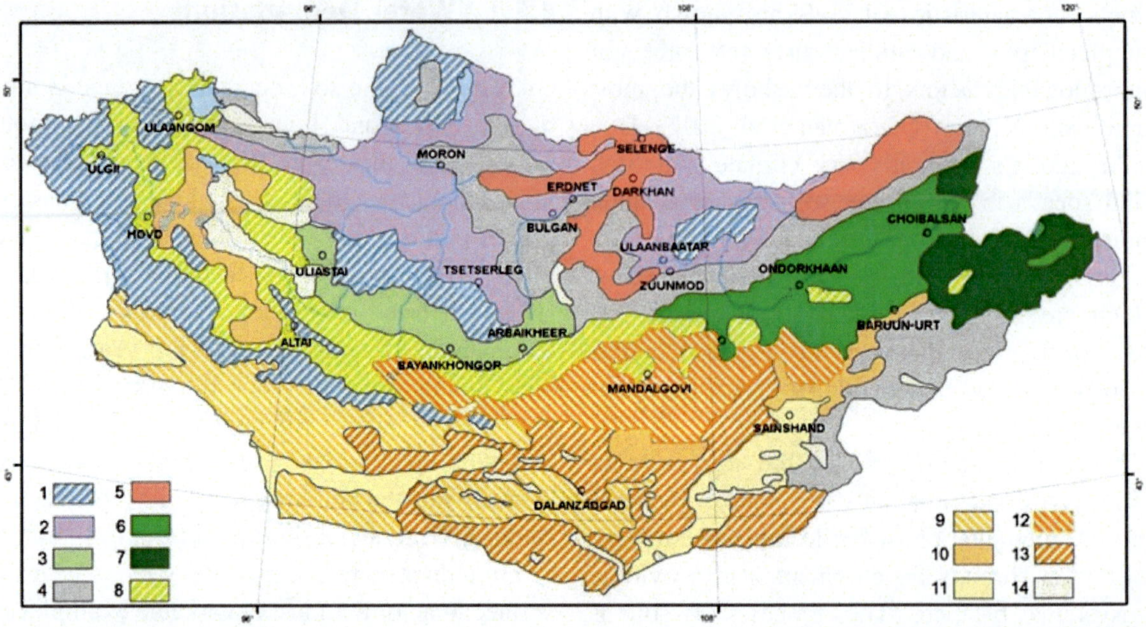


**Fig. 8.11** Water erosion vulnerability map (based on National Atlas of Mongolia 2009). 1. Glacier and permafrost-related hazards; 2. Lower rank due to relatively slope run-off; 3. High rank due to relatively slope run-off; 4. Lower rank due to relatively flood; 5. High rank due to relatively flood; 6. Low risk rank due to relatively water; 7. No risk rank due to relatively water; 8. High erosion by break-stone; 9. Low erosion by break-stone; 10. Lakes; 11. High surface run-off; 12. Low surface run-off; 13. Low effect surface run-off erosion; 14. High effect surface run-off erosion

combination with desertification, especially through soil erosion. The last research used the ArcGIS wind erosion equation model to assess wind erosion throughout Mongolia (Mandakh et al. 2016). Due to the fact that the driest and windiest periods of the year coincide, soil erosion intensifies in the spring. In the Gobi Desert region, sand cover is often formed on the surface of the soil where sand movement occurs in some places due to strong winds. Sandy soils cover more than 3 million hectares in the western and southern parts of the country. Depending on the thickness of the sand carried by the wind, the properties of the soil will change and affect the vegetation cover (Batkhishig et al. 2018).

In general, the process of wind erosion is more pronounced in semi-desert and desert areas in Mongolia (Gunin 2019). High-grade wind erosions are noticed in Gobi regions, Great Lakes depression of and along the lakes valley from the soil mapping. Specially there are some high grades that are recorded at the Baruun Khuurai depression, Trans-Altai Gobi, Ulaan lake and near Mandal-Ovoo. Plants shrouds are short and scarce in these places and there are fewer slopes and obstacles on the surface. In other words, every indication estimated by us has high grade in these places. Steppe and desert steppe zones of our country, especially, eastern prairie, Zamiin Uud and Sainshand of eastern Gobi are included in moderate grade. The changes of wind erosion processes are clearly noticed in valleys and gorges of Sharga, Nomin, Ingen khuuvur, Galba and Borzon Gobi in southern sides of Gobi-Altai, Bayankhongor and Umnugovi aimags and there is an estimation released by comparing 2000 and 2010, that 165.7 t/ha of soil is moved annually from these regions (Fig. 8.12). However, soil erosion by wind is 57.7 t/ha a year in the western part of the Great lake depression or in Khar Us lake and along the Buyant river basins and reduced in other places.

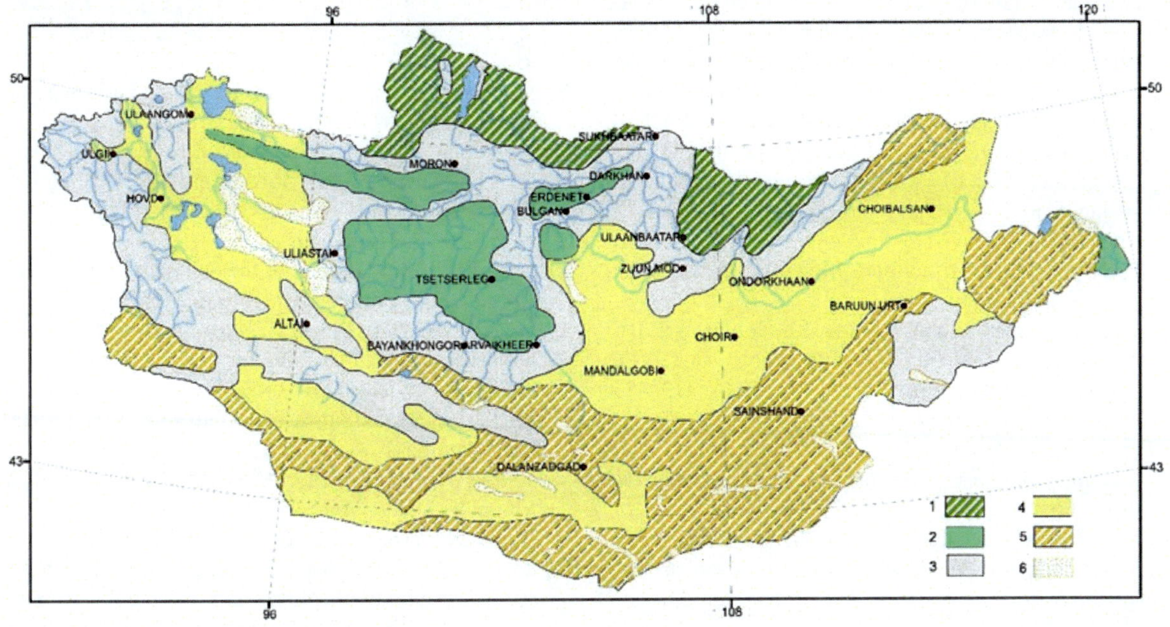


**Fig. 8.12** Soil wind erosion vulnerability map (based on National Atlas of Mongolia 2009). 1. No risk, 2. Low, 3. Moderate, 4. High, 5. Very high, 6. Sands

## References


Avaadorj D (2018) Mongolian soil study overview. Mongolian J Soil Sci 3:7–15. Admon printing, Ulaanbaatar (in Mongolian)

Batkhishig O (2016) Soil classification of Mongolia. Mongolian J Soil Sci (01):18–31

Batkhishig O et al (2018) Soil and soil cover. In: Nyamdavaa G (ed) Environment of Mongolia: the pasture, soil, use and conservation. Publishing house ministry of environment and tourism, Ulaanbaatar, pp 71–81 (in Mongolian)

Bazha SN, Danzhalova EV, Drobyshev Yu I, Khadbaatar S (2018) Transformation of land ecosystems southern part of the basin Baikal. Tvorishestvo Science Press, Moscow (in Russian)

Bespalov ND (1951) Soils of Mongolian People's Republic. USSR Academy of Sciences Publishing House, Moscow (in Russian)

Dorjgotov D (1974) Soil map of the Mongolian people's Republic. Moscow State Institute of Geodesy and Cartography

Dorjgotov D (1976) Soil geographical zoning of Mongolia. Publishing house Mongolian Academy of Science, Ulaanbaatar (In Mongolian)

Dorjgotov D (1981) Soil map of the Mongolian people's Republic. Moscow State Institute of Geodesy and Cartography

Dorjgotov D (1986) Soil map of the Mongolian people's Republic. Moscow State Institute of Geodesy and Cartography

Dorjgotov D (1992) Soils of Mongolia (genesis, systematics, geography, resources and use). Moscow State University Publishing, Moscow (in Russian)

Dorjgotov D (ed) (2009) National Atlas of Mongolia. Institute of Geography. Mongolian Academy of Sciences, Ulaanbaatar (in Mongolian)

Dorjgotov D (2003) Soils of Mongolia. Admon Printing, Ulaanbaatar (in Mongolian)

Dorjgotov D, Batbayar D (1986) Soil classification of the Mongolian People's Republic. State Publishing House, Ulaanbaatar (in Mongolian)

Garidkhuu J (1975) Soils of the basin Orkhon and Selenge. Academy Publishing House, Novosibirsk (in Russian)

Jahn R, Blume HP, Asio VB, Spaargaren O, Schad P (2006) Guidelines for soil description food and agriculture organization of the United Nations

Gunin PB (2019) Ecosystems Atlas of Mongoia, Gunin P, ed. Mongolian Academy of Science, RAS, MNUE. Admon Publishing, Ulaanbaatar (in English)

Gunin PD, Vostokova EA (eds) (2005) Ecosystems of Mongolia. Atlas. Mongolian-Russian Joint Biological Expedition, Moscow (in English)

IUSS Working Group WRB (2006) World reference base for soil resources 2006 - A framework for international classification, correlation and communication, Rome, Food and Agriculture Organization of the United Nations

IUSS Working Group WRB (2015) International soil classification system for naming soils and creating legends for soil maps. World soil resources report No. 106. FAO, Rome

Khadbaatar S (2010) Landscape-ecological features of soil degradation in the central part of the Selenga basin. Doctoral dissertation of Geography. Publishing House MGPU, Moscow. SPU (in Russian)
Khudulmur S, Tsogtbaatar J, Dash D, Batjargal Z, Mandakh N (2013) Desertification Atlas of Mongolia, Tsogtbaatar JK (ed). Admon Printing, Ulaanbaatar (in Mongolian)
Krasnoshchekov K (2013) Soil cover and the soils of mountain forests in northern Mongolia. Novosibirsk. Nauka (in Russian)
Mandakh N, Tsogtbaatar J, Dash D, Khudulmur S (2016) Spatial assessment of soil wind erosion using WEQ approach in Mongolia. J Geog Sci 26(4):473–483. https://doi.org/10.1007/s11442-016-1280-5
National Atlas of Mongolian People's Republic (1990) Academy of Sciences Publishing House, Moscow (in Russian)
National Atlas of Mongolia (2009) Institute of Geography, Mongolian Academy of Science, Ulaanbaatar
Nogina NA et al (1980) Soil map Mongolia. Moscow State Institute of Geodesy and Cartography
Nogina NA, Dorjgotov D (1982) Soil-geographical zoning of Mongolia. Soil Sci (4):23–32. Moscow (in Russian)
Nogina NA et al (1984) Soil cover and soils of Mongolia. Nauka, Moscow (in Russian)
Nyamsambuu N (2004) Wind erosion of arable soils and its consequences. In: Environment of Mongolia. Publishing house Munkhiin Useg, Ulaanbaatar, pp 62–70 (in Mongolian)
Pankova EI (1992) Genesis of the soil salinity in desert. Uprpoligraph printing, Moscow [in Russian]
Tsegmid Sh (1969) Physical geography of Mongolia. Publishing House, Ulaanbaatar [In Mongolian]
Undarmaa J et al (2018) Rangeland ecosystems of Mongolia. Munkhiin useg publishing, Ulaanbaatar [in Mongolian]
Vostokova EA, Gunin PD (eds) (2005) Ecosystems of Mongolia Atlas. Joint Russian-Mongolian Complex Biological Expedition, Moscow (in Russian)
Vostokova EA (2019) Soil salinization map, Gunin PS (ed). Institute of Ecology and Evolution. Russian Academy of Science, Ulaanbaatar

# 9 Biogeographical Characteristics of Mongolia

Khurelbaatar Tsogbadral

**Abstract**

Due to Mongolia's geographical location between the Siberian taiga and the desert of Central Asia, geological and topographical features, and extreme climatic conditions, the distribution of vegetation cover is very unique. These natural conditions contribute to the diversity of fauna ecosystems: snow tiger and sheep in the high mountains, reptiles in the Gobi region. Mongolia's fauna includes 141 mammal species, 502 bird species, and 79 fish species. More than 3127 species of vascular plants, 574 species of mushrooms, 1056 species of lichens, 2003 species of algae, and 580 species of mushrooms have been found in Mongolia. Among vascular plants, 14% are endemic to Mongolia. Biodiversity of Mongolian fauna and flora correlates with various biogeographic roots. The chapter consists of two sub-parts: the phytogeographical and zoogeographical features of Mongolia, which briefly describes the geographical distribution of Mongolia's flora and fauna with different physical and geographical conditions, as well as the brief history of the researches. The chapter concludes with a short description of the main wildlife representatives.

**Keywords**

Biogeography · Phytogeographical zones · Zoogeographical regionalization · Flora · Fauna

## 9.1 Phytogeographical Features and Studies

The vegetation of Mongolia ranges over intersection of Siberian taiga and Central Asian desert. From north to south, a cold and humid climate gradually shifts to a warm and dry climate. This results in the gradual replacement of vegetation in plains and from the foot of mountains to their peaks, thereby forming belts. The distribution of vegetation corresponds mostly to the distribution of precipitation and temperature. The vast territory of Mongolia with the high mountain ranges, depressions, forest steppes and Gobi Desert, and extreme continental climate and its factor influences to the vegetation cover (Nyambayar 2009).

The average annual temperature for most of Mongolia is just below 0 °C and only in the southern Mongolia is the annual positive temperature. The average vegetation period of Mongolia is very compressed and increases by more than 90–120 days. Frosts very often occur in May and are repeated in August. Dry climate in combination with light and well-drained soils creates little favorable for the existence of

K. Tsogbadral (✉)
Department of Geography, Mongolian National University of Education, Ulaanbaatar, Mongolia
e-mail: tsogbadral@msue.edu.mn

© Springer Nature Switzerland AG 2021



B. Yembuu (ed.), *The Physical Geography of Mongolia*, Geography of the Physical Environment,
https://doi.org/10.1007/978-3-030-61434-8_9

vegetation water regime of soils. The harsh environmental conditions of vegetation development lead to relatively low rates of vegetation accumulation. Mongolia's climate regime has a significant impact on plant renewal processes. This diversity is also reflected in changes in precipitation and distribution of plants, starting from the foot hills to the top of the mountain ranges, in vertical belts.

Plants start to grow when daily mean temperatures exceed 5 °C in spring, and plants withered when they fall below 5 °C in autumn. Daily mean temperatures exceeding 10 °C are associated with high plant growth and development rates. In spring, daily average temperatures exceed 5 °C around the 6th of May in the high mountains, the 13th of May in the high mountain forests, and the 29th of April in the mountain forest steppes and steppes, the 19th of April in the desert steppes and the 6th of April in the deserts. In autumn, temperatures fall below 5 °C on the 26th of September in the high mountains, the 22nd of September in the mountain forests, and the 3rd of October in the forest steppes and 5th October in steppes, the 13th of October in the desert steppes, and the 23rd of October in the deserts. The dates when daily average temperature exceeding 10 °C are about 20 days later in spring and also same number of days earlier in autumn compared to consequent dates of 5 °C in all vegetation zones and belts. The annual sum of daily mean air temperatures exceeding 5 °C ($\sum T > 5$ °C growing degree days) indicates heat accumulation for vegetation. During the growing season, the annual sum is 900–2400 °C over the whole vegetation zones and belts of Mongolia. Generally, the annual sum increases with decreasing latitude. However there is less heat accumulation in the high mountain and mountain forest belts (Table 9.1) (Undarmaa et al. 2018).

The first studies of Mongolian flora and fauna were started by Russian scientists in the 1820s. Since then, reference collections, literatures, and species descriptions were added by Russian, Mongolian, and European botanical and zoological researchers. In 1830, the Russian botanist N.S. Turchaninov studied the flora of northern Mongolia, which is the first scientific researches of plants in Mongolia. However, more complete researches of Mongolian vegetation were carried out at the end of the nineteenth century by the famous researcher N.M. Przevalsky (1870–1873, 1879–1880, 1883–1885), who traveled and collected 702 specimens of mammals, 5010 birds, and so on (Yunatov 1950).

P.N. Krylov (1903) determined a brief botanical description of the northern shore of Uvs lake (northwestern Mongolia) and the Tagna-uul mountains during the research in Uriankhai region. V.L. Komarov (1905) gave a more detailed description of the Alpine area of the Eastern Sayan Mountains (Munkh-Saridag, bordering Russia) and

**Table 9.1** Main vegetation types by the natural zones and belts (based on Gomboluudev et al. 2018)

| Natural zones and belts | Main plant formations | Dates of daily mean temperature exceed 5 °C | Dates of daily mean temperature exceed 10 °C |
|---|---|---|---|
| High mountain (alpine) belt | Festuca lenensis, Poa attenuata, Festuca kryloviana | 902 | 308 |
| Mountain taiga belt | Pinus sibirica, Larix sibirica | 821 | 275 |
| Forest steppe | Larix sibirica, Betula platyphylla, Pinus sylverstris | 1299 | 608 |
| Steppe zone | Stipa krylovii, Stipa grandis, Cleistogenes squarrosa | 1392 | 689 |
| Semidesert zone (Gobi) | Stipa gobica, Stipa glareosa, Allium polyrrhizum | 1800 | 1006 |
| Desert zone | Anabasis, Haloxylon, Artemisia | 2417 | 1505 |

the lake Khuvsgul. He collected plants and described several species from Mongolia (*Caragana, Nitraria*), published the first textbook, "Introduction to the Flora of China and Mongolia" (1908) (Urgamal and Bayarkhuu 2018). In addition to the Mongolian vegetation map compiled and defined the structure and composition of the community (Ulziikhutag 1989), numerous national and foreign researchers contributed significantly to the study of Mongolian flora, for example, Grubov (1955, 1982), Davaajamts (1983), Buyan-Orshikh (1977, 1988), Ulziikhutag (1989), Banzragch et al. (1978), Sanchir (1974), Jamsran (1971), Gubanov (1996), Ganbold (2010), Dariimaa (2010, 2014), Hilbig (2010), Tuvshintogtokh (2014), Urgamal et al. (2013, 2014, 2016), and so on.

### 9.1.1 Phytogeographical Regions

The phytogeographical regionalization issues of Mongolia were included of the researches of G.N. Potanin (1876–1877, 1879, 1886) and M.V. Pevtsov (1899–1901, 1907–1909, 1923–1926), V. A. Komarov (1905), I.V. Pavlov (1929), E.M. Murzaev (1952), A.A. Yunatov (1950, 1954), V.I. Grubov (1955, 1963, 1984), and E.M. Lavrenko (1965, 1970, 1978). Mongolia's first phytogeographic zoning scheme conducted by Pavlov, based on the study of the Khangai mountain flora, includes the following provinces: (1) Mongol Altai; (2) Khentii mountainous area; (3) Mongol Daguur; (4) Gobi; (5) Depression of Great Lakes; and (6) Sayan province (Tsegmid 1969).

The comprehensive four three-tiered phytogeographic regionalization of Mongolia was described by A.A. Yunatov in 1950, which was based on holistic biogeographical features of the country. He singled out the boreal type of zonality, which in Mongolia is expressed in the form of the Sayan variant and is typical for the western Khuvsgul and Khentii, as well as the arid type, which includes three variants: Khangai, Altai, and Gobi. The vegetation cover of the mountains is very diverse, which is connected with different zonal provincial positions of mountain ranges, as well as with the history of formation of mountainous vegetation. In this regard, the vertical zonality in the mountains is as important concerning distribution of vegetation as the latitudinal zonality in the plains. Since 1950s, many works have been devoted to the study of mountain vegetation and the typology of zoning. The high-altitude zonality of Khangai (Bannikov and Khudyakov 1976); (Banzragch and Munkhbayar 1978) mountain forest steppe of the Mongolian and Gobi-Altai Banzragch D et al. (1978); Volkova (1988) was studied in detail.

The 16 geobotanical regions of Mongolia, defined by Grubov (1955, 1982) and Ulziikhutag (1989), are based on the geographical features of the country and the characteristics of the vegetation cover (Shagdar and Yadamsuren 2017), which is still used as the main source of flora regionalization. These 16 phytogeographical regions (Grubov 1954) are strongly differ in the number of recorded vascular plant species (Fig. 9.1). The mountainous areas in the west (Mongol Altai, Khovd) and the north (Khangai, Mongol Daguur, Khentii, Khuvsgul) of Mongolia show the highest richness of vascular plant species with a range of 1078 to 1636 species per region. The lowest species numbers are found in some of the dry phytogeographical regions, the Alashan Gobi, Trans-Altai Gobi, valley of lakes and depression of Great lakes with a range of 272–481 species per region (Urgamal et al. 2016).

The regional differences in species richness are in part due to differences in habitat characteristics and diversity between the individual regions. The main part of specific richness is concentrated in the middle altitudinal belts of great mountain systems of Mongol Altai, Gobi-Altai, Khangai, Khentii, southern Sayans basin, as well as in the area of these landscape is not large but the Flora is rich. However, there are differences in the state of floristic research, by high increases in vascular plant species compared to the conspectus of Gubanov (1996) in the Mongol Altai and the Khentii mountain Urgamal et al. (2016). The composition of the vegetation

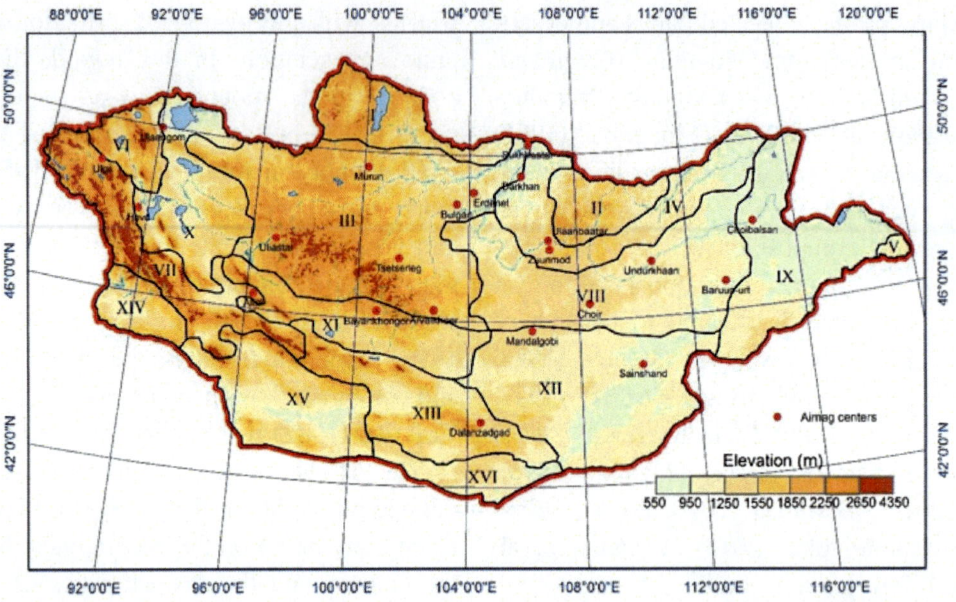


**Fig. 9.1** Phytogeographical regions of Mongolia (modified based on Grubov, 1954). I. Khuvsgul Mountain Taiga; II. Khentii Mountain Taiga; III. Khangai Mountain Forest Steppe; IV. Mongol Daguur Mountain Forest Steppe; V. Khyangan mountain meadow steppe; VI. Khovd desert steppe; VII. Mongolian Altai mountain steppe; VIII. Middle Khalkh Steppe; IX. Eastern Mongolian Steppe; X. Desert Steppe of Depression of Great Lakes; XI. Desert Steppe of Valley of Lakes; XII. Desert Steppe of Eastern Gobi; XIII. Desert Steppe of Gobi-Altai mountains; XIV. Zuungar Gobi; XV. Trans-Altai Gobi; XVI. Alashaa Gobi

cover of the Mongol Altai, Khangai, Khentii, Khuvsgul and Mongol Daguur regions is comparatively richer than other regions (Table 9.2) (MEGD and UNDP 2014).

The research on florogenesis of Central Asia and Altai mountains conducted by R.V. Kamelin (2010) was distinguished floristic zoning and vegetation classification. In the new version of the zoning, it is crucial that Mongolia's territory be considered as the region where the botanical data show the borders of the three sub-kingdoms: *Holarctic-Boreal, Ancient Mediterranean (2 provinces: Zuungar and North East Gobi) and East Asian* with 8 provinces and 23 districts (Kamelin 2010). There are 4 provinces (Altai-Sayan, Baikal-Juggur, Altai Zuungar and Tuva-Mongol) of the Boreal sub-kingdom, 2 provinces (Zuungar and North East Gobi) of the Ancient Mediterranean, and also 2 provinces (Daguur Mongol-Manchur Khalkh-East-Mongol transition territory) of the East Asian sub-kingdoms on the territory of Mongolia (Fig. 9.2).

### 9.1.2 Distribution Patterns of the Flora

Mongolian vegetation presents those special features which have developed through time because of local landscape forms, the organic environment, and extreme continental climate. Therefore, Mongolian flora characterized by a mixture of elements of central Asian, Mongol Daguur, Manjuur, Tuva-Mongol, central Kazakhstan-South Altai-Zuungar, and Zuungar desert plants (MEGD and UNDP 2014). Water is a main limiting factor for vegetation growth in southern part, while energy is a major constraint for the north of Mongolia (Chu and Guo 2012). Most of the territory is covered by steppe vegetation, which is supported by pasture fund V.I. Grubov (1955). Yunatov (1950, 1952) used the Russian classification system, and plant communities were named by one or more dominant species. In the combined names of plant communities, the dominant species was identified as the dominant species (Tuvshintogtokh 2014).

**Table 9.2** Vegetation regions, area percentages and number of species

| Regions | Area percentage | Number of species |
|---|---|---|
| Khuvsgul Mountain Taiga | 4.96 | 1081 |
| Khentii Mountain Taiga | 3.05 | 1276 |
| Khangai Mountain Forest Steppe | 17.59 | 1551 |
| Daguur Mountain Forest Steppe | 6.6 | 1310 |
| Khovd Mountain Steppe | 1.98 | 1045 |
| Mongol Altai Mountain Steppe | 7.02 | 1662 |
| Khangai Mountain Steppe | 0.87 | 847 |
| Middle Khalkh Steppe | 11.54 | 793 |
| Eastern Mongolian Steppe | 3.94 | 972 |
| Desert Steppe of Depression of Great Lakes | 6.11 | 900 |
| Desert Steppe of Valley of Lakes | 3.18 | 482 |
| Desert Steppe of Gobi-Altai mountains | 5.02 | 889 |
| Desert Steppe of Eastern Gobi | 9.35 | 480 |
| Alashaa Gobi | 6.43 | 272 |
| Trans-Altai Gobi | 5.72 | 383 |
| Zuungariin Gobi | 1.62 | 879 |

*Source* Biological diversity in Mongolia 2017

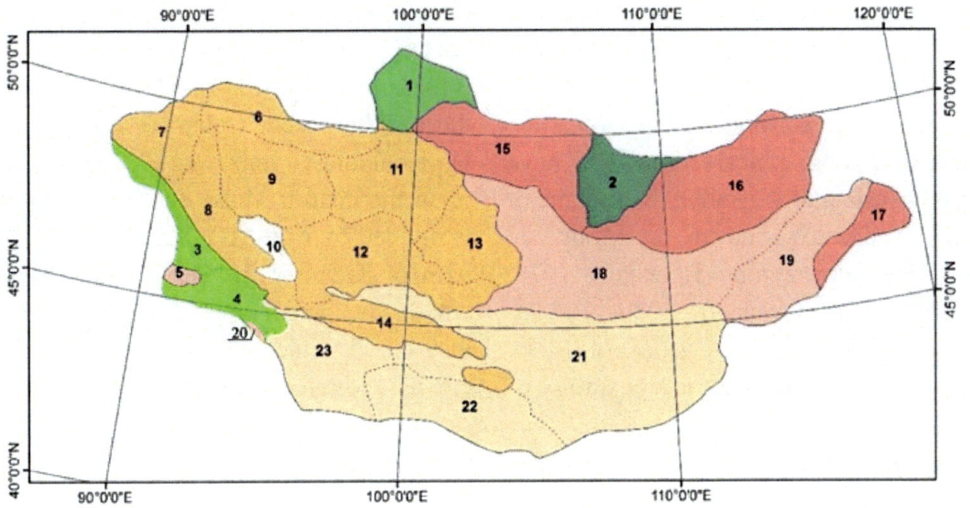


**Fig. 9.2** Geographical regions of vegetation in Mongolia (R.V. Kamelin 2010). *Source* Kamelin R.V. (2010). **I. Boreal Subkingdom**: Altai-Sayan Province (green color): 1. South Sayan district (3 subdistricts); Baikal-Juggur province (colored dark green): 2. Khentii district (2 subdistricts); Altai Zuungar province (light green): 3. Zuungar-Altai district (2 subdistricts); 4. Southern Mongol-Altai district (2 subdistricts); 5.Baruun Khuurai district; Tuva-Mongol Province (colored sandy brown): 6. Uvs lake district (5 subdistricts); 7. Chui-Achit lake district (4 subdistricts); 8. Khovd district (5 subdistricts); 9. Valley of Lakes district (2 subdistricts); 10. Shargiin-Gobi exclave (Gobi suburb); 11. North-Khangai district (2 subdistricts); 12. South Khangai district (2 subdistricts); 13. East Khangai transitional district (1 subdistrict); 14. Gobi-Altai district (2 subdistricts); **II. East Asian sub kingdom**: Daguur Mongol-Manchur province (pink color): 15. Selenge-Daguur district (5 subdistricts); 16. Onon-Kherlen-Daguur district (4 subdistricts); 17. Central Khyangan district (2 subdistricts); Khalkh-East-Mongol transition territory (colored light pink): 18. Khalkh district (3 subdistricts); 19. Dariganga-Eas t Mongolian district (3 subdistricts); **III. Ancient Mediterranean sub kingdom**: Zuungar province (Wheat color): 20. Nomiin district; North-East Gobi province: 21. North Gobi district (3 subdistricts); 22. North-Alashaa-Gobi district (4 subdistricts); 23. Trans Altai Gobi district (4 subdistricts)

Over the past 30 years, the flora has been enriched with 2,539 new species, including 232 species of mushrooms, 1,366 species of algae, 274 species of lichens, 176 species of mosses, and 684 species of vascular plants. In 1989, a total of 4,500 species were registered in Mongolia's flora, and also a total of 7,315 plant species and subspecies (new 2,815 species added over the past 30 years), which is an increase of 40% in diversity (Urgamal and Bayarkhuu 2018). Since 1998, the number of vascular plants has increased by 10 families, 20 genera, and 130 species displacement of zonal vegetation types on this territory. At present, 3160 species and subspecies of vascular plants, distributing over 684 genera, 133 families, and 39 orders are recorded from Mongolia (Fig. 9.3 Comparative study in Mongolian vascular plants Fig. 9.3). The priority families by species number in Mongolian flora are *Asterasea* (451 species), *Fabaceae* (349), *Poaceae* (257), *Rosaceae* (156), *Brassicaceae* (149) (MEGD and UNDP 2014) and their species number compared to the 1998 flora study has increased as follows: Asteraceae by 104 species, Fabaceae by 80 species, Poaceae by 45, Rosaceae by 39, and Brassicaceae by 18 species (Urgamal and Bayarkhuu 2018).

The botanical-geographical researches have specified and considerably detailed the scheme of regions of a vegetative cover of Mongolia, especially in the Gobi regions (Beresneva 1978; Evstifeev 1976; Rachkovskaya 1989; Buyan-Orshikh 1977, 1992; National Atlas of MNR 1990). The high hyposometric position of Mongolia combined elements of latitudinal and vertical zonality in the vegetation cover (Grubov 1963, 1982). It described the displacement of zonal vegetation types on its territory.

When comparing zonal and subzonal types of vegetation of the communities, a southward shift of the zonal site is noted, determined by high mountain ranges of Southern Siberia and Northern Mongolia. In northwest Mongolia, arid vegetation types, including desert vegetation, are moving northward in depressions. The configuration of the bands and their latitude in Mongolia has its own peculiarities. If on their main territory the zonal boundaries have latitudinal extent, then on its eastern margin it acquires a meridian character. This can be clearly seen in the steppe region of Eastern Mongolia, as well as in the southeastern part of Gobi. Reason of this condition is the monsoon type of precipitation distribution in the eastern part of the Asian continent. The zonal series of vegetation begins on low hilly plains in the northeast of Mongolia with a strip of herbaceous cereals and richly scattered cereals, mainly on dark chestnut steppes, partly on chernozem soils. The main dominant vegetation here is the coarse-grained grass (*Stipa baicalensis)* and the turf herbaceous plant (*Filifolium sibiricum)* Urgamal and Oyuntsetseg (2017).

In the central Mongolia the latitudinal region is divided by extensive Khangai plateau. Further west of Khangai, the Great Lakes basin begins, where the vegetation cover is concentrically distributed under the influence of the "basin effect", which is to increase the aridity and

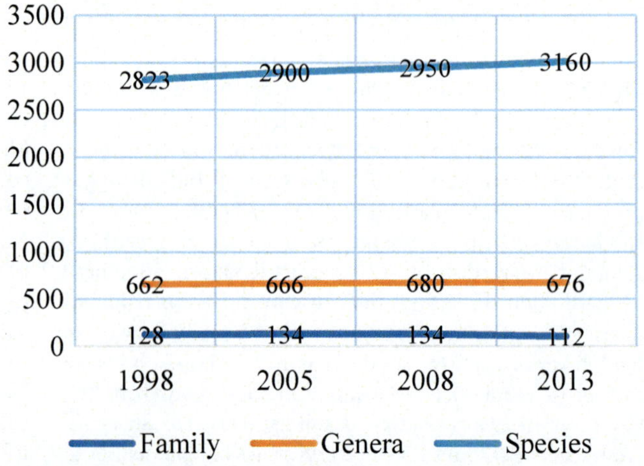


**Fig. 9.3** Comparative study in Mongolian vascular plants. *Source* (MNET 2017)

reduce the absolute height to the center of the basin. At the same time, the zonal types of plant communities are at the maximum levels of depression. In the northern part of the Great Lakes basin (Uvs lake depression) there are specific types of dry steppe with western species that are not typical of Mongolia—*Stipa capillata, Festuca valesiaca.* In most of the territory, the dry steppe subregion characterized by the predominance of soddy herbaceous communities were formed mainly by the *Stipa krylovii*, as well as rhizome grains. Mongolian dry steppes are also characterized by the presence of caragane (*Karagan microphylla, C. stenophylla, C. pygmaea*) and semi-shrubby steppe *Artemisia frigida.*

Communities of heterogeneous and dry steppes belong to the subtype of real steppes. In the south, they are replaced by desert steppes, which are dominated by *Stipa gobica* and *S. Klemenzil.* Desert steppes are communities of unique composition and structure, and are distributed throughout the Central Asian desert region. Desert steppe types are endemic to Central Asia and are unparalleled in other arid areas of the Palaearctic. A.A. Yunatov (1950, 1974) described the desert steppe communities are dominated by fine-woven feather grasses *Stipa gobica* and *S. glareosa*, densely woven onions (*Allium polyrrhizum*) and (*Gabbian*) semi-shrubs (*Anabasis Brevifolia, Salsola xominerina, Artemisia*)—fruit bushes (*A. Passerine*).

The steppe deserts form a rather wide strip and territory in the west of the southern foothills of the northern Gobi-Altai, and then occupy vast areas in the central part of the Gobi Desert. Deserts in the south are characterized by complete dominance in the prices of Central Asian semi-shrubs and shrubs. This strip contains a large number of dominant species and a significant diversity of the communities they form. Deserts with *Haloxylon ammodendron* are most widespread throughout Gobi: *Anabasis brevifolia, Zygophyllum xanthoxylon, Reaumuria songorica.* Western Gobi is characterized by communities such as *Ephedra przewalskii, ljinia regelii, Sympegma regelii.* The East Siberian types of these deserts include *Brachanthemurn gobicum, Salsola passerina, Potaninia mongolica, Nitraria sphaerocarpa*, and *Amygdalus mongolica.*

The latitudinal position of the country's territory and regularities of vegetation distribution is related to solar zoning, and the influence of such a factor as the degree of continental climate is on the vegetation. The steppe of Western Mongolia is still a transition zone from the west Siberian steppes. Here, along with Mongolian plant communities, more western elements also participate. The deserts of Western Mongolia are characterized by maximum aridity and the poorest composition of dominants. Vegetation of eastern Mongolia is characterized by the influence of flora of Eastern Asia. Here, special types of Daguur-Manchurian steppes are present in the extreme peculiar southeast and east Siberian deserts. Laws of altitude zonality in the Khuvsgul mountains were studied by Irkutsk botanists. The analysis of latitudinal and longitudinal changes of mountainous vegetation in Mongolia was carried out by Z.V. Karamysheva (1988). These data allowed to develop the system of Mongolian vegetation altitude zonality developed by A.A. Yunatov. At the same time, we adhered to the idea of zonal consideration of the belt systems and the associated typology of vegetation zoning with botanical-geographical zoning, which was expressed by many scientists (Schiffers 1946) (Sochava 1952).

There is no endemic genus in Mongolia, but more than 149 endemic and 140 subendemic species have been recorded. These species include genera of *Oxytropis, Astragalus, Saussurea* and *Potentilla*, families of *Fabaceae, Asteraseae* and *Brassicaseae.* The endemic plants have been divided by their origin, times and age to the paleo-endemics and neo-endemics (Fig. 9.4).

Those originating to the paleo-endemics belong to ancient desert and steppe species, whereas the Mongol-Altai and Khangai mountains species are neoendemics. For example, ancient

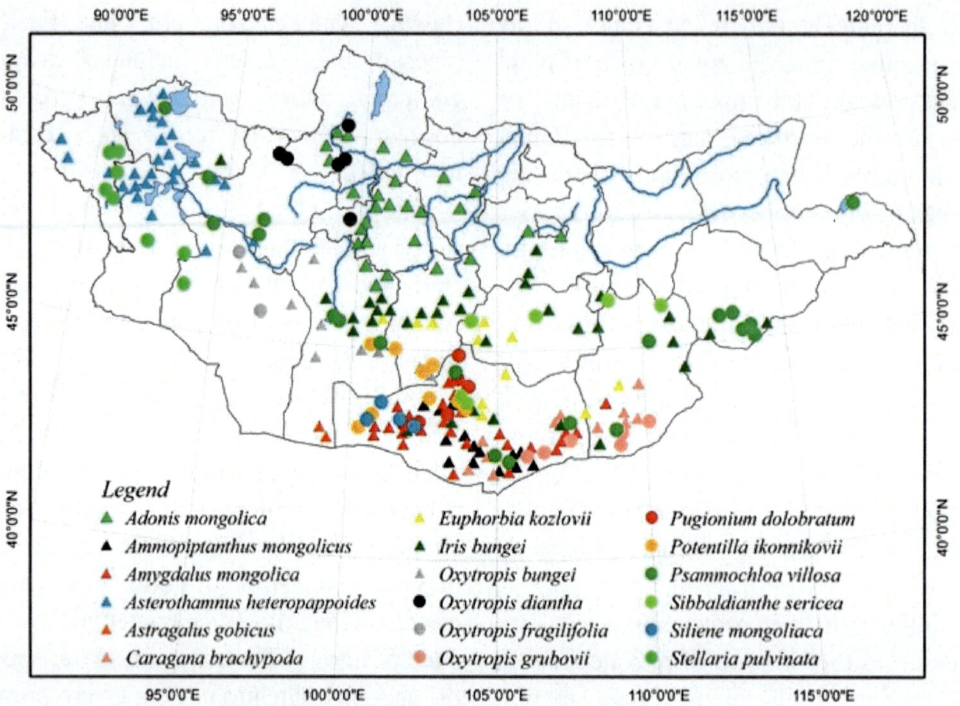


**Fig. 9.4** Distribution of some endemic species in Mongolia. Endangered and threatened plant species (Gunin et al. 2019)

endemics are *Sympegma regelii Bge*, *Brachanthemum gobicum* Krasch, *Amygdalus mongolica* Maxim, *Potaninia mongolica* Maxim, *Ammopiptanthus mongolicus* (Maxim.ex Kom) Cheng f, *Incarvillea potaninii* Batal and neoendemics are *Potentilla khenteica*, *Potentilla ikonnikovii* Juz, *Festuca venusta* St.ves, *Alchemilla khangaica* (Fig. 9.5) (Urgamal and Sanchir 2015).

Mongolia's pasture ecosystem is undergoing change due to climate and human-induced impacts. Thus, these issues are becoming the primary theme of the study, focusing on the rangeland vegetation community, covering changes (Undarmaa and Sasaki 2018), grazing pressure and management (Kakinuma et al. 2013).

## 9.2 Zoogeographical Features and Studies

### 9.2.1 Zoogeographical Regions

The zoogeographical research in Mongolia began in the 1920s with the participation of the Russian Academy of Sciences. For example, the expeditions of P.K. Kozlov, E.V. Kozlova (1923–1926), A.N.Formozov (1929), S.A.Kondratev (1930) and A.G.Bannikov (1947, 1953, 1954) were very valuable in the development of zoogeographical research of Mongolia. An attempt at zoogeographical regionalization in Mongolia

*Ajania grubovii Muldashev* *Astragalus gubanovii ulzij* *Papaver saichanense Grubov* *Potentilla schmakovii*

**Fig. 9.5** Some endemic plants of Mongolia (photo on Gundegmaa V)

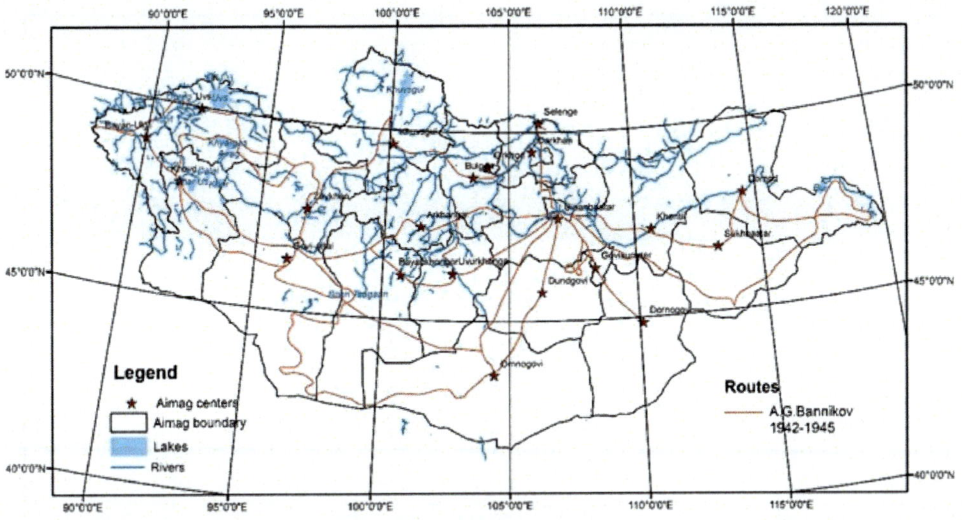


**Fig. 9.6** Routes of A.G. Bannikov 1942–1945 (Bannikov 1954)

began in the 1930s, which was based on specific fauna, such as Sushkin (1925) and Tugarinov (1929) based on the distribution of birds across the country (Tsegmid 1969). Also, some general considerations expressed by A.I. Formozov in 1929, and the point about the zoogeographical regions of the territory of Mongolia is described on the basis of comprehensive data, in brief remarks in the introductory chapter of G. Allen in 1938. In 1942–1945, Bannikov's expedition spent six times traveling in all geographical regions of Mongolia and its total routes reached 19,600 km (Fig. 9.6).

On the territory of Mongolia, as P.P.Sushkin (1925) pointed out, there are two subregions: North and Mountain of Asia. P.P.Sushkin referred to the northern area as taigi, forest steppe and steppe regions. Attempts of zoogeographical regions of the whole territory of Mongolia is reported on the basis of the comprehensive data, except for brief remarks in the introductory chapter of G. Allen (1938), as well as some general considerations expressed by A.I. Formozov (1929).

The comprehensive zoogeographical zoning of Mongolia by Bannikov in 1954, which was based on the long research of fauna throughout the country. He divided the territory of Mongolia into 10 zoogeographical sections and subregions, which were included into Mongol-Tibetan (Gobi and forest steppe subregions) and taiga regions (Fig. 9.7). The advantage of the Bannikov's regionalization is more related to the biogeographic characteristics of the country and is being used as the main source of fauna zoning in Mongolia.

### 9.2.2 Distribution Patterns of the Fauna

The vast territory of Mongolia is characterized by a wide variety of wildlife populations. The animal life of steppes, semi-deserts, and mountains of Mongolia is characterized by a high degree of endemism, especially among invertebrates, mammals, and reptiles. Rare, threatened, and endangered species include animals from various habitats (Munkhbayar 1976).

It is clear that the presence of mountain ranges, which determine the vertical zone, causes diversity of habitats. As a result of the latitudinal belt zoning, the fauna of Mongolia contains all physical and geographical groups of mammals: from highland tundra animals to the inhabitants of shrubby deserts. Within each zone, the rugged terrain creates a greater diversity of habitats than the plains.

Of the 141 species of mammals in Mongolia, 16% are regionally threatened, of which 2% are critically endangered (Gobi Bear, Przhevalsky's Horse and Red Deer), and 11% are endangered (Mongolian Marmot, Eurasian Beaver, Alashan Ground Squirrel, Small Five-toad Jerboa,

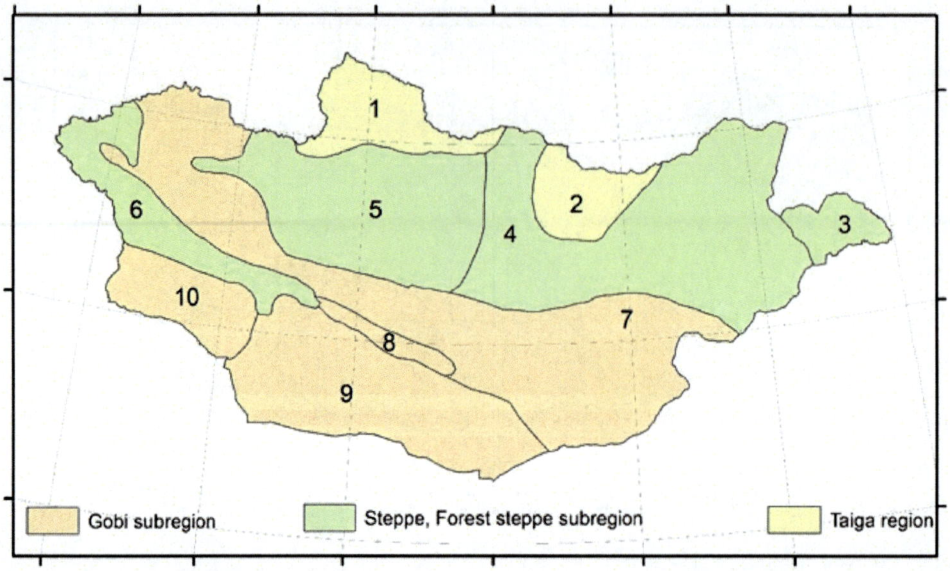


**Fig. 9.7** Zoogeographical regions and sections of Mongolia (Bannikov 1954): 1. Khuvsgul, 2. Khentii, 3. Western Khyangan, 4. Mongol Daguur, 5. Khangai, 6. Northwest Mongolia, 7. Northern Gobi, 8. Gobi-Altai, 9. South Gobi, 10. Western Gobi

Mongolian Three-toad Jerboa and others) while 3% are vulnerable (Long-eared Jerboa, Sable, Black-tailed Gazelle, Reindeer) (Fig. 9.8).

Large herbs of migratory ungulates are observed in steppes and desert steppes. In addition, some endangered and endemic species inhabit mountains and deserts of Mongolia. In the Mongolian Red List of mammals and birds (Seidler et al. 2011), 21 (16%) of the 128 mammals were assessed and 20 (4%) of the 506 birds were assessed, including data-deficient species (DD) were classified as regionally threatened [critically endangered (CE), endangered (EN), and vulnerable (VU)].

Mongolia's sandy desert fauna, which occupies a tiny area (about 4% of total area), consists mainly of widespread forms. These Asian forms are currently difficult to relate more precisely to any desert and sandy fauna formation. Fauna of stony deserts and semi-deserts of Mongolia has a different appearance and a slightly different history. Ecologically much more vulnerable forms of stony deserts are close to semi-desert forms, and the latter to the forms of zonal and partly mountain steppes. Fauna of the steppe zone of Mongolia, as well as other areas, was probably formed mainly by desert fauna (Podtyazhkin 1986).

The faunas of mountain steppes and highlands of Mongolia are also peculiar. The geographical distribution of mammals of these belts are connected with the steppe zones and semi-desert species. These faunas, which evolved before the

**Fig. 9.8** Species composition of Mongolian fauna. *Source* (MEGD & UNDP 2014)

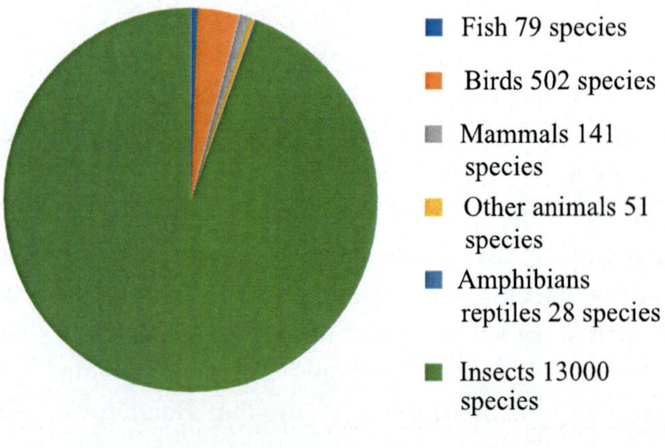

conditions of harsh continental, changing climate, turned out to be very vulnerable and gave adaptive forms of mountain steppe and highlands, in general, close to the original steppes and deserts. Argali (*Ovis ammon*), ibex (*Capra sibirica*), and Eurasian lynx (*lynx lynx*) distributed from deserts to high mountain tundra are different subspecies from each other. As far as taiga fauna is concerned, the issue is extremely complex and requires special consideration.

***Subalpine, alpine and mountain tundra highlands***. The height of mountains, the importance of this complex in the formation of living cover is especially great in the west (Mongol Altai) and consequently, in the east is rapidly decreasing: while Khangai, to a lesser extent, the mountains of Khuvsgul, still retains it almost completely. In the same sequence, afforestation in the mountains increases, which results in the isolation of forestless highlands from steppe and other open landscapes of plains and foothills.

Mongolian highlands are inhabited by low grass meadows, often steppe meadows, mountain tundra regions, and shrub subalpines (subalpine), all in different combinations. In addition, there are also steppe communities, including desert and steppe petrophyte communities. Rock falls and rocky massifs almost everywhere complicate the mosaic of habitats. The mountainous animal population is generally characterized by low biodiversity and relatively simple structure. A common feature of communities that can be attributed to different taxonomic groups of animals is their oligo and even monodomination. For example, buckwheat dominates weeds among birds, and the mountain range inextricably dominates the steppe meadows and remains in some places as the only inhabitant of the alpine belt. Small mammal populations are usually dominated by only one species of voles, either large or Mongolian, and one of the brown teeth, the tundra (Shvetsov 1986). They do not differ in the diversity of the mountainous community and their ecological appearance, for example, phytophages (rodents, hares, ungulates) are represented here almost only by greenery.

Regional differences in the composition of mountain fauna within the country are mainly due to impoverishment—in the general scheme from northeast to southwest—of the fauna of the tundra-holtz group, which flowed from the north. Thus, wild reindeer (*Rangifer tarandus*) penetrate only in Khuvsgul, polar birds—in Khangai, white and tundra partridges—in Khangai and Mongol Altai, but all of them are absent in the trans-altai Gobi. Accordingly, in the fauna and price groups of vertebrates of this southernmost of the largest mountain systems of the country, the mountain-tundra hue practically disappears, which indicates that the population of animals of the "extra-zonal" upper belts of the mountains also experiences a certain influence of latitudinal-zonal structures.

***Mountain taiga***. Mongolia's taiga forests are distributed over the northern mountainous area, which is direct continuation of the world's largest taiga region of North Asia. This also applies to lava plateaus, domes, and flat tops, as the analogs of flat taiga are mainly located at absolute altitudes approaching 2000 m and do not occupy a significant area. Nevertheless, the Mongolian taiga occupies a worthy place in the series of zonal shifts (Bannikov 1954; Tsegmid 1969). The taiga forest in Mongolia is the southern periphery of the boreal forest zone in the northern hemisphere. Although the taiga forest inside the country's territory is scarcer than the core of the Siberian taiga, Mongolia is home to most taiga species of fauna, such as squirrels, sable, bear, deer, wild boar and most flying bird species. In other words, the taiga within the country, being very limited in terms of territory, is quite representative as a fragment of the palaeogeographic taiga zone as a whole. The depletion of their fauna compared to the neighboring regions of Russia is insignificant or not expressed at all. Especially, for mammals the difference is reduced to the absence of separate species in the north of Mongolia Sokolov et al. (1982).

***Forest steppe***. Fauna of taiga "flows" to the forest islands of larch and forest steppe of Khangai very widely and completely; negative signs revealing the zoogeographical specificity of forest steppe Mongolia in this case give very little. This conclusion can be referred to bird species, although among this group there are more taiga species inhabiting the Khangai mountain forest. The penetration of forest fauna to the shoreline varies slightly depending on the mountain system. In Khentii, the pine-birch forest steppe directly frames large areas of taiga and is inseparable from them. In the Great Khyangan region of Mongolia, the influence of taiga fauna extends far beyond the limits of the taiga itself. Here, clinging to the oppressed birch forests and floodplain willow, among the vast steppe plateaus, there are not only moose but also Tolai hare *(lepus tolai)*, Eurasian lynx *(Lynx lynx)*, Mongolian gazelle *(Procapra gutturosa)*; middle and large tooth chipmunk, 1980, chipmunks, gray-headed partridges, 1986; chipmunks, winged chipmunks, and so on (Vostokova et al. 1995).

***Steppe***. Mongolia's steppe communities are the systematic and biological animal groups characteristic of this zone. The middle class of birds is mainly larks, large terrestrial birds from the crane of the left leg (bustards, cranes), and predatory myophages (the Tengmalm's mound of the steppe eagle). Among mammals there are green voles, squirrels, and forage plants. Previously, the Asiatic wild ass or Khulan (*Equus hemionus)* and, possibly, Przhevalsky's horse or Takhi *(Equus ferus)* penetrated widely into the steppe, then they "separated" from Siberian roe deer *(Capreolus pygargus)* and red deer *(Cervus elaphus or C. canadensis)* steppe, and everywhere in the melkosopochniks and intermountain depressions lived argali (*ovis ammon*). Mongolian gazelles *(Procapra gutturosa)*, as a typical steppe inhabitant, which are widely distributed in the north-east (Dornod plain) of the country. However, the area of Mongolian gazelles has also significantly degraded, which has survived only in the east of Mongolia, where it is primarily associated with the subzone of dry steppes (Sokolov et al. 1982).

***Desert Steppe***. The semidesert zonal type of communities under consideration is clearly expressed in the deep interpenetration of the actual steppe and desert elements of the fauna, which can be traced for different groups of animals, albeit with significant fluctuations in the ratio of both. Mongolia's desert steppe or Gobi is a region of species composition, a combination of fauna elements penetrating from surrounding areas. The zone under consideration is characterized by many insect characteristic forms, and in such families gravitating to arid areas as black bodies and takhins, it is the fauna of desert steppes that is characterized by the greatest wealth "both in terms of its overall composition and the number of endemics, including clans" (Yemelyanov et al. 1986). This zone limits the habitat of about 20% of rodent species in the Mongolian open zone habitat. At the same time, the fauna of desert steppes and deserts has 12 common species, while that of desert and real steppes is twice as small (Podtyazhkin 1986).

***Desert***. Animals in the deserts of Mongolia are largely a community for all deserts, but they are found here in a special Central Asian refraction. Three typical dendrophiles (*grey prawns, saxaul Sparrow, desert Rambler, and Mongolian saxaul Jay*) are also found in the bush area. Changes in the predominant biological groups and a sharp increase in the diversity of the bird population are particularly noticeable: the number of nesting species increases almost fivefold. A number of small mammal species, including the hygrophilous form of lake voles, live only in oases, but their margins are optimal for sandy spears (Vostokova et al. 1995). Among the large animals holding on to these peculiar tugai, there is a wild boar and a strict desert, Bactrian camel *(Camelus ferus)*, Gobi bear *(Ursus arctos gobiensis)*, and Snow leopard *(Uncia uncia)* (Sokolov and Zhirnov 1986). As a rule, monotonous plains are unattractive for the inhabitants of ungulate deserts; Mongolian gazelles *(Procapra gutturosa)*, Asiatic wild ass or Khulan *(Equus hemionus)* and Bactrian camel *(Camelus ferus)* gravitate to foothills and small meadows and penetrate into mountains, and mountain ungulates, especially argali *(ovis ammon*).

### 9.2.3 Some Faunas

***Mammals***. Mongolia has 141 species of mammals, belongs to 73 genus and 23 families, 8 orders, which includes 13 species of insectivores; 12 species of chiropteras; 6 species of lagomorphs; 69 species of rodents; 24 species of carnivores; 2 species of perissodactyls; 1 species of tylopoda; and 11 species of artiodactyls (Fig. 9.9).

A detailed research on the diversity of Mongolian mammals has not been completed; only a few studies on some subspecies of mammals have been conducted. According to recent studies, 57 subspecies of 30 mammal species were identified. For last 100 years, 6 subspecies of mammals, belonging to 3 species and 1 genus were vanished from Mongolian fauna, caused by direct and indirect human activities (Fig. 9.10).

***Birds***. The distribution of threatened bird species in Mongolia varies depending on the landscapes and regions. The main bird habitats in Mongolia comprised grassland steppe, semi-desert and desert, mountain steppe, high mountains, forested mountains, wetlands, and riparian areas. Characteristic of birds in grassland steppe include Upland Buzzard *Buteo hemilasius*, Steppe Eagle *Aquila nipalensis*, Saker Falcon *Falco cherrug*, Mongolian Lark *Melanicorypha mongolica*, Crested Lark *Galerida cristata*, Eurasian Skylark *Alauda arvensis*, Demoiselle Crane *Anthropoides virgo*, Mongolian Plover *Charadrius mongolus*, Great Bustard *Otis tarda*, Northern Wheatear *Oenanthe oenanthe*, Isabelline Wheatear *Oenanthe isabellina*, Lesser Short-toed Lark *Calandrella rufescens*, and Horned Lark *Eremophila alpestris*. Birds adapted to desert and semi-desert habitats include Pallas's Sandgrouse (Ocock 2006). Currently, about 502 bird species have been recorded in Mongolia, belonging to 61 families and 19 orders (Gombobaatar et al. 2012). About 81 of them are resident birds and 391 species are migratory birds. 254 species of migratory birds breed in Mongolia, 10 species are winter visitors from Siberia, 8 species are summer visitors, and 64 species are vagrants. Four main global migratory routes have been recognized in Mongolia: the East Asia-Australasia flyway; the Central Asia flyway; the West Pacific flyway; and the Africa-Eurasia flyway (MNET 2017).

Dornod Mongol and Great Lakes Valley are strictly protected and vital areas for rare birds. Rare species found in these habitats include the Reed Parrotbill *Paradoxornis heudei*, which is dependent on a very few wetland sites in eastern Mongolia, and Dalmatian Pelican *Pelecanus*

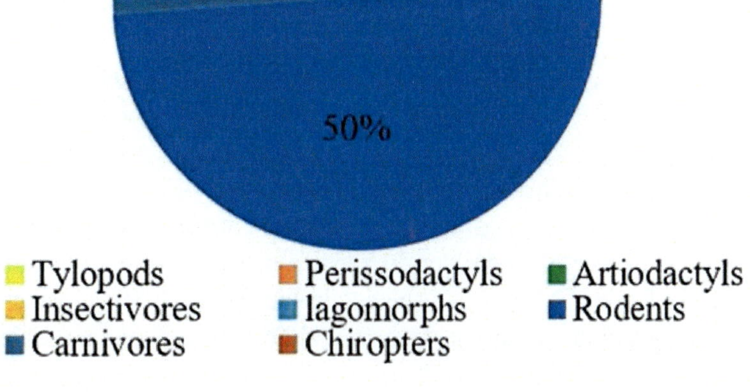


**Fig. 9.9** Mammals diversity of Mongolia Inventory of Mongolia's mammal diversity

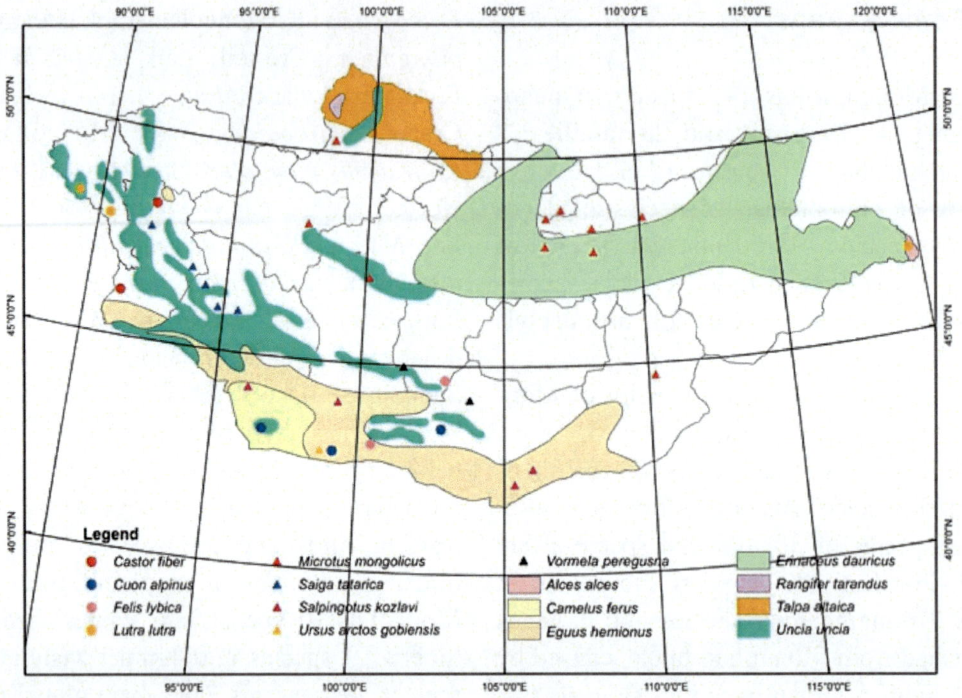


**Fig. 9.10** Distribution of rare mammals (Vostokova and Gunin 2005)

*crispus*; the eastern population breeds only at a few lakes in Western Mongolia. Special protected areas such as Numrug, Altan els, Mongol Daguur, Onon-Balj, Khugnu Khaan, Otgontenger, Uvs, Khar-Us lake, Tsambagarav, Toson Khulstai, Khar Yamaat and Ikh Nart are high in species richness Gombobaatar et al. (2011, 2012).

*Endemic bird areas*. Endemic bird areas (EBAs) are areas with at least two restricted-range bird species (species with a total global breeding range of less than 50,000 km$^2$) and are entirely confined, while secondary areas (SAs) are areas that support one or more restricted-range species but to which less than two species are entirely confined (Schiffers 1946). In common with other countries in north-east Asia, Mongolia has low levels of bird endemism, with most of the species occurring in the country having wide breeding ranges (Fig. 9.11).

***Fish***. Mongolia has 79 species of fish belonging to 46 genera and 14 families reported in three main watersheds. There are 29 species in Arctic Ocean Drainage Basin, 43 species in the Pacific Drainage, and 10 in the Central Asian Inland Basin. According to the book on the "Fishes of the Mongolian People's Republic"(1983) a total of 59 fish species have been recorded, later 74 species were described by G.Baasanjav, Ya. Tsend-Ayush in 2001, and 76 fish species were listed in the biodiversity assessment and conservation plan in 2002 (Baasanjav and Tsend-Ayiush 2001).

***Amphibians and reptiles***. Mongolian herpetological fauna seems to be poorer than the vast territory of the country, but their species are relatively well studied. They are divided into two classes, three orders, 10 families, and 18 genera consisting of 6 reptile species (Munkhbaatar and Munkhbayar 2012; Terbish et al. 2006). The geographical distribution of amphibians is more widespread in northern Mongolia, where there are more lakes, rivers, and ponds. Reptiles are mainly found in arid regions, particularly in the Gobi Desert. However, due to climate change and anthropogenic impacts, the distribution of some amphibian and reptile species has declined (Terbish 2006). The population of Siberian salamander in the Tuul River basin near Ulaanbaatar has declined, and Mongolian toad has not been recorded in recent years.

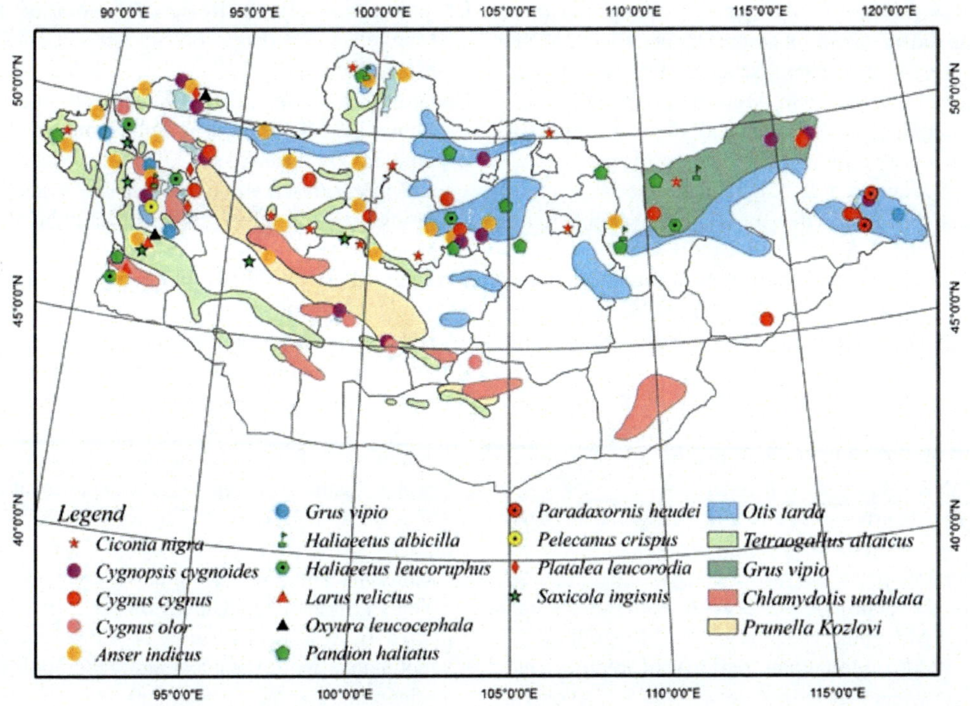


**Fig. 9.11** Distribution of rare birds: modified from Ecosystems of Mongolia, Atlas. (Vostokova and Gunin 2005)

## References


Baasanjav G, Tsend-Ayiush Y (2001) Fishes in Mongolia. Admon Co Ltd, Ulaanbaatar (in Mongolian)

Bannikov IA, Khudyakov OI (1976) Soil-vegetation subbelt of the forest belt of south-eastern Khangai. Moscow (in Russian)

Bannikov AG (1954) Mammals of the People's Republic Mongolia. Publishing House of RAS, Moscow (in Russian)

Banzragch D et al (1978) Vegetation of the Atlas-Bogdo-Ula middle mountain massiva in Zaaltaiskaya Gobi. Moscow (in Russian)

Banzragch D, Munkhbayar S (1978) On the borders of the high-mountainous vegetation belt in the Hangai Mountains. Geography and dynamics of flora and fauna of MNR, Moscow (in Russian)

Chu T, Guo X (2012) Characterizing vegetation response to climatic variations in Hovsgol, Mongolia using remotely sensed time series data. Earth Sci Res 1:279–290

Gombobaatar S, Sumiya D et al (2012) Bird Red List and its future development in Mongolia. In: Enforsch. biol. Ress. Mongolei (Halle/Saale), pp 169–182, Ulaanbaatar (in Mongolian)

Gombobaatar S et al (2011) Regional red list series, vol 7. Birds. Zoological Society of London, National University of Mongolia and Mongolian Ornithological Society (in Mongolian)

Grubov VI (1954) Vegetation regionalization of Mongolia. Probl Geogr 35 (in Russian)

Grubov VI (1982) Key to the vascular plants of Mongolia (with an atlas). Probl Geogr 28 (in Russian)

Grubov VI (1963) Botanical-geographical zoning of Central Asia. Moscow (in Russian)

Gunin PD, Bazha SN, Khadbaatar S et al (2019) Ecosystems Atlas of Mongoia, Gunin P, ed. Mongolian Academy of Science, RAS, MNUE. Admon Publishing, Ulaanbaatar

Kakinuma K et al (2013) Rangeland management in highly variable environments: Resource variations across the landscape mediate the impact of grazing on vegetation in Mongolia. Grassl Sci 59(1):44–51

Kamelin RV (2010) Mongolia on the map of the phytogeographical division of palaeoarctics. Turczaninowia 13(3):5–11 [in Russian]

MEGD & UNDP (2014) Convention on biological diversity. The 5th national report of Mongolia. Ministry of Environment and Green Development, Ulaanbaatar

MNET (2017) Biological diversity in Mongolia. Ministry of Environment, Ulaanbaatar (in Mongolian)

Munkhbaatar M, Munkhbayar K (2012) Herpetological diversity of Mongolia and its conservation issues. Erforsch.biol. Ress. Mongolei (Halle/Saale), pp 203–212

Munkhbayar Kh (1976) Reptiles and amphibians of Mongolia. Publishing House of Ministry of Education, Ulaanbaatar (in Mongolian)

Nyambayar D (2009) Flora of Mongolia. Bembi fund, Ulaanbaatar (in Mongolian)

Ocock JB (2006) Mongolian red list of fishes, regional red list series, vol 3. Zoological Society of London, London

Podtyazhkin O (1986) To the question of the independence of rodent fauna of desert steppes of the MNR and its place in the general scheme of zoogeographic division of the Palaearctic. Moscow (in Russian)

Schiffers E (1946) On geobotanical regionalization of mountainous countries// owns botany. Moscow

Shagdar D, Yadamsuren O (2017) Plant species composition and vegetation cover of Kherlen Toono Mountain, Mongolia. J Asia-Pacific Biodiv 10:132–136

Shvetsov YI (1986) Faunal complexes of mammals in the highlands of western and northern Mongolia. In: Zoogeographic zoning of the MNR. Moscow, pp 109–123

Sochava VB (1952) Main provisions of geobotanical zoning. Bot J 3. Moscow (in Russian)

Sokolov VE et al (1982) Modern distribution of gazelles in the MNR. Moscow (in Russian)

Sokolov VE, Zhirnov LV (1986) Large mammals: science. In: Deserts of Zaaltaiskaya Gobi. Science, Moscow, 350 pp

Terbish K et al (2006). A guide to the amphibians and reptiles of Mongolia. Ulaanbaatar (in Mongolian)

Terbish K (2006) Mongolian red list of reptiles and amphibians, regional red list series, vol 5. Zoological Society of London, London

Tsegmid S (1969) Physical geography of Mongolia. Institute of Geography and Permafrost. Mongolian Academy of Sciences. State Publishing House, Ulaanbaatar, 405 p (In Mongolian)

Tuvshintogtokh I (2014) Steppe vegetation of Mongolia. Institute of Biology. Mongolian Academy of Sciences. Bembi san publishing, Ulaanbaatar (in Mongolian)

Ulziikhutag N (1989) Review of flora of Mongolia. Mongolian Academy of Sciences, Ulaanbaatar (in Mongolian)

Undarmaa J et al (2018) Rangeland ecosystems of Mongolia. Munkhiin useg Publishing, Ulaanbaatar

Urgamal M, Oyuntsetseg B (2017) The relict plant species and their conservation status to the vascular flora of Mongolia. Biodiv Res Mongolia:41–43 (in Mongolian)

Urgamal M et al (2016) Flora and fauna of Current Mongolia. Mongolian Academy of Science News, Ulaanbaatar (in Mongolian)

Urgamal M, Snachir C (2015) Preliminary analysis of the vascular flora of mongolia. International Conference of Ecosystems of the Central Asia under current conditions of socio-economic development. Ulaanbaatar: Institute of General and Experimental Biology, Mongolian Academy of Sciences, pp 262–264

Urgamal M, Bayarkhuu S (2018) Plant diversity of Mongolia and their quantitative changes in Last 30 Years (1989–2017). Global Biodiv:2–3

Volkova E (1988) Vegetation of the South Mongolian mountains: main botanical and geographical patterns: sciences. Moscow (in Russian)

Vostokova EA, Gunin PD, Rachkovskaya EI (1995) Mongolian Ecosystems: sciences. Moscow

Vostokova EA, Gunin PD (eds) (2005) Ecosystems of Mongolia Atlas. Joint Russian-Mongolian Complex Biological Expedition, Moscow

Yunatov A (1950) Main features of the vegetation cover of the People's Republic of Mongolia. Publishing House of RAS, Moscow (in Russian)

# 10 Division of the Physiographic and Natural Regions in Mongolia

Dash Doljin and Batchuluun Yembuu

**Abstract**

Mongolia has a complex landscape and unique terrain that has been formed through internal and external processes. Due to the different landscapes and physical conditions of the country, the geographical distribution of physical patterns is differing. This chapter begins with a brief history of physiographic regionalization and an updated map of Mongolia's physical-geographical regions. In addition, this chapter provides a detailed description of natural zonalities (altitudinal belts and latitudinal zones) and the principles of their development. Most of Mongolia's natural zone sources are based on biological approaches, mainly referred to as vegetation cover, and also have different names and different data on areas in each zone. This chapter provides information on Mongolia's natural zones and zoning in terms of physical geography. The updated author's map presented in this chapter has been used as the main source of information on Mongolia's natural zones since the 2000s. A detailed description of the six natural zones (altitudinal zones and latitudinal zones) and the principles of their development is also presented in this chapter. Mongolia has six main natural zones and belts (with sub-divisions), such as alpine and mountain taiga, mixed and deciduous forests, forest steppe, steppe, Gobi (desert steppe), and desert zones.

**Keywords**

Physiographic region · Natural zone · High mountain belt · Mountain taiga belt · Steppe zone · Desert zone

## 10.1 History of the Geographical Regionilization in Mongolia

### 10.1.1 Physiographic Regions

A physiographic region is a dividing landform into distinct regions and morphological unit with an internal coherence in its landform characteristics appropriate to the level of sub-division. Most common classification of the physiographic regions is based upon Nevin Fenneman's classic American three-tiered approach of divisions, provinces and sections (Fenneman 1914, 1916). Even though the classification of the physiographic region was followed by Russian scientists, the concepts are the basis for similar of other countries.

The different classifications of physiographic regions of Mongolia have been developed by

D. Doljin (✉) · B. Yembuu
Department of Geography, Mongolian National University of Education, Ulaanbaatar, Mongolia
e-mail: dash@msue.edu.mn

B. Yembuu
e-mail: batchuluun@msue.edu.mn

© Springer Nature Switzerland AG 2021

B. Yembuu (ed.), *The Physical Geography of Mongolia*, Geography of the Physical Environment,
https://doi.org/10.1007/978-3-030-61434-8_10

several researches (Table 10.1) since 1930s, by Kondratiev (1930), Simukov (1934), Murzaev (1948), Sh.Tsegmid (1969) (Dash 2002). In 1930, the first classification of the physiographic regions of Mongolia created by Russian scientist S.A. Kondratiev was working at the Geography Department of the Institute of Science of Mongolia. According to his classification, Mongolia is divided into 11 physiographic regions: *(1) Khentii mountain ranges, (2) Orkhon Depression, (3) Khangai and Khuvsgul mountain ranges, (4) Depression of Western Lakes, (5) Mountain Tagna, (6) Mongolian Altai, (7) Central Depression of the Lakes, (8) Gobi-Altai and Trans-Altai Gobi, (9) Eastern Gobi of Mongolia, (10) Eastern Steppe of Mongolia, and (11) Khyangan mountain range*

The second map of the physiographic regions of Mongolia was included in the "Atlas of Mongolian Peoples of Republic" in 1934, which was published by the Institute of the Mongolian Science. According to this classification by A.D. Simukov, Mongolia is divided into 12 physiographic regions: *(1) Altai Mountain Area, (2) Depression of Southern Lakes, (3) Khangai Mountain Ranges, (4) Area of around Khuvsgul, (5) Central Depressions, (6) Gobi-Altai, (7) The Western Gobi, (8) Middle Platea of Khalkha, (9) High Plataea of Shanghai, (10) Khentii Mountain Ranges, (11) Eastern Platea of Mongolia, (12) Hilly plains of eastern Gobi.*

The third classification of the physiographic regions developed in 1937, by Geography Department of the Institute of Science of Mongolia, was published in description of the "The Geography of Mongolian Peoples of Republic". According to this classification, Mongolia is divided into 10 physiographic regions: (1) Altai Mountain Ranges, (2) Depression of Western Lakes, (3) Khangai mountain ranges, (4) Khentii mountain ranges, (5) Eastern Platea of Mongolia, (6) Middle Platea of Khalkha, (7) Central Depressions, (8) Gobi-Altai Говь Алтай, (9) High Plataea of Shanghai, and 10) Hilly plains of eastern Gobi.

Although these physiographic classifications which were compiled in 1930s contained some disadvantages, they had a lot of advantages such as a first attempt for the classifying of Mongolia's vast territories based on its physical differences of the country, as well as some independent regions

**Table 10.1** The physiographic regionalization of Mongolia

| | Name of the scholars | Region's name | Hierarchy of the classification | Number of regions, provinces and sections |
|---|---|---|---|---|
| 1 | S.A. Kondratiev (1930) | Physiographic regions | Single | 11 Regions |
| 2 | A.D. Simukov (1934) | Physiographic regions | Single | 12 Regions |
| 3 | E.M. Murzaev (1948) | Physiographic regions | Two-tiered | 5 Regions, 27 Provinces |
| 4 | Sh.Tsegmid (1969) | Physiographic regions | Three-tiered | 4 Regions, 12 Provinces, 27 Sections |
| 5 | N.Y. Phadeeva, Kh. Tulgaa, J. Natsag et al. (1981) | Landscape-typological | Five-tiered | 4 Regions, 14 Provinces, 34 Sub-provinces, 354 Sections, 92 Sub-sections |
| 6 | N.Y. Phadeeva, Kh.Tulgaa et al. | Natural | Five-tiered | 3 Regions, 14 Provinces, 34 Sub-provinces, 418 Sections, 21 Sub-sections |
| 7 | E.A. Vostokova and P.D. Gunin (1995) | Landscape-Typological | three-tiered | 6 Regions, 14 Provinces, 36 Sections |
| 8 | Sh. Tsegmid (1995) | Байгалийн | Two-tiered | 4 Regions, 9 Provinces |

were corresponded to the physical features of the certain area (Tsegmid 1969). Moreover, the schemes of these physiographic regions were a basic background for the future classification of the physiographic regionalizing of Mongolia.

A more scientific map and two-tiered scheme of the physiographic regions of Mongolia was created by Russian scientist E.M.Murzaev in 1948. According to this map, Mongolia was divided into five regions (Altai mountain, Depression of Great Lakes, Khangai and Khentii mountains, Dornod plain and the Gobi region) and 27 sub-regions (Murzaev 1948). This classification contains plenty of advantages based on landforms and its geomorphological patterns of Mongolia. Also, Murzaev used the latitudinal features of some regions for his classification. So, the physiographic regions of Mongolia considered the basic fundamental of the following classifications and valuable resources from his decedents of physical geographers.

Comprehensive three-tiered physiographic regionalization of Mongolia was created by Tsegmid in 1969, based on the landforms pattern. According to this classification, Mongolia was divided into 4 distinct physiographic regions (Khangai and Khentii mountains, Altai mountains, Eastern Mongolian Plain and The Gobi), 11 physiographic provinces and 27 physiographic sections.

cThe principle of background is the physiographic divisions with three-tiered approaches by Tsegmid. The largest unit of his classification (regions) was based on combined features of geomorphology and latitudinal characters, while second unit of physiographic provinces and sections (third unit) was based on geomorphological features of the areas. Furthermore, since 1970s, plenty of results were accumulated in the researches of physical geography of Mongolia, as well as different types of serial maps were compiled within these periods, such as Map of Mongolian Landscape, Map of Geomorphology, Soil Map of Mongolia, and maps of natural conditions.

Also, detailed maps such as Map of Landscape and Ecology of Mongolia, Map of Natural Zonation were compiled as a result of the last researches of physical geography of Mongolia.

One of the significant scientific works is Map of Natural Realms of Mongolia compiled by Tulgaa with Russian scientists V.L. Lvov, E.L. Smirnova, N.V. Fadeeva in 1990, which is based upon the five-tiered hierarchies: 3 divisions, 14 provinces, 34 sub-provinces, 418 sections and 21 sub-sections of natural landscapes in Mongolia. In addition, the landscape features of Mongolia, as well as dominant and non-dominant physico-geographical characteristics, were determined in detail on this map. The methods of mapping approaches of this map were considered a most significant and it was a main advantage of the map of natural realms, compared with previous maps. The basic principle of the natural realms of this map was landscape distinctions in Mongolia.

One of the serial wall maps (at scale 1:1,000,000) were compiled by Russian and Mongolian researchers based on satellite map and data (1981), the Map of Landscape-Typology Realms in Mongolia. In 1995, Russian scientists E.A. Vostokova and P.D. Gunin created a "Map of Landscape-Ecological classification in Mongolia" which was included in the book "The Ecosystems of Mongolia" (Gunin et al. 1995). The fundamental basics of this map were the features of flora and fauna as well as a holistic feature of the landscape and its similarities with different places in the country.

The systemic approach of the classification of ecosystems distinction on the Mongolian territory was a three-tiered mapping, with 6 regions, 14 provinces and 136 sections of the landscapes. Although the above-mentioned classifications were based on certain approaches and named differently, all are innovative and modified versions of the physiographic regions by Tsegmid (1969) and results of the researches of physical geography of Mongolia. In particular, the covering area of regions and provinces and borders became more detailed (Dash 1999).

The new classification of the physiographic regions by Dash is based on the synthesis of previous physiographic regions by several scientists and based of their advantages. In addition, this is revised version of the classification of the physiographic region by Tsegmid in 1969, and is also based on the feature of physical geography and landscapes differentiation of Mongolia. On the other hand, the background of these revised physiographic regions includes ecological-landscape approaches. The updated physiographic regionalization is characterized by the geotecture and morphostructural features, climatic condition, the distribution of the ecosystem in Mongolia and based on the landscape principle. Consequently, there are 6 physiographic regions *(Altai-Sayan mountains; Trans-Baikal mountain Taiga; Khangai mountains; Daguur-Mongol steppe; Great Khyangan mountians; Semi-desert and desert of the Central Asia)*, 16 provinces and 45 sections (Fig. 10.1). The areas of the physiographic regions are different, the *semi-desert and desert region is largest* (42.0%) (Table 10.2).

A. ***Region of Altai-Sayan mountain: I.*** *Mongol Altai high mountain; 1.Kharkhiraa-Turgen mountains; 2.Central part of Mongol Altai; 3. Munkhkhairkhan mountains; II. Eastern Sayan high mountains: 4. Western of Lake Khuwsgul; 5. Eastern of Sangilen mountain; III. Tagna high mountain and depression: 6. Uvs lake depression.*
B. ***Region of Trans-Baikal Mountain Taiga: IV.*** *Eg-Selenge mountain taiga; 7. Eastern part of Khuvsgul lake; 8. Buren-Buteel mountains; 9. Confluence of Orkhon-Selenge river s;* ***V.*** *Khentii mountain Taiga; 10. Northwestern part of Khentii mountain range; 11. Central part of Khentii; 12. Eastern part of Khentii; 13. Southern part of Khentii;*
C. ***Khangai Mountain Region: VI.*** *Khangai high mountain; 14. Western part of Khangai; 15. Central part of Khangai; 16. Southern part of Khangai; 17. Northeastern part of Khangai; 18. Tarvagatai mountains; 19. Bulnai mountains;* ***VII.*** *Mountains of the Orkhon and Tuul; 20. Orkhon-Khanui river basin; 21. Orkhon-Yeruu river basin; 22. Tuul-Tarna river basin;*
D. ***Daguur-Mongol Steppe Region: VIII.*** *Daguur mountain steppe; 23. Ulz river basin;* ***IX.*** *Steppe of Eastern Mongolia; 24. Plateau of Middle of Khalkh; 25. Plain of northern part of Kherlen river; 26. Undurkhaan-Baruun-Urt hummocky; 27. Pediplain of Menen;* ***X.*** *Dariganga-Erdenetsagaan mountain steppe; 28. Dariganga upland; 29. Erdenetsagaan low mountains;*

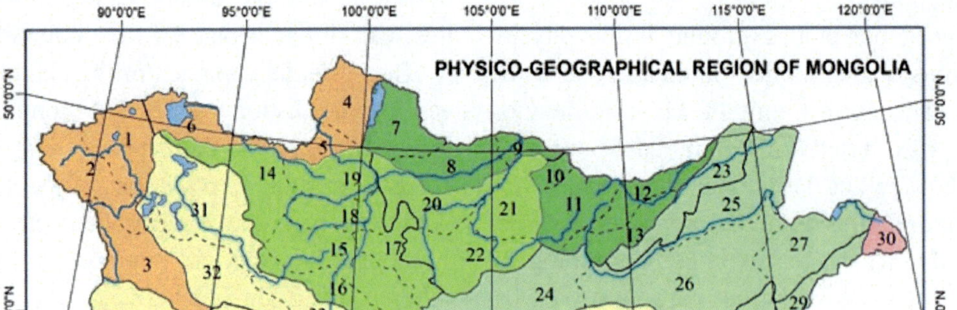


**Fig. 10.1** Physiographic Regions of Mongolia (by Dash 2002)

**Table 10.2** The physiographic regions by areas

| | The physiographic regions and provinces | Area (km$^2$) | Percentage |
|---|---|---|---|
| | ***A. Region of the Altai-Sayan Mountains*** | **167960.4** | **10.7** |
| I. | Mongol Altai high mountain | 104478.5 | 6.7 |
| II. | Eastern Sayan high mountains | 40299.7 | 2.5 |
| III. | Tagna high mountain and depression | 23182.2 | 1.5 |
| | ***B. Region of Trans-Baikal Mountain Taiga*** | **144639.4** | **9.2** |
| IV. | Eg-Selenge mountain taiga | 68718.7 | 4.4 |
| V. | Khentii mountain Taiga | 75920.7 | 4.8 |
| | ***C. Khangai Mountain Region*** | **287530.4** | **18.3** |
| VI. | Khangai high mountain | 185314.3 | 11.8 |
| VII. | Mountains of the Orkhon and Tuul | 102216.1 | 6.5 |
| | ***D. Daguur-Mongol Steppe Region*** | **300157.0** | **19.2** |
| VIII. | Daguur mountain steppe | 32965.7 | 2.1 |
| IX. | Steppe of Eastern Mongolia | 239728.1 | 15.3 |
| X. | Dariganga-Erdenetsagaan mountain steppe | 27463.2 | 1.8 |
| | ***E. Great Khyangan Mountian Region*** | **8598.3** | **0.6** |
| XI. | Forest Steppe of the western part of Khyangan | 8598.3 | 0.6 |
| | ***F. Semi-desert and Desert Region of the Central Asia*** | **657614.5** | **42.0** |
| XII. | Depression of Great Lakes | 139104.1 | 8.9 |
| XIII | Gobi-Altai hilly mountains | 75731.7 | 4.8 |
| XIV | Zuungar gobi | 27778.9 | 1.8 |
| XV. | Trans-Altai Gobi | 134838.0 | 8.6 |
| XVI. | Dornod Gobi | 280161.8 | 17.9 |

E. ***Great Khyangan Mountian Region: XI.*** *Forest Steppe of the western part of Khyangan; 30. Mountains of Degee-Numrug River Basin;*

F. ***Semi-desert and Desert Region: XII.*** *Upper Gobi of the Altai; 31. Khar-Us and Khyargas lake depression; 32. Depression with Altai sub mountains; 33. Valley of Lakes*; ***XIII.*** *Gobi-Altai hilly mountains; 34. Khar Azarga and Gichgene mountains; 35. Mountain depression around the Ikh Bogd; 36. Gurvansaikhan and Shanghai mountains;* ***XIV.*** *Zuungar gobi; 37. Baruun Khuurai depression*; ***XV.*** *Trans-Altai Gobi*; *38. Extreme arid desert of the western part of the Trans-Altai Gobi; 39. Eastern part of the Trans-Altai Gobi*; ***XVI.*** *Dornod Gobi; 40. Hilly plain of the Mandal-ovoo and Khuld; 41. Mandalgovi-Ulziit hummocky; 42. Delgerekh- Altanshiree plain; 43. Low mountains of Khatanbulag-Khuvsgul; 44. Borzon Gobi; 45. Galba Gobi.*

### 10.1.2 The Natural Zones and Belts

Since 1950, regional research in Mongolia has expanded, and analytical and systematic researches were initiated throughout the country. In particular, the Russian scientists A.A. Yunatov, E.M. Murzaev and Bespalov had contributed extensively in the research studies of the natural zones of Mongolia. At that time, the new definition of natural zones was formulated. The first classification of the vegetation of Mongolia was developed by Yunatov in 1948, based on the vegetation diversity of Mongolia (Yunatov 1948). According to Yunatov's classification, the vegetation zones are classified by vegetation of high mountain (alpine), mountain taiga, mountain steppe and forest zone, steppe zone, desert steppe, and desert zone. Meanwhile, E. Murzaev (1952) defined the natural zones of an alpine (mountain meadow) vegetation, mountain taiga, mountain steppe and forest steppe; high steppe, desert steppe (Gobi) and desert based on geographical features of the country.

Traditionally, Mongolian nomads have extensive experience in observing the surrounding nature, weather conditions and physical features of their places of residence. Therefore, they have identified northern mountainous areas with high precipitation and forests called Khangai. The southern area with less moisture and precipitation, with arid and sparse grasslands, is called Gobi Desert. This nature recognition experience and lessons are the main contributions of Mongolians to natural science. The foreign explorers who studied Mongolia and Central Asia considered that the "Khangai" and "Gobi" is the name of certain regions, which is the "single-sided" of the understanding. However, the terms of "Khangai" and "Gobi" even represent the certain regions in Mongolia, but these names represent the natural conditions and landscape features for Mongolians. On the other hand, the basics of natural zone classification in Mongolia was developed by nomads of Mongolia, who experienced their traditional life since ancient period (Dash 2002).

Sh. Tsegmid (1969) had determined the two-altitudinal belts and four latitudinal zones based on Russian scientists' research and that classification is introduced into school textbooks and still is being used. The natural zones and belts are updated and compiled as a wall map "Map of Natural zones and belts in Mongolia" (scale 1:2 500 000) by D.Bazargur in 1972 (Bazargur 1972) and this map is presented in several atlases of physical geography for schools. The second map of Natural zone of Mongolia (1:2,500,000) was developed by the National Agency of Geodesy and Cartography and published (edited by S. Tserenbaymbaa) based on the zonation of the vegetation cover.

The map of Mongolia's vegetation zones and detailed descriptions created by N. Ulziikhutag in 1989 was based on Yunatov's classification scheme of vegetation. There were several updates in this map, which classified the depression belts and altitudinal belts on some places in Mongolia (Ulziikhutag 1989). The natural zone map and description (introduced by Ulziikhutag) was in the several official documents such as "National report on biological resource" and "The action plan on the conservation of biological diversity of Mongolia". In addition, the short description of the natural regionalization had included in research works by D. Dorjgotov (1976) and S. Jigj (1979).

The updated Map of Natural Zones of Mongolia (editors D. Dash and Kh. Tulgaa) developed by Institute of Geography and Permafrost (former name) of the Academy of Science of Mongolia in 1990. However, this map was not introduced for the public; the classification of natural zones had presented as a research paper. The updated map of "Natural Zones of Mongolia" (scale 1:1,500,000) introduced by Dorjgotov in 2006 is still the main teaching and learning material for all levels of education. There were four altitudinal belts (high mountain meadow, tundra, mountain taiga and high mountain steppe) and three zones (steppe, desert steppe and desert) and subzones in this map. He has divided the steppe zone into four subzones (meadow typed steppe, semi-arid steppe, arid steppe and extreme dry steppe). The desert steppe zone is divided into two subzones (deserted steppe, steppe-dominated desert) and the desert zone is divided into two subzones—true desert and extreme dry desert (Dorjgotov 2006). The advantages of this map were that the scale is relatively large and boundaries between the zones were detailed; however, the internal structure of the zones were not much based on the landscape feature corresponding to the names.

### 10.1.3 Development and Principles for Updated Classification of the Natural Zones

As mentioned before, some variations of the classification of natural zones were introduced and maps were produced. However, the boundaries of the natural zones and belts were not clear, names were inappropriate, and it was critical that where forest steppe is either natural zone or the belt. Moreover, the size of the coverage areas was inconsistent. Those factors were the reason to produce new classification of the natural zones. Especially, the relief is one of the main factors that influence a spatial origin of the different landscapes, through climate impact on the latitudinal distribution. Thus, the complicated issue is that the northern part of the country belongs to which natural zone.

The natural pattern in the Mongolian territory has been characterized by the spatial distribution of the landscapes and its features. Therefore, the landscape regionalization principle should be the classification of the natural zones. There are several maps of natural zones compiled based on the landscape principles, but boundaries of the zones were not much clear, the names of the zones are not corresponding with these features and different views are related to the forest steppe.

The newly developed classification of the natural zones and belts in Mongolian territory was created by Dash et al. in 2002, based on updated researches since 1990. The justification of the newly updated classification of the natural zone of Mongolia has reduced the disadvantages of the previous one. For instance, there was a high generalization of the natural zone map and the boundary of zones, the titles of the zones were inappropriate in certain areas, as well as the boundary between some natural zones needed to change on the research.

The followings are the frameworks for the development of the updated classification of the natural zones in Mongolia and its mapping:

- One of the main influencing factors on the spatial variations of the landscape is the landforms, which also influenced the latitudinal distribution of climatic parameters. This is one of the pushing factors for updating the classification of natural zones in Mongolia. In particular, due to the complex topography of the northern part of Mongolia, classification of natural zones and belts in mountainous areas is still under discussion.
- The structure of the altitudinal belts of the mountainous landscape of Mongolia depends much on the location (geographical latitude, direction), the absolute height of mountains and orientations of slopes.
- Therefore, the number of altitudinal belts in high mountains and its components are different in different mountain ranges.
- In the depression's influence is many clear intermountain valleys and its intensity of influence very much depends on size, form and deep of the depressions. The central and southeastern plains are characterized by the latitudinal variations of the natural zones and become a bio-climatic factor for the landscape components in Mongolia (Dash 2015).

The general patterns of natural zones and belts in Mongolian territory were determined by these above-mentioned landscape features. Therefore, in the updated classification of the natural zones in Mongolia, the altitudinal and depression's belts were considered along with the latitudinal zones and subzones. On the other hand, it is landscape-ecological, holistic and systematic approaches of the classification issues of the natural zones and belts (Table 10.3.).

The map of the natural zones and belts presents a general view of nature. However, in terms of classification and mapping, it cannot be too generalized, because of the needs of mapping that generally characterized in every detail. The updated map of natural zones and belts was compiled based on the previous maps: "The landscapes of Mongolia" (1:3,000,000), "The map of Ecosystems of Mongolia" (1:1,000,000). The first version of the Map of Natural Zones in Mongolia had created at a scale of 1:1,000,000, and the detailed calculation of the coverage area and percentage of each zone and belt had been based on the digitized map by GIS (Table 10.4).

The updated map of the natural zone of Mongolia by Dash (1999) contains more detailed and the parallel of latitudinal and altitudinal subzones and depression belts; it also has a less generalization (Fig. 10.2). According to the principles of this map, the landscape type with relatively common distribution in high

**Table 10.3** The classification of natural zones and belts (Dash 2016)

| | Latitudinal zones and subzones | | Altitudinal belts | The depression belts |
|---|---|---|---|---|
| I | | | High mountain belt | |
| II | | | Mountain taiga belt | |
| III | Forested steppe zone | 1.Mountain forest steppe | Mountain forest steppe ($III_1'$) | Steppe with meadow-($IV_1''$) Steppe ($IV_2''$) |
| IV | | 1. Meadow steppe | | |
| | Steppe zone | 2. Steppe | Mountain steppe ($IV_2'$) | Semi desert ($V_2''$) |
| | | 3. Dry steppe | | |
| V | Semi-desert (Gobi) zone | 1. Desert steppe | Dry Steppe ($IV_3'$) | Steppe-desert ($VI_1''$) |
| | | 2. Semi-desert | | True desert ($VI_2''$) |
| | | 1. Steppe desert | Desert type: steppe ($V_1'$) | Extreme arid desert ($VI_3''$) |
| VI | Desert zone | 2. True desert | Semi-desert ($V_2'$) | |
| | | 3. Extreme arid desert | | |

**Table 10.4** The natural zones and belts (area and percentage)

| | Natural zones | Area ($km^2$) | Percentage |
|---|---|---|---|
| 1. | High mountain belt | 56,394.0 | 3.6 |
| 2. | Mountain taiga belt | 70,492.5 | 4.5 |
| 3. | Forested steppe (Khangai) zone | 238,108.0 | 15.2 |
| 4. | Steppe zone | 535,743.0 | 34.2 |
| 5. | Semi-desert (Gobi) zone | 366,561.0 | 23.4 |
| 6. | Desert zone | 299,201.5 | 19.1 |
| | Total | **1,564,500.0** | **100.0** |

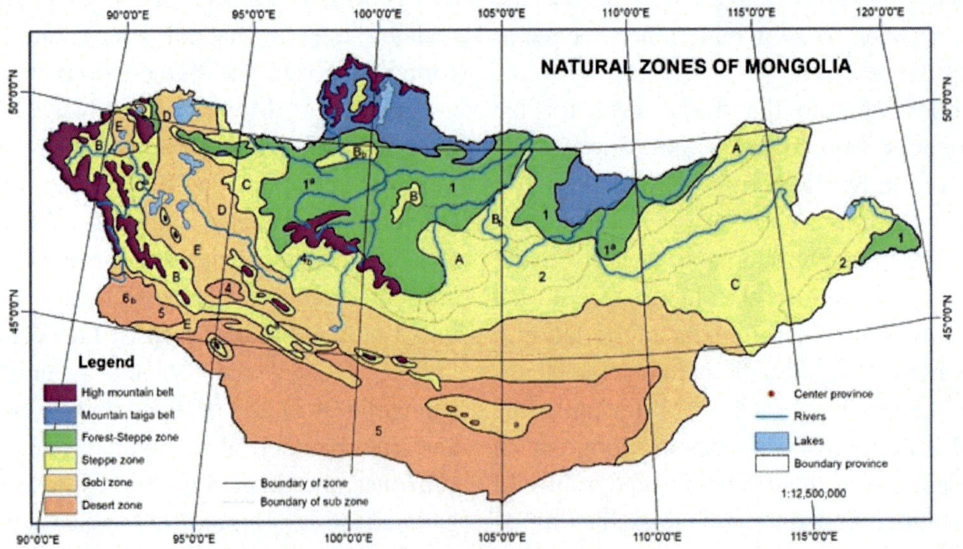


**Fig. 10.2** Natural zones and belts in Mongolia (Dash 2002)

mountains is named a "*belt*" which cannot establish a zone in the territory of Mongolia. The belt is divided into two types in Mongolia: High mountain belt and Mountain taiga belt. As a distribution, areas of these belts belong to the forest steppe zone (Dash 2015).

In the map of "Natural zone of Mongolia", which is a commonly used, the coverage area of the taiga belt does not correspond to the landscape type definition. D. Dorjgotov used the names of "high mountain meadow" and "tundra", while "high mountain belt" is shown in the map of the vegetation of Mongolia, which is more suitable for the landscape type of this area (Dorjgotov 2006). Because of the parts, it was named as "*alpine belts*" (by Jambaajamts 1989) in the high mountains such as Altai, Khuvsgul and Khangai, have a specific feature of combined landscape types: mountain meadow, snow cover, nival zone, and naked peaks of the mountains. Therefore, the name of "alpine belt" cannot represent the whole area of the top of high mountains (Dash 2002).

## 10.2 Main Characteristics of the Natural Zones and Belts

The general features of the various natural zones on the Mongolian territory are determined by the different type of landscapes and its spatial distribution caused by the latitudinal variations, incoming solar radiation, and climatic parameters patterns (Tables 10.5 and 10.6; Fig. 10.3). Therefore, the natural zone classification should be based on the landscape's features. The territorial and landscape differentiation in Mongolia is characterized by the several types of natural zones and belts: Four latitudinal zones (forest steppe, steppe, Gobi and desert zone) and two altitudinal belts in the mountains (high mountain and mountain taiga belt). Furthermore, the zones are divided into subzones, while the belt is divided into latitudinal and depression belts. The steppe zone is divided into three subzones (steppe with meadow, true steppe and arid steppe), Gobi is divided into two subzones (desert steppe and semi-desert), while desert zone is divided also into three subzones (steppe-desert, true desert and extreme arid desert). All these subzones are characterized by exceptional components of nature and terrestrial patterns.

### 10.2.1 High Mountain Belt

The high mountain belt occurs above the tree line in Khuvsgul, Khentii, Khangai, and the Mongol Altai and Gobi-Altai mountain ranges and as of area 56,394.0 km$^2$ or 3.6% of the total land area of Mongolia (0.8% in Khuvsgul mountains, 0.9% in Khangai and 1.9% in Altai mountain) (Dash 2015). According to Ynatov (1950) and Ulzi-ikhutag (1989) the high mountain belt is divided into four sub-belts: summits, lower summits, alpine and sub-alpine. The summits consist of perpetual snow-capped peaks, barren rocks and gravel.

The high mountain in the Khangai Mountains is represented mostly by mountain meadows, while in Khuvsgul it is represented by tundra and by glaciers in the Altai. The altitude of the lower limit of the high mountain belt differs for each mountain range depending on its geographical location. For example, the lower limits of the high mountain belt occur in the Khuvsgul mountains and in the Central Khentii at 2000–2200 m a.s.l., in the eastern Khentii at 2000–2050 m, in the western Khentii at ~2200 m, in the western Khangai at 2350–2400 m, in the northern Mongol Altai at 2300–2400 m, in the central Khangai at 2500–2560 m, in the southern

**Table 10.5** Monthly average temperature in each natural zones and belts (MEGD 2014)

| Monthly mean temperature, °C | High mountain belt | Mountain taiga belt | Forest steppe zone | Steppe zone | Gobi zone | Desert zone |
|---|---|---|---|---|---|---|
| Maximum | 14.5 | 14.3 | −17.6 | 15.9 | 20.5 | 23.8 |
| Minimum | −21.1 | −27.8 | −22.2 | −21.2 | −20.1 | −14.9 |

**Table 10.6** Dates and annual sum of daily mean temperature exceed 10 °C (based on (Gomboluudev et al. 2018)

| Natural zones and belts | Annual sum > 10 °C | Periods (from spring to autumn) |
|---|---|---|
| High mountain belt | 300–310 | 1 June–1 Sep |
| Mountain forest belt | 280–300 | 10 June–25 Aug |
| Forest steppe zone | 600–610 | 20 May–15 Sep |
| Steppe zone | 690–700 | 20 May–20 Sep |
| Gobi zone | 1000–1005 | 10 May–25 Sep |
| Desert zone | 1500–1510 | 25 Apr–10 Oct |

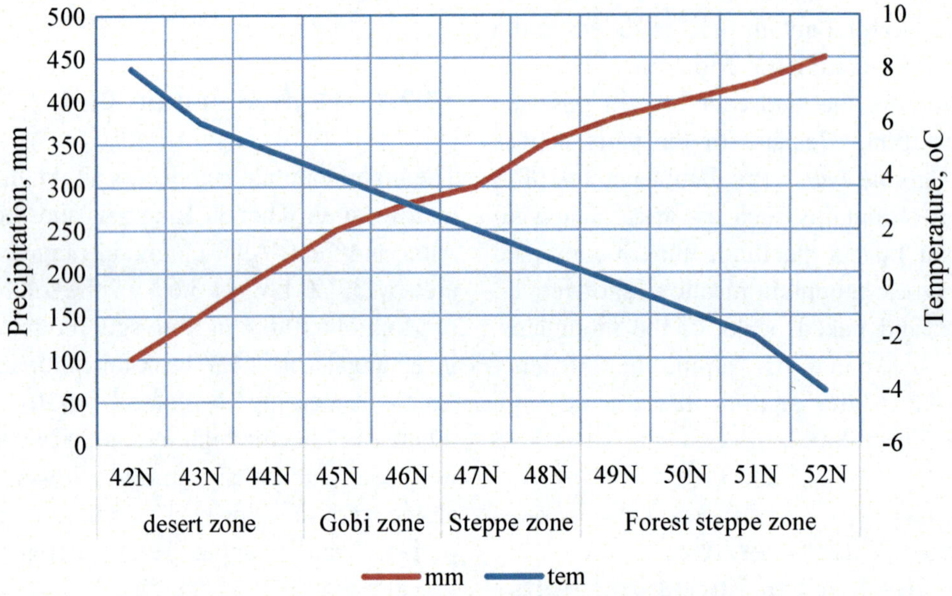


**Fig. 10.3** Annual mean temperature and precipitation variations by the natural zones

Mongol Altai at 2700 m, in the southern Khangai at 2750–2800 m and in the Gobi-Altai mountain range at 2700–2900 m (Ulziikhutag 1989). The internal structure of the high mountain belt is very complicated, with a mountain meadow in Khangai mountains, mountain tundra in Khuvsgul mountains, and glaciers with the naked peak in Altai mountains (Dash 2015).

The annual precipitation in the high mountain belt exceeds 350–400 mm. The growing season is short and occurs for 60–80 days due to low temperature. Mountain peat meadow soils occur in flat and low parts of the mountain top areas where soil surface water accumulated. The upper layer of the soil has a thin peat-accumulated layer, which is dark, abundant in roots and compact. One of the predators in the high-mountain belt is the snow leopard (*Panthera uncia*). Snow leopards in Mongolia are distributed in the Altai mountain ranges. The main threats to the survival of snow leopards include habitat degradation of illegal hunting (Takehiko et al. 2018). The high mountain belt is characterized by unique vegetation, such as high mountain steppe with short herbaceous plants, thickets of shrubs, and lichens and mosses (Tuvshintogtokh 2014).

### 10.2.2 Mountain Taiga Belt

The mountain taiga belt spreads over southern Baikal's large mountain taiga forest regions, including the Khentii, Khuvsgul mountains, the Zed-Buteel ranges in the eastern part of Khuvsgul and in the southern Khangai occupies only the northern slopes. The lower limit of this belt lies at 1700–1800 m a.s.l. in these mountains. On the other hand, it spreads in areas where the mountains become lower and latitudinal zonation factors prevail. The main feature of the mountain taiga belt is that forests cover the slopes of mid-elevated mountains and the middle parts of high mountains. The mountain taiga belt comprises 4.5% of the country, 3.0% belongs to Khuvsgul region and 1.5% to the Khentii mountain range. Especially, it is well developed in the central and northern part of Khentii mountain, caused by the location of southerly latitude compared to the Khuvsgul mountains. Therefore, the northern slopes of the Khentii mountains are covered by pine-birch (*Pinus sylvestris and Betula platyphylla*) and larch-birch (*Larix sibirica* and *Betula platyphylla*) (Undarmaa et al. 2018). Due to the arid climate conditions, the mountain taiga belt do not occur in the Mongol Altai mountain range (Yunatov 1948; Ulziikhutag 1989; Dash 2002). Scattered mountain taiga is also distributed on the northern slopes of the Tarvagatai and Khan Khukhii mountains. Though it has a humid climate with yearly average precipitation of 400–500 mm, the vegetation founded there have a short growing season, for around 90 to 110 days due to low temperatures (Table 10.4). Taiga is boreal coniferous forest, comprising primarily of Siberian larch *(Larix sibirica),* (70%) and Siberian pine (*Pinus sibirica),* rich in mosses and lichen (Fig. 10.4).

### 10.2.3 Forested Steppe (Khangai) Zone

Forested steppe (Khangai) zone occupies 15.2% or 238,108.0 km$^2$ area (Dash 2002). This zone is characterized by a unique combination of forests in the northern slopes and mountain steppes in the southern slopes. According to some research ("mountain forest steppe belt" by Ulziikhutag 1989; Gresimov and Lavrenko 1952), forested-steppe is still on the discussion which is a latitudinal zone or as an altitudinal belt. However, an author of this chapter recommended to use the name of "forest and mountain steppe belt" instead of forested-steppe zone, in the article "the forest-steppe landscape and its features". However, the general term "forested-steppe" is still used in many researches. Otherwise, the areas

**Fig. 10.4** Coniferous forest in the Darkhad depression (own photo)

**Fig. 10.5** Khangai mountain (own photo)

that belong to the forest steppe are the latitudinal character. Moreover, according to the most common approach of the landscape, the region of the forest steppe zone is a combined feature of the forest and mountain steppe (Fig. 10.5). Thus, the combination is characterized by latitudinal zone and altitudinal. However, the landscape features of the forested steppe are more dominated as latitudinal characters and altitudinal as well (Dash 2002). In high elevated areas (above 2000 m), the forest cover is located mostly on the northern slopes of mountains, while the steppe vegetation dominates on the other slopes. The lower limit of the mountain forest steppe lies at 850–950 m a.s.l. in the northeastern Khentii, 1650 m in the Khuvsgul mountain region, 1300–1400 m in the southwestern Khentii, 1000–1200 m in the northern Khangai, 1400–1500 m in the southeastern Khangai, 1950–2000 m in the central Khangai, and 1800–2000 m Mongol Altai (Ulziikhutag 1989).

With its comparatively cool and wet climate conditions, the landscape can be regarded as a mountain forest steppe zone. The boundary of the forested steppe zone corresponds with the isotherm- 2 °C of the mean annual temperature; +16 °C in July mean temperature or −24 °C in January mean temperature (Shagdar et al. 2006). Annual precipitation in the northern part of the forested steppe zone is appr. 300 mm and in the southern part is 200–300 mm. The growing season continues for around 135 to 150 days.

### 10.2.4 Steppe Zone

The steppe zone covers the eastern Mongolian steppe and the middle Khalkh plain steppe reaching out to the western tip of the southern slopes of the Khan-Khukhii mountains. There is steppe zonality in the southern parts of the Altai and Khangai mountains, as well as in Khasagt Khairkhan, Ikh Bogd, and Baga Bogd mountains. Steppe zone areas in depressions spread in the Orkhon-Tuul river valley, at the end of the Khanui river, the Delger river valley and the Shishigt watershed. The steppe zone in Mongolia is the largest in Eurasia (Tuvshintogtokh 2014). It occupies 32.4% of Mongolia and encompasses 4.3% of meadow steppe, 10.1% of the real steppe, 14.3% of dry steppe and 5.5% of mountain steppe subzones. Typical plants found in the steppe zone is several types of caaragana shrubs, dry bushes and undershrub. Steppe zone is divided into three sub-zones: *meadow steppe, steppe, and dry steppe*. The distinguished boundary of these sub-zones is very clear from the Dornod plain to the eastern part of Khangai mountains, while part of the Khan-Khukhii mountains is dominated by the dry steppe.

According to the altitudinal character of the regionalization, the steppe occurs in the southern part of the Khangai, Mongol Altai, Khasagt khairkhan, Ikh Bogd and Baga Bogd mountains, which are located in the southern part of Mongolia. As depression zones of the steppe, it also,

**Table 10.7** Climatic parameters in selected places in the steppe zone

| Stations | Elevation, m | Mean temperature, °C | | | Total annual precipitation, mm |
|---|---|---|---|---|---|
| | | Annual | January | July | |
| Mandalgobi (45° 46′N) | 1939 | 10 | −18 | 20 | 159.7 |
| Undurkhaan (47°51′N) | 1027 | −0.8 | −22 | 18 | 255.7 |
| Baruun-Urt (46°41′N) | 981 | 0.6 | −21 | 20 | 203.7 |
| Choibalsan (48°04′N) | 747 | 0.5 | −21 | 21 | 250.6 |

steppe occurs in the Orkhon and Tuul river basins, the low stream of the Khanui river, Delger river basin and Shishkhed river basin in the Darkhad depressions (Dash 2015). The majority of reliefs are hilly in the steppe zone. The majority of relief of the steppe zone occupies by plains and rolling plains of central and eastern Mongolia and becomes narrower to the west as it passes the foothills of low mountains of the Khangai in the south, to continue to the Great Lakes depression and ending in the south of the Khan Khukhii mountains. The annual precipitation is 200–250 mm and the growing season occurs for 150–170 days (Table 10.7).

### 10.2.5 Semi-desert (Gobi) Zone

The Gobi zone, known in Western countries as "semi-desert", is a unique zone located at the intersection between the steppe and desert zones in southern and southwestern Mongolia, and in the northern parts of the Central Asian desert. The Gobi zone is an area of 366,561.0 $km^2$, representing 23.4% of Mongolia and consisting of two main features of landscape or subzones; desert steppe (9.1%) and semi-desert (14.3%). It can be regarded as a separate zone and as an intermediate transition zone from steppe to desert zone. This zone extends to the Great Lakes depression, Valley of Lakes, Gobi-Altai mountains and low elevations of the eastern Gobi. The northern edge of the Gobi zone, which the Mongolian nomads determine by the marmot distribution, coincides with our version of the zone's northern boundary. Annual precipitation is 100–125 mm and less; the growing season occurs for 170–190 days.

By the broadest definition, Gobi is a long stretch of desert in Asia. Traditionally, "Gobi" is not only a place name for Mongolians, but they also consider "Gobi" as sandy depressions with a scarce vegetation cover and saline soils. In terms of this point, there are 33 different Gobi in Mongolia. However, this is not fixed; it depends on the criteria defined by different researchers. There are more than 370 Gobi according to the sand researcher Baasan, based on the features of physical geography and sand surfaces (Baasan 1988). These places include both natural zones of Gobi or semi-desert zone and desert zone (Fig. 10.6). Mainly, the intermountain depressions are named as the "Gobi" such as Galba Gobi, Sharga Gobi, Zakhui Gobi and Zarman Gobi that related to the desert type of landscapes. This zone has dry steppe plants, including *Artemisa frigida* and *Caragana*. At the same time, it also includes the plant species of transitional zones which are influenced by the Central Asian desert vegetation cover (Evstifeev and Rachkovskaya 1985).

### 10.2.6 Desert Zone

The desert steppe zone occupies 19.1% of the total land area of Mongolia or 299,201.5 $km^2$, consisting of three subzones: steppe desert (6.2%), true desert (9.2%) and extra-arid desert (3.7%), which covers appr. from 44°30′N to the country border in the south. Desert zone occupies the south of Mongol Altai and Gobi-Altai mountain ranges. The landforms of the desert zone are flat, plains and pediments of the mountains (Fig. 10.7). Most of the parts of the Gobi-Altai mountain ranges and Tyan-Shan include the

**Fig. 10.6** Khongor sand dunes (own photo)

semi-desert and steppe zones. According to the zonation of depression, the Baruun Khuurain Khotgor (Western Dry-Depression) is a unique place, which includes the desert zone. The eastern edge of the desert zone reaches the Gurvan Bayan Khooloi, and furthermore, it continues to the Khyangan mountain, which is affected by the difference in geographical longitude (Stolz et al. 2012). The isolated mountains are very common in the Trans-Altai Gobi, due to physical weathering in arid climate condition. The Aj Bogd, one of these mountains, is characterized by the altitudinal belts. The different types of vegetation depend on the elevation and the slopes (Fig. 10.8. Altitudinal belts in the Aj Bod mountain (Vostokova 1986; Gunin 2019).

Few plants are found in this zone, due to dry land and very harsh climate, 40 °C in summer to −40 °C in winter. Desert zone belongs to the dry warm climate zone, where precipitation is less than 100 mm (Table 10.8), strong wind frequency is 50–60%, and drought frequency is 70–80%. Desert gray-brown type of soils is widely distributed in the desert zone. Some extra-arid areas of this zone occurred borzon and takyr type of soils. Gobi light-brown soils are formed under the condition with low soil moisture, where evaporation is more than precipitation (Vostokova and Gunin 2005). Because of low moisture, carbonate and salts easily accumulated on the soil surface. Mongolian desert is the homeland of the dinosaur's tracks and footprint sites.

### 10.2.7 Climate Change and Anthropogenic Impacts

Because of the dry climate conditions, drought occurs throughout the country; its frequency increases from north to south and is consistent with the pattern of moisture distribution. According to the results of the drought calculation based on the satellite data and the NDVI, it has increased significantly in the Gobi Desert region, especially in the Great Lakes depression and Valley of Lakes. Recent study shows that the drought is likely to occur in the high mountainous belt—forest steppe and steppe zones: 1–2 times during 10 years; in the desert steppe zone: once every two years; and in the middle part between the steppe and desert steppe zones: once every three years (MEGD 2014).

Almost the entire Mongolian steppe zone has experienced a significant reduction in biomass of vegetation. About 60% of this decrease can be related to climate-related trends: in particular, a decreasing amount of precipitation and an

**Fig. 10.7** Khermen Tsav, Gobi Desert

increasing temperature. While water is a main limiting factor for vegetation growth in southern part, energy is a major constraint for the north of Mongolia (Chu and Guo 2012). In addition, the sharp increase in the number of goats and the degradation of pastures is the reason for most areas continuing to experience a decline (MET 2017).

According to the latest research on main climate parameters, temperature and precipitation change for various different natural zones. For instance, in taiga zones NPP > 296 C h/m$^2$ (Net Primary Productivity) does not change temperature when precipitation changes (MEGD 2014; Nyamtseren et al. 2018). The research using the HadCM3 model shows that the forest steppe is likely to turn into a steppe and steppe zone, which is likely to be pushed by the semi-desert zone by 2080 (MEGD 2014; MET 2017). In

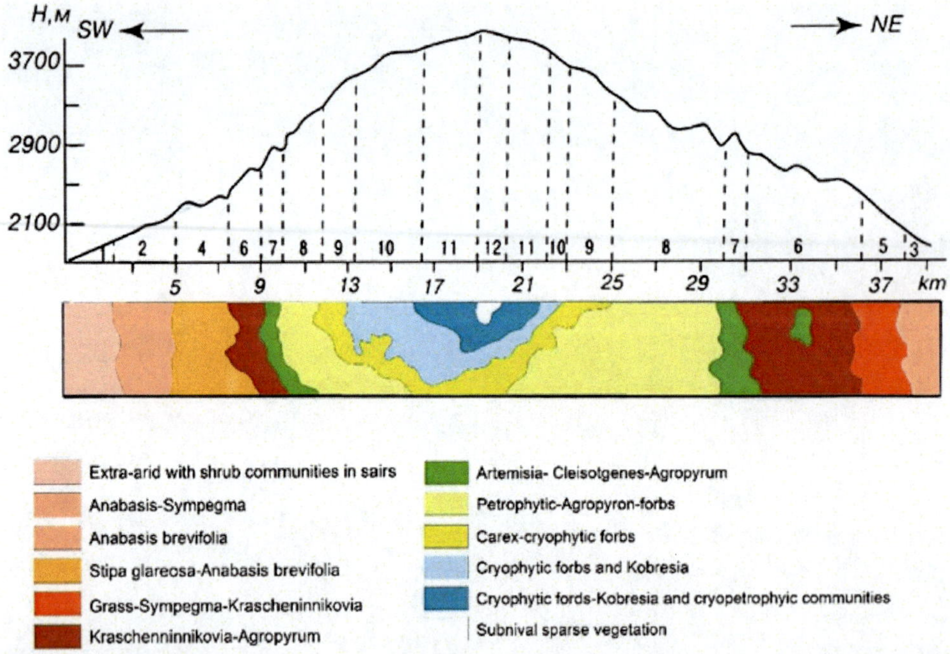


**Fig. 10.8** Altitudinal belts in the Aj Bod mountain (Vostokova 1986; Gunin 2019)

**Table 10.8** Climatic parameters in selected places in the semi-desert and desert zones

| Stations | Elevation, m | Mean temperature, °C | | | Total annual precipitation, mm |
|---|---|---|---|---|---|
| | | Annual | January | July | |
| Dalanzadgad (43°34′N) | 1465 | 4 | −16 | 20 | 129.5 |
| Ulaangom (49°59′N) | 939 | −3.7 | −32 | 16 | 138.8 |
| Sainshand | 938 | 3.5 | −18.4 | 24 | 115.9 |
| Ekhiin gol (43°15′N) | 1000 | 8.3 | −11.8 | 26.2 | 50 |

addition, the results of studies have highlighted significant differences in the plant phenological timing and length for each natural zone. According to the seasonal change in soil moisture (Nandintsetseg and Shinoda 2011), in the northernmost forest steppe zone, the summer recharge was longer than in the southern steppe and the desert zone, while in the forest steppe zone, autumn and spring drying phases were shorter. On analysis of long-term remote sensing data, in the mountainous area (about 31% of the country) NOAA/NDVI values have increased and in rest of the parts the NDVI values have decreased (Erdenetuya et al. 2016).

## References


Baasan T (1988) What is the Gobi and how many Gobi in Mongolia? Sci Life 4:15–25 (in Mongolian)

Bazargur D (1972) Map of the natural zone and the animals of Mongolia (1:2500 000). Map. The Institute of Geodesy and Cartography, Ulaanbaatar (in Mongolian)

Chu T, Guo X (2012) Characterizing vegetation response to climatic variations in Hovsgol, Mongolia using remotely sensed time series data. Earth Sci Res 1:279–290

Dash D (1999) Issues of the new physiographic regionilization of Mongolia. In: Proceeding of the international conference on Mongolian Plateau. Institute of Geography, Mongolian Academy of Sciences, Ulanbaatar, pp 25–33 (in Mongolian)

Dash D (2002) Natural zones and Regions of Mongolia. Extended Abstracts. In: International conference "Geographical study of central Asia and Mongolia. Institute of Geography, Mongolian Academy of Sciences, Ulaanbaatar, pp 28–33

Dash D (2015) The landscape and ecological issues of Mongolia. Munkhiin Useg LLT, Ulaanbaatar (in Mongolian)

Dash D (2016) The issues of the natural zones and belts of Mongolia. Geoforum Mongolia (1):4–12 (in Mongolian)

Dorjgotov D (2006) The map of natural zones and belts in Mongolia. The National Agency of Geodesy and Cartography, Ulaanbaatar

Erdenetuya M et al (2016) Mongolian rangeland changes with a changing climate. In: 10th International rangeland congress, p 924

Evstifeev YG, Rachkovskaya EI (1985) Main zonal types of Gobi desert. Mongolian Geogr Iss (21) (in Russian)

Gomboluudev P et al (2018) Climate of Mongolia. In: Undarmaa J (ed) Rangeland ecosystems of Mongolia. Munkiin Useg Co. Ltd, Ulaanbaatar, pp 76–100

Gunin PB (2019) Ecosystems Atlas of Mongolia, Gunin P, ed. Mongolian Academy of Science, RAS, MNUE. Admon Publishing, Ulaanbaatar

Gunin PD et al (1995) Ecosystems of Mongolia. Moscow (in Russian)

MEGD (2014) Mongolia second assessment report on climate change-2014, Dagvadorj D, Batjargal Z, Natsagdorj L, eds. Ministry of Environment and Green Development, Ulaanbaatar

MET (2017) Mongolia's initial biennial update report under united nations framework convention on climate change, Batjargal Z, ed. Ministry of Environment and Tourism, Ulaanbaatar. https://unfccc.int/files/national_reports/non-annex_

Murzaev EM (1948) The classification of the physiographic regions of Mongolian Peoples of Republic. Iss Phys Geogr 13 (in Russian)

Nandintsetseg B, Shinoda M (2011) Seasonal change of soil moisture in Mongolia: its climatology and modelling. Int J Climatol 31(8):1143–1152

Nyamtseren M et al (2018) A comparative study of temperature and precipitation-based aridity indices and their trends in Mongolia. Int J Environ Res 12 (6):887–899

Shagdar Sh, Batchuluun Ye (2006) The physical geography of Mongolia. Munkhiin Useg Publishing, Ulaanbaatar

Stolz C, Hülle D, Hilgers A, Grunert J, Lehmkuhl F, Dasch D (2012) Reconstructing fluvial, lacustrine and aeolian process dynamics in Western Mongolia. Zeitschrift für Geomorphologie 56(3):267–300

Takehiko Y, Ito et al (2018) Chapter 4. Wildlife of Mongolia. In: Undarmaa J (eds) Rangeland ecosystems of Mongolia. Munkhiin Useg Publishing, Ulaanbaatar, pp 142–205

Tsegmid S (1969) Physical geography of Mongolia. Institute of Geography and Permafrost. Mongolian Academy of Sciences. State Publishing House, Ulaanbaatar, 405 p (In Mongolian)

Tuvshintogtokh I (2014) The Steppe vegetation in Mongolia. Bembi San Printing, Ulaanbaatar (in Mongolian)

Ulziikhutag N (1989) The overview of the vegetation of Mongolia. State Publishing House, Ulaanbaatar (in Mongolian)

Undarmaa J et al (2018) Chapter 3. Vegetation of Mongolia. In: Undarmaa JE (ed) Rangeland Ecosystems of Mongolia. Munkhiin Useg Co Ltd, Ulaanbaatar, pp 113–115

Vostokova EA, Gunin PD (eds) (2005) Ecosystems of Mongolia Atlas. Joint Russian-Mongolian Complex Biological Expedition, Moscow

Vostokova EA (1986) Alpine vegetation of the mountain ridges of Southern Mongolia: high mountains vegetation. Nauka, Leningrad (in Russian)

Yunatov AA (1948) On zonal-belt dismemberment of the vegetation cover of the Mongolian People's Republic. Izvestiya Vses Geogr Soc 80(4)

# 11 Land Use and Nature Conservation in Mongolia

Bat-Erdene Tsedev

**Abstract**

This chapter provides information about Mongolian land use and some conservation. The first part of this chapter relates to land use in Mongolia. Thus, it can be seen that they focused on issues related to the integrated route and stationary survey of different land use types in Mongolia. Mongolia has a wide and vast territory and a few population, therefore land use is specific. Nomadic livestock is primarily economic sector, so the types of land use occupies the center of focus. Nowadays, the agricultural land is diminishing, at the same time, urban lands are expanding when the socialist period of agricultural land use intensified. Further specific land use is growing by the policy of nature conservation. Pasture lands are decreasing at a faster rate due to economic needs for urban expansion, mines, agricultural land and necessary infrastructure. The tradition of nature conservation, protected areas, as well as the protected areas covered by the International convention heritage for nature conservation in Mongolia, have been examined in the following chapter. Nature conservation has been a much challenging task in the present state of affairs where the economy prevails over ecology. Historically, the tradition of the nature conservation of feelings and beliefs has been used as a powerful tool for nature conservation. As the data sources, data of the Ministry of Nature, Environment and Tourism, the National Statistics Office of Mongolia, Ministry of Food, Agriculture and Light Industry and other study are used in this chapter. Also, land use in Mongolia map was available.

**Keywords**

Land use · Nature conservation · Protected area · Ramsar convention

## 11.1 Land Use and Resources

Land is one of the essential and non-renewable resources as well as it is a resource of people's livelihood and social existence, means of production and classic form of property. Thus, land is considered quite an important thing that could be a kind of ownership and resource of livelihood, property and business. Land affairs have specific features in every nation, which have been developed by different historical stages. From the revolution of national freedom until the present, different types of land use have emerged as a result of several changes of politics, society and economics. Consequently, coordinating a range of

B.-E. Tsedev (✉)
Department of Geography, Mongolian National University of Education, Ulaanbaatar, Mongolia
e-mail: baterdene@msue.edu.mn

© Springer Nature Switzerland AG 2021
B. Yembuu (ed.), *The Physical Geography of Mongolia*, Geography of the Physical Environment,
https://doi.org/10.1007/978-3-030-61434-8_11

land affairs was expanded and refined, and a legal document which coordinated those in due course was approved. In addition, this implementation has been organized. By 2002, it was being rearranged and approved in accordance with the situation and requirements of society and economics. Land use information for Mongolia has been collected from various sources (e.g. state statistical book, local and regional maps, satellite imagery) and has, subsequently, been evaluated in order to analyze land use changes.

Over the last 70 years, there is an overall trend of land use intensification in Mongolia, although this varies regionally and is set against a slow decline in the proportion of Mongolia's land area used for pasture land. Mongolia considered that we need land of special usable and included lands are as follows: State specially protected land; land of boundary line; land designed to defend and secure safety of the state; land of diplomatic agent of foreign country and consulate, land given for agency of international organization, regular observing field of weather condition, experiment and nature environment, scientific and technological trial; inter-aimags pasture for fattening livestock; hay land of state fodder storage; petroleum contract land which utilized to explore under the allocation of product and land of free zone.

The Law of Mongolia on Land describes, regardless of the form of ownership, that all lands within the borders of Mongolia constitute a unified land territory. The unified land territory shall be classified based on the general purpose of its use and the need for its use (The Great Khural of Mongolia 2002). According to the Law of Mongolia on Land, the objectives of land use classification can be categorized into the following.

Basic Classification of the Unified Land Territory (The Great Khural of Mongolia 2002): Agricultural land; Land of cities, villages; Land under roads and networks; Land with forest resources; Land with water resources; Land for special needs. Mongolian territory covers 0.31% of earth surface and 1.16% of land resource, and area of each land use of Mongolia (Fig. 11.1).

Agriculture land including pasture, hay land, arable land, fallow land, and construction ground of farm and other manufacturing purpose of agriculture comprises 115232.6 thousand hectares (ha) −73.6% of all land use of Mongolia. In

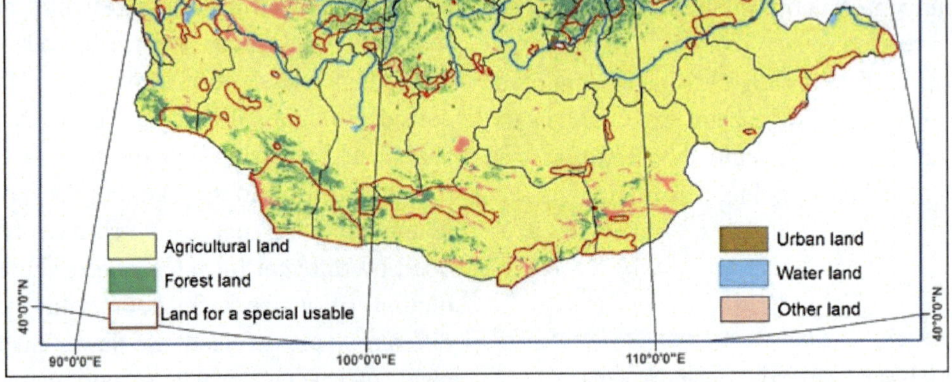


**Fig. 11.1** Land use in Mongolia, (updated Bat-Erdene, Ts)

**Table 11.1** Land use structure in Mongolia, (NSO 2017)

| Form of land use | Agriculture land | Cities, village, and other settlements land; | Transportation and network land | Forest resource land | Water resource land | Special usable land | Total |
|---|---|---|---|---|---|---|---|
| Size of area (thousand/ha) | 115232.6 | 466.0 | 355.4 | 14748.1 | 967.6 | 24641.8 | 156411.6 |
| Rate (%) | 73.6 | 0.3 | 0.23 | 9.45 | 0.62 | 15.7 | 100 |

addition to this, pasture occupies more area than others as Mongolians lead agriculture (Table 11.1).

The total area of Mongolia is 15,641,000 ha, of which 73.6% of the territory of Mongolia is agricultural land, 16.1% is land for special needs, 9.2% is with forest resources, 0.5% is urban area, 0.4% is with water resources, and 0.3% is under roads and networks, as of 2016 (Fig. 11.2) (MNET 2017). Agricultural land includes plowed fields, perennial crops, fallow and grazing lands. The area of pasture lands has decreased considerably in recent decades.

The size and purpose of each land database have been changed due to society and the economic development of Mongolia (Table 11.2). The political and socio-economic changes that have occurred in Mongolia since 1990 have had a considerable impact on land use structure due to the increasing rate of land conversion.

### 11.1.1 Agricultural Land Use Activity

Agriculture land in Mongolia comprises pasture, hay, arable land, fallow land, construction ground of farm and other manufacturing purposes of agriculture. Agricultural land occupies about one-third (114982.8 thousand ha) of the area of Mongolia (73.5%). Of the total area of agricultural land, 110493.8 thousand ha is pasture (96.1%), 1742.4 thousand ha is hay land (1.5%), 1067.7 thousand ha is arable land (0.9%), 260.6 thousand ha is fallow land (0.2%), 110.6 thousand ha is land for agricultural buildings (0.1%), and 1259.7 thousand ha is land of not suitable for agricultural (1.1%) (MNET 2017).

**Pasture land use**: Pasture land means rural agricultural land covered with natural and cultivated vegetation for grazing of livestock and animals. Mongolia's pasture land is in the form

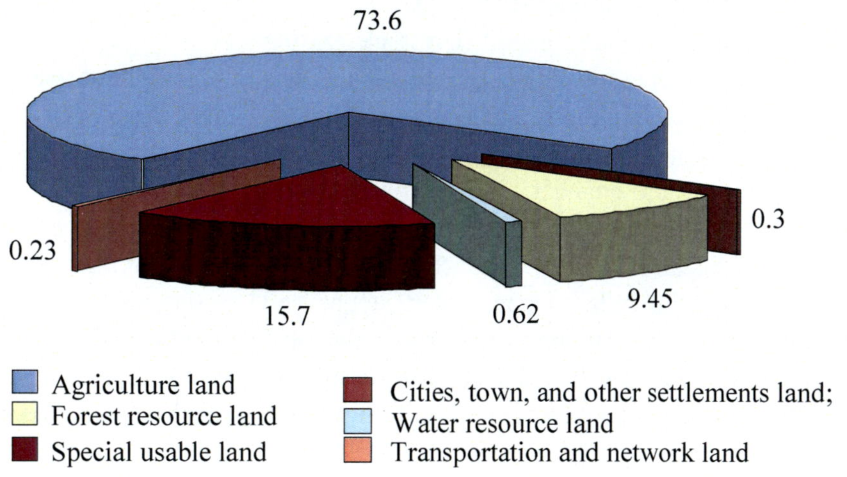


**Fig. 11.2** Land use structure in Mongolia (shown by percentages)

**Table 11.2** Area changes of an integrated land database (MNET 2017)

| Land use type | 2014 | 2016 |
|---|---|---|
| Agricultural land: | 115008.6 | 114931.1 |
| Arable land | 704.5 | 835.7 |
| Hayland | 1823.0 | 1823.3 |
| Pasture | 111679.8 | 11485.2 |
| Land of cities, villages and other settlements | 712.1 | 753.6 |
| Land under road and networks | 454.8 | 473.8 |
| Hayland with forest resources | 14320.5 | 14334.3 |
| land with water resources | 686.7 | 686.1 |
| land for special needs | 25228.9 | 25232.7 |
| Total area | 156411.6 | 156411.6 |

**Fig. 11.3** Arid regions pasture, Uu Durulj, Govi-Altai aimag, Western Mongolia, (own photo, 2017)

of a high mountain tundra steppe, mountain tundra and steppe fodder (Fig. 11.3).

Before the socialist period, livestock grazing involved seasonal movement of herds and flocks between upland (in the summer) and lowland (in the winter) pastures, the so-called transhumance. Frequent pasture changes allowed the regeneration of their productivity. These practices have proven sustainable over centuries. Pasture is combined with hills, steppe, desert steppe and Gobi. The infrastructure of urban, mine and agriculture land are transferring from pasture to different rate, the reason being to unfold pasture. Today's pasture land has decreased by 9.0 million ha compared to 1964 (MNET 2017).

At present, state special needs of 783.3 thousand hectares in nine inter-provincial aimags ha and 5.4 million hectares of grazing land in the inter-soum pasture reserve land (Table 11.3).

**Arable land use**: Arable land use: Prior to 1991, most arable land in Mongolia was held in 54 large state farms (Tim and Jennifer 2001). In 1991, the state determined what portion of the ownership of each farm it would retain, and the value of the rest of the farm's infrastructural assets was divided among workers in the form of vouchers. The market value of arable land currently appears to be very low (based in large part on the high availability of arable land) in areas that have previously been under cultivation. The Green Revolution Program in Mongolia, established in 1997 has provided up to 100,000 families with 40 small plots of arable land for

**Table 11.3** Agricultural land use, (ha) (MNET 2017)

| | Region | Pasture | Hayland | | Region | Pasture | Hayland |
|---|---|---|---|---|---|---|---|
| 1 | Western region | 28.76 | 56.2 | 4 | Central region | 34.9 | 193.53 |
| 2 | Khangai region | 25.17 | 132.4 | 5 | Ulaanbaatar | 0.22 | 5.5 |
| 3 | Eastern region | 21.39 | 1216.3 | | | | |

cultivating vegetables for sale and household sustenance (Tim and Jennifer 2001). According to the Land Law, participants who are granted possession rights to small garden plots (the average to be one ha per household, although those families were entitled to two hectares) are entitled to 25-year possession rights to this land (The Great Khural of Mongolia 2002).

A net decrease in arable land under productive use is from 787,000 hectares in 1991 to 323,000 hectares in 1999 (Tim and Jennifer 2001). About 200 thousand tons of grain has been harvested for feed in Mongolia in 2018, which may be attributable to several factors, including lack of operating capital, the virtual absence of a rural credit market, and inadequate processing and marketing infrastructure. Sustainable use of natural and social resources in Mongolian agriculture should bring essential economic benefits and save natural resources from exhaustion. On the other hand, ecologically sound agriculture should help to reconcile social needs with the requirements of the environment. Large areas of original arable land are located in the central regions (Selenge, Bulgan, Central, Khuvsgul and Arkhangai aimag) of Mongolia (Table 11.4).

### 11.1.2 Land of Cities, Villages and Other Settlements Land Use Activity

Land of cities, villages and other settlements land occupy about one-third (466.9 thousand ha) of the area of Mongolia (0.5%), of which 77.6 thousand ha is common tenure lands (10.3%), 43.5 thousand ha is factory land (5.8%), 236.1 thousand ha is mining land (31.3%), 71.7 thousand ha is ger (Mongolian traditional dwelling) area (9.5%) (MNET 2017). Built-up lands reach a maximum in the Ulaanbaatar region (Fig. 11.4). Other biggest urban land areas are located in the Darkhan, Erdenet, Choibalsan and other aimag centers.

The urban land use depute to change in land use of Ulaanbaatar. According to the survey is to compare the changes that occurred in the main urban land cover classes of Ulaanbaatar, the capital city of Mongolia during centralized economy (i.e. 1969–1990) with the changes occurred during market-based economy (i.e. 1990–2011) and describe the socio-economic reasons for the changes. For the analysis, multi-temporal high-resolution optical and microwave

**Table 11.4** Arable land use, (ha) (MOFA 2018)

| Region | Arable Land | Used area | Fallow land | Unused Land/% |
|---|---|---|---|---|
| Central | 821208 | 569145 | 252063 | 30.7 |
| Khangai | 139768 | 44098 | 95670 | 68.4 |
| West | 126393 | 22302 | 104091 | 82.4 |
| East | 181753 | 104811 | 76942 | 23.6 |
| Govi | 376 | 376 | 0 | 0 |
| Total/State | 1269498 | 740732 | 548766 | 51.9 |

**Fig. 11.4** The built-up area of Ulaanbaatar city (own photo, 2017)

remote sensing (RS) images, as well as geographical information system (GIS) and census data sets are used.

Researchers have selected Ulaanbaatar, the capital city of Mongolia, as a test site. Ulaanbaatar is situated in the central part of Mongolia, in the Tuul River valley, at an average height of 1350 m above the sea level. By 2018 statistics, the population of the city was counted at about 1,460,000 inhabitants (NSO 2017).

The study area chosen for the study covers an area of 28 km × 20 km. It covers the majority of the area belonging to the capital city, although there are some areas extending outside of the selected image frame. For the selected area, it is possible to define such classes as the built-up area, ger area (Mongolian traditional dwelling), forest, grass, soil and water. The built-up area includes buildings of different sizes, while ger area includes mainly gers surrounded by fences.

The data from the source Landsat TM of 20 September 2011 with a spatial resolution of 30 m were used (Fig. 11.5). In addition, a topographic map of 1969, scale 1:50,000, and a general urban planning map were available.

For several decades, single-source multispectral remote sensing data sets have been successfully used for land cover mapping, and for the generation of thematic information, diverse supervised and unsupervised classification methods have been applied (Fig. 11.6).

Initially, to create the primary historical GIS data, the classes: building area, ger area, forest and water were digitized in a UTM map projection from a topographic map of 1969, scale 1:50,000 using ArcGIS (Fig. 11.7).

The total areas related to each class defined from the classified multitemporal RS images as well as a digitized map are shown in Table 11.5. Although we have census data of Ulaanbaatar city, it is not possible to directly relate it to the current analysis, because our study area does not cover all the areas from where the final census data was collected. However, as the test area covers the majority of the area belonging to the capital city, it is possible to use the census data for a comparison of the general population increase with the actual urban expansion process.

In 1969, in Ulaanbaatar city, the building area and ger area covered 2498.92 and 978.86 ha, respectively, whereas in 1990 these two urban classes covered 3996.71 and 2719.65 ha, respectively. Moreover, it is seen that the forest and water resources had been reduced. The available census data indicated that in 1969 the population of the capital city was 267400, while in 1990 it had become 574900. As observed, within the 21-year period of the centralized

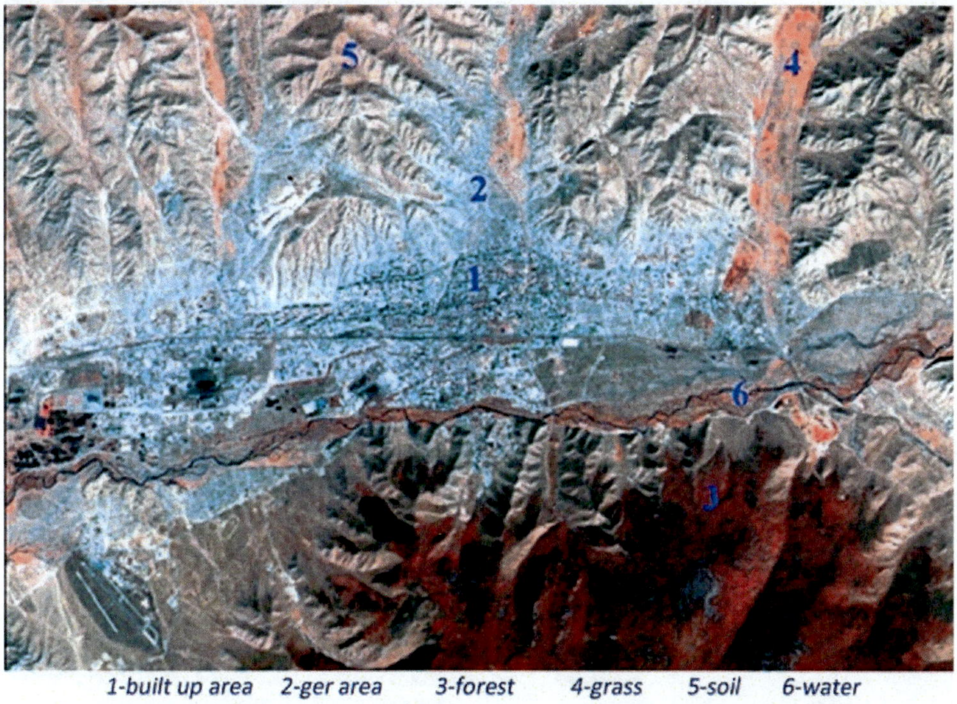


**Fig. 11.5** 2011 Landsat TM image of the selected part of Ulaanbaatar city (Bat-Erdene 2014)

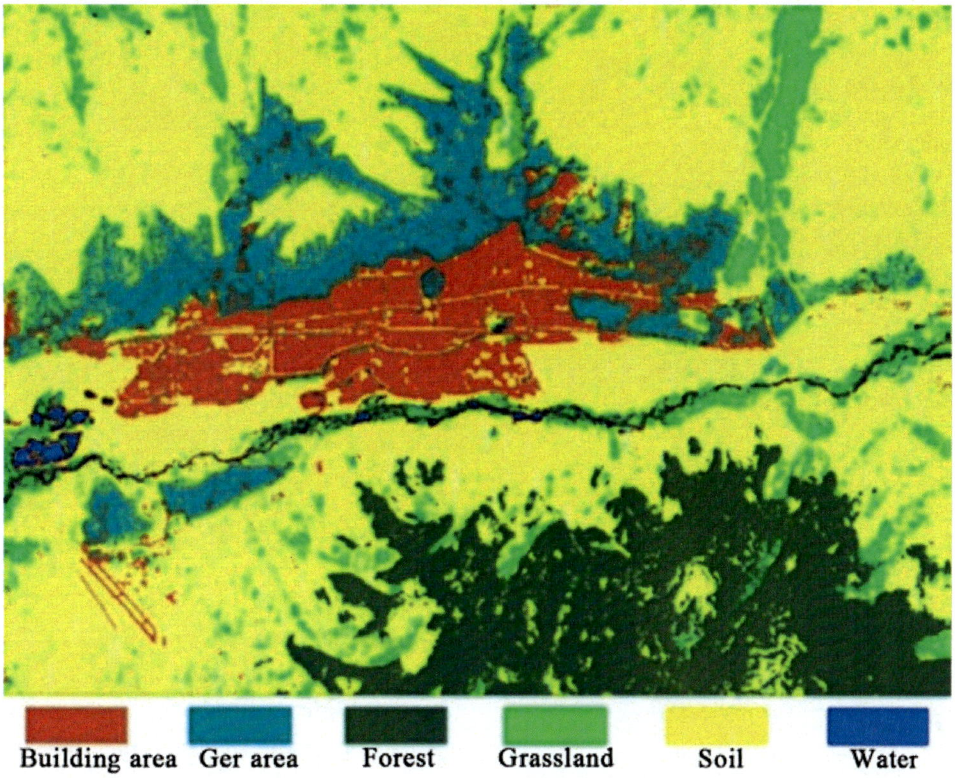


**Fig. 11.6** Classified image using 2011 data sets (Bat-Erdene 2014)

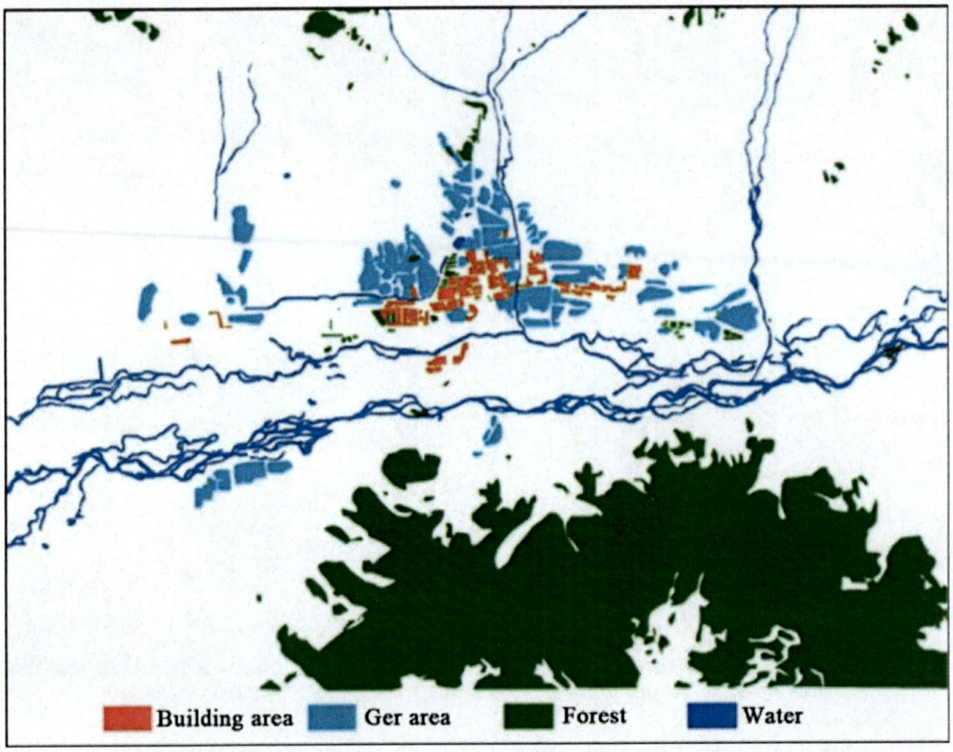


**Fig. 11.7** Digitized topographic map of Ulaanbaatar area of 1969 (Bat-Erdene 2014)

**Table 11.5** The total areas for each class in different years, evaluated from multitemporal GIS and RS data sets

| Land Use Type | 1969 (ha) | 2011 (ha) |
|---|---|---|
| Building Area | 2498.92 | 5236.09 |
| Ger Area | 978.86 | 8613.16 |
| Forest | 11896.12 | 8209.97 |
| Grass Land | Not available | 3104.65 |
| Soil | Not available | 30687.74 |
| Water | 1078.84 | 556.58 |
| Total | 56408.19 | 56408.19 |

economy, the building areas were increased by only 59.94%, whereas the ger areas were increased by more than two-fold (Bat-Erdene 2014).

### 11.1.3 Forest Resources Land Use Activity

Forests occupy about one-third (14.3 million ha) of the area of Mongolia (9.2%). Of total area of forest resources land, 12189.2 thousand ha is forest land (85%), 142.5 thousand ha is clear-cut forest area (1%), 57.9 thousand ha is tree nursery (0.4%), 743.6 thousand ha is forest land under regeneration (5.2%), and 1201.2 thousand ha is other areas of forest land (8.4%). Forest resource land includes forest, felling field, glade, tract and land dedicated to breed forest and provides a chance to grow. Large areas of original forests are located in the mountainous area in the north (Table 11.6). The dominant type of forests comprises conifers (71.4% is larch, 11.1% is birch, 9.5% is cedar, 6.3% is pine, 1.7% is other) and shrubs (MNET 2017).

**Table 11.6** Land with forest resources by region, aimag and the capital (thousand hectares, (MNET 2017)

| Region | Total | Forest land | Clear-cut area | Tree Nursery | Forest land under regeneration | Other areas of forest land |
|---|---|---|---|---|---|---|
| Western region | 2,312.0 | 2,084.6 | 15.9 | 13.6 | 33.5 | 164.4 |
| Khangai region | 7,490.3 | 6,103.8 | 57.3 | 0.5 | 538.7 | 790.1 |
| Central region | 3,153.1 | 2,858.7 | 25.8 | 40.4 | 59.7 | 168.5 |
| Eastern region | 1,304.8 | 1,077.4 | 43.4 | 1.3 | 111.6 | 71.1 |
| Ulaanbaatar | 74.1 | 64.7 | – | 2.1 | 0.2 | 7.2 |

### 11.1.4 Special Needs Land Use Activity

The total area of special needs land is 25229.0 thousand ha; 16% of the territory of Mongolia. Special needs land consists of 21140.8 thousand ha or 84.1% protected area (see heading 11.2), 3112.0 thousand ha is border strip land (13%), 124.1 thousand ha is lands provided for ensuring national defense and security (1%), 0.033 thousand ha is land specified for foreign diplomatic missions and consulates, as well as resident offices of international organizations (0.001%), 22.9 thousand ha is land for scientific and technological tests, experiments and sites for regular environmental and meteorological observation (0.009%), 691.4 thousand ha is inter-aimag reserve pasture rangeland (2.3%), 110.9 thousand ha is hay land (0.4%), 24.5 thousand ha is contracted oil exploration sites to be utilized in compliance with the production sharing agreements (0.1%), 2.1 thousand ha is a free trade zone (0.008%) (NSO 2017).

### 11.1.5 Other (Under Roads and Networks, Water Resources) Land Use Activity

The total area of under roads and network land occupies 473.8 thousand ha, its 0.3% of the territory of Mongolia. Of which roads land is 335.9 thousand ha (70.9%), for railway land is 32.5 thousand ha (6.8%), for air transport land is 9.5 thousand ha (2.0%), for networks land 95.9 thousand ha (20.1%) (MNET 2017).

The road network occupied 335.9 thousand hectares (70.9%), the railway station 32.5 thousand hectares (6.8%), the air transportation network accounts for 9.5 thousand hectares or 2.0%, and the landline network is 95.9 thousand (NSO 2017). In recent years, the square of roads and networks land has been increasing rapidly due to the growth of the infrastructure sector.

Water bodies cover 0.4% of the country's surface and include lakes, rivers, streams and artificial reservoirs. The land with water resources occupies 686.1 thousand hectares, or 0.4% of the total land area. The land with water resources including rivers and streams there are 228.5 thousand hectares or 33.3% in rivers, streams and lakes, 443.6 thousand hectares or 64.7% in lakes and ponds, 12.4 thousand hectares or 1.8% in streams and springs, 1,6 thousand hectares or 0.2% respectively (MNET 2017). The above land with water resources information is different from the results of regular surface water monitoring results. For example, the lake area is 15372 $km^2$ or 1% of the territory of Mongolia. Large areas of water bodies' cover are located in the central, north and west region of Mongolia.

### 11.1.6 The Nature Conservation

The Mongolian land's current situation is the probability of degradation has high risen why our country located in Central Asia and highland, semi-dry, have low forest because natural conditions are vulnerable, and climate change. So about 30% of the territory is protecting and implementing a policy on protection of state policy. Currently, 17.8% of Mongolia's area or 27,892,781 hectares are under protection (MNET 2017). Environmental issues are boundless, in other words, the most global issue on our planet. Therefore, it is best to resolve the environmental issues in the country and to work with other countries. Mongolia is a signatory to the following international conventions on conservation.

### 11.1.7 The Tradition of the Nature Conservation

The tradition of protecting nature, fauna and flora has a long history in Mongolia. Mongolia was probably one of the first countries in the world to realize the importance of conservation. There were closed seasons for hunting rabbits, deer, antelope, saiga and gazelles in the time of Marco Polo; later the laws of "*Khalkh Juram*" between 1709 and 1799 set aside 16 mountains that were to be protected from hunting, cultivation and timber felling (Samiya and Mühlenberg 2006). Bogd Khan Mountain has been protected since the twelfth or thirteenth century as a holy mountain. It was established as Mongolia's first official protected area in 1778.

In the thirteenth century, the Laws of Chinggis Khaan's "*Great Khural*" and the sixteenth century law enforcement letter "*Khalkh Juram*" were written on ancient Mongolian laws. "*Khalkh Juram*" prohibited hunting, land disturbance, and cutting of trees in 14 beautiful mountain ranges such as Bogd Khan, Khan Khentii, Khugnukhaan and Tuvkhunkhaan in 1778, Bogd Khan. In 1818, Otgontenger and Bulgan were officially protected by government, in 1911, the Bogdkhan mountain Protected Area Administration was first established by Mongolian Government policy (Samiya and Mühlenberg 2006).

The impact still remains and it is a valuable area in terms of biodiversity and recreation for the citizens of Ulaanbaatar. A further two sites were accorded protected area status in 1957 and another five in 1965; the Great Gobi strictly protected area was established in 1976. Since 1990, environmental protection has been given high priority by the government, and a total of 26 protected areas covering 12.6 million hectares, 8% of the country, have been established to date.

Mongolians have ancient traditions and customs to respect for nature greatly. There are many different ways of the tradition of nature conservation. In terms of attitudes, content and principles of environmental conservation, the tradition of the nature conservation can be divided into five main categories: charity, quarantine, doctrine, beliefs, observation and cognition (Tsendeekhuu 2001).

***The tradition of respect*** is often taught to protect the land, mountains, forests, wildlife, and plants and to train the way they use it to fit their lives. Among the customs that have been honored in nature, the universal, most respectful ritual was a mountaineering ceremony.

***The tradition of thoracic*** has been seen as a multi-faceted activity in understanding the nature of abusive behavior and maintaining the natural ability to self-regeneration.

***The tradition of the doctrine*** was a way of convincing individuals to understand the natural phenomena of nature, to respect the nature and to admit it, but to be corrupted and abusive.

***The traditions of the devotional*** are considered to have been a method of confusion. One of the more important traditions of religious traditions is worship in the land and in the mountains, praying for the sake of sincerity and fortune.

***The observation and cognition tradition*** is a way of keeping track of nature, weather, phenomena, and changes in life experiences, and have a life-changing way of life.

### 11.1.8 The Protected Areas

The Mongolian Government's central piece of environmental legislation is the Environmental Protection Law of Mongolia, 1995. It provides a legal framework for protecting and managing flora, fauna, ecological communities and heritage places that are defined as matters of national environmental significance. Since 1778, the protected areas system of Mongolia has designated into 61 protected areas (MNET 2017), covering 17.8% (27.9 million hectares) of the Mongolian territory (Fig. 11.8, Table 11.7). In the National Program on protected areas adopted by the Mongolian Parliament (1998), the government plans to increase the amount of land in protected areas to 15% of the country's territory by 2001 and 30% by 2030. This extensive expansion of protected areas is likely to come at a significant cost to current land users and possessors, to economic production, and possibly to sustainable land use in unprotected areas (MNET 2017). Mongolian protected areas have become the main tourism destinations in Mongolia (Oyungerel 2004).

Much of the land currently in protected areas and planned for protected area expansion is traditional grazing land. Even with full implementation of the Law on Land, much of this land use (summer and fall pastures, and in many cases winter pastures outside of winter camps) will not be legally protected by possession contracts. Every country has its own legal framework for the selection and management of protected areas. These are divided into four categories of protection (Table 11.7). Strictly protected areas represent the strongest level of protection, followed by National Parks. Nature Reserves are established for the purpose of ecological, biological, paleontological and geological conservation, and Natural Monuments are classified as natural and or historical/cultural (Oyungerel 2011). Mongolia's protected areas represent the country's very best landscapes, ecosystems, wildlife habitats, watersheds and forests. International and domestic visitors are attracted to these protected areas because they offer the opportunity to enjoy wide open spaces, solitude, adventure, to connect to nature and the chance to experience some of the best hospitality in the world. Globally, these values are becoming increasingly rare and sought after. Protected areas are important carbon sinks and provide buffers against the effects of human-induced global warming (Oyungerel 2016).

Representation of rare and rare animals is highest in the areas located at the junction of large natural provinces. In terms of the number of genera and species of rare animals the superiority belongs to the Great Gobi strictly protected area, as well as National parks Govi Gurvansaikhan, Khan Khentii, Numrug and Khuvsgul (Oyungerel 2011) (Fig. 11.9).

Mongolia has decided that each strictly protected areas or National Park shall be allocated a differentiated protection level or "zoning" in accordance with modern conservation principles (Table 11.8). The strictly protected areas have the highest protection status in the Mongolian context (Oyungerel 2004) (Fig. 11.10).

A number of reports and plans approved by the Mongolian Parliament or government have stated that the current protected areas system needs improvement in quantity (area, number) and in quality.

**Table 11.7** Network of protected areas (2017) (MNET 2017)

| Protected area category | Numbers of sites | Area (ha) | Percent of territory |
|---|---|---|---|
| Strictly Protected Area | 12 | 10.554.523 | 6.75 |
| National Park | 22 | 9.229.905 | 5.9 |
| Nature Reserves | 19 | 2.006.270 | 1.28 |
| Natural Monuments | 8 | 102.083 | 0.07 |
| Total | 67 | 27.892.781 | 17.80 |

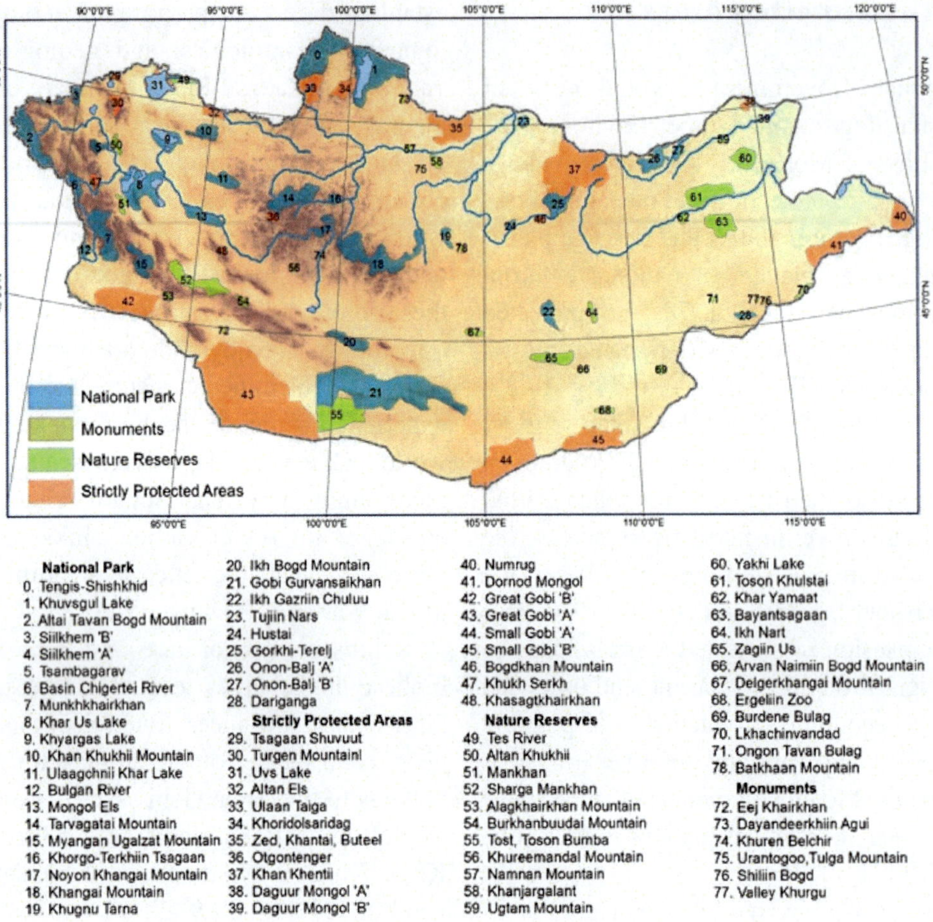


**Fig. 11.8** Location of protected areas (updated Bat-Erdene, Ts)

**Fig. 11.9** The protected area of the National Park Gorkhi-Terelj, central Mongolia (own photo, 2007)

**Table 11.8** Protection levels in protected areas, (MNET 2017)

| Protection Level | Category of protected area | Protection Zone | Area (ha) |
|---|---|---|---|
| Level I | Strictly Protected Areas | Pristine Zone | 1,910,725 |
| Level II | Strictly Protected Areas | Conservation Zone | 3,140,599 |
| | National Park | Special Zone | 1,122,819 |
| Level III | Strictly Protected Areas | Limited Use Zone | 5,827,444 |
| | National Park | Travel and Tourism Zone | 1,406,836 |
| Level IY | National Park | Limited Use Zone | 6,741,339 |
| Level Y | Nature Reserve | N/A | 2 017 495 |
| | Natural Monument | N/A | 103 230 |

**Fig. 11.10** The protected area of the Natural Monument Eej khairkhan, Western Mongolia (own photo, 2007)

## 11.1.9 Ramsar Sites in Mongolia

Environmental issues are boundless; in other words, the most global issue on our planet. Therefore, it is best to resolve the environmental issues in the country and to work with other countries. Mongolia joined the Ramsar Convention (Convention on Wetlands of International Importance especially as Waterfowl Habitat) on 8 April 1998. There are currently 11 wetlands designated as Ramsar sites (Table 11.9, (Ramsar Convention 2019)). Although much of Mongolia is arid and the dominant habitats are desert and steppe, there are some rich wetlands that support important breeding and migratory populations of water birds. Mongolia has the only breeding population of the threatened Dalmatian Pelican Pelecanus crispus in East Asia, and significant breeding populations of the threatened Swan Goose Anser cygnoides, White-naped Crane Grus vipio and Relict Gull Larus relictus (Birdlife International 2018).

## 11.1.10 World Heritage, Man and Biosphere Sites

Mongolia has joined the World Heritage Convention, since its commitment to the convention. The Government of Mongolia has designated 5 sites as World Heritage Sites (UNESCO 2017).

**Table 11.9** Ramsar sites

| | Name | Area ($km^2$) | Features |
|---|---|---|---|
| 1. | Ugii lake | 25.1 | Maximum length 7.9 km, maximum width 5.3 km, surface area 25.7 $km^2$, average depth 6.6 m, maximum depth 15.3 m, volume water 0.17 $km^3$ |
| 2. | Valley of the Lakes | 456 | Area 45.600 ha, 500 km long, has a width of about 100 km, 1000-1400 m above sea level |
| 3. | Uvs lake and surrounding wetlands | 5850 | The largest lake in Mongolia, at 3,350 $km^2$, surface elevation of 759 m, averages a depth of just 6 m, the remnant of an ancient lake that covered 92,000 $km^2$ and plunged 740 m deep |
| 4. | Terkhiin Tsagaan Lake | 61.1 | Stunningly beautiful lake in a rugged volcanic area. The lake is at 2060 m elevation, extending 16 km east to west, and ranging between 4 and 10 km across |
| 5. | Mongol Daguur (Daur) | 2100 | Area 2,100 $km^2$, 260 bird species use the site for staging, breeding or wintering, including six species of cranes of which two are threatened. |
| 6. | Lakes in the Khurkh, Khuiten rivers valley | 429.4 | Use the site for summering of the world's white cranes 11%, 3% of the black cranes, 1% of the great cranes, and of the region's black storks 16 percent. Area 429.4 $km^2$ |
| 7. | Khar-Us lake National Park | 3213.6 | Maximum length 72.2 km, maximum width 36.5 km, surface area 1.578 $km^2$, average depth 2.2 m, maximum depth 4.5 m, volume water 3.432 $km^3$ |
| 8. | Ganga lake and surrounding wetlands | 32.8 | Maximum length 2.1 km, maximum width 1.6 km, surface area 2.2 $km^2$, is a unique landscape of lakes, steppes, and sand dunes |
| 9. | Buir lake and surrounding wetlands | 1040 | Maximum length 40 km, maximum width 21 km, surface area 1.040 $km^2$, average depth 6–10 m, maximum depth 50 m |
| 10. | Airag lake | 450 | Average length 16 km, width 13 km, lake surface area 143 $km^2$, average depth 5.7 m |
| 11. | Achit lake and surrounding wetlands | 737.3 | Site area 737.3 $km^2$, average length 30 km, maximum width 16 km, lake surface area 311 $km^2$, average depth 5 m |

*Source* Ramsar Convention on Wetlands

The Great Gobi strictly protected area was the first to be designated followed by the Bogd Khan mountain strictly protected area, Uvs lake, Khustai mountain, Dornod Mongol and Mongol Daguur (Table 11.10, Fig. 11.11).

Almost all of the World Heritage Sites and the Man and Biosphere Reserves are part of the national protected areas system (UNESCO 2017) (Table 11.11).

### 11.1.11 Important Bird Areas

Important bird areas identifies 70 areas in Mongolia which could thus qualify for inclusion in the national protected areas system. The defined area covers 8.358.313 ha (5% of Mongolia), of which 70% is already covered by protected area regulation. There are, however, 41 important bird areas that currently lack any form of protection (Fig. 11.11).

### 11.1.12 Transboundary Protected Areas

A transboundary protected area is a clearly defined geographical space that consists of protected areas that are ecologically connected across one or more international boundaries and involves some form of cooperation (Maja, et al. 2015).

Mongolia is an important step in working with neighbors to protect the environment, one of which is to establish a trans-border protected area.

**Table 11.10** Properties inscribed on the World Heritage

| № | Name | Area (ha) | Type |
|---|---|---|---|
| 1 | Uvs lake basin | 1,068,854 | Natural |
| 2 | Orkhon valley | 121,967 | Cultural |
| 3 | Petroglyphic complexes of the Mongol Altai | 22,000 | Cultural |
| 4 | Great Burkhan Khaldun mountain and surrounding sacred landscape | 443,739.20 | Cultural |
| 5 | Landscapes of Dauria (Daguur) | 912,624 | Natural |

*Source* UNESCO

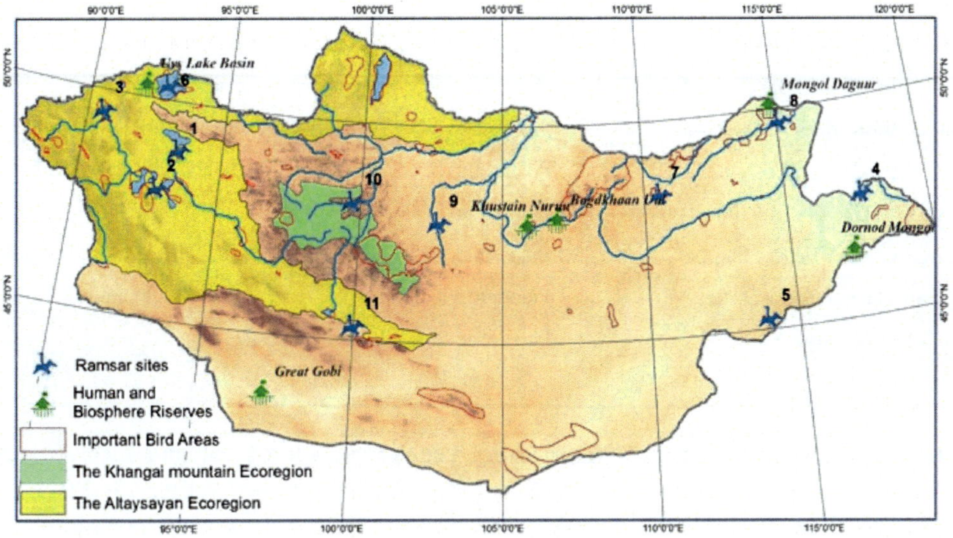


**Fig. 11.11** The protected areas covered by the international convention, (updated Bat-Erdene)

**Table 11.11** Man and Biosphere site

| № | Name | Location | Area (ha) |
|---|---|---|---|
| 1 | Great Gobi Strictly Protected Area | Govi-Altai, Bayankhongor and Khovd aimag | 5,560,412 |
| 2 | Bogdkhan mountain | Tuv aimag, Ulaanbaatar capital city | 41,322 |
| 3 | Uvs lake basin | Uvs aimag | 771,700 |
| 4 | Khustai Nuruu National Park | Tuv aimag | 48,889 |
| 5 | Dornod Mongol Strictly Protected Area | Dornod aimag | 589,906 |
| 6 | Mongol Daguur Strictly Protected Area | Dornod aimag | 108,154 |

*Source* UNESCO

The establishment of a trans-border protected area creates a transboundary protection corridor and a favorable environment for migratory movements, collective management of transboundary forests and steppe fires, joint ecosystem baseline studies, and regional conservation activities. Also, transboundary tourism routes and centers will be established and green tourism will be developed, when do invent protected area's administration. Daurian (Daguur) international protected area (Russia-China-Mongolia), Uvs lake basin strictly protected area (Russia-Mongolia), Onon Balj,

**Table 11.12** Ecoregions in Mongolia (TNC)

| | Eco name | Realm name | Major habitat types Name |
|---|---|---|---|
| 1 | Khangai Mountains Alpine Meadow | Palearctic | Montane Grasslands and Shrublands |
| 2 | Khangai Mountains Conifer Forests | Palearctic | Temperate Conifer Forests |
| 3 | Mongolian-Manchurian Grassland | Palearctic | Temperate Grasslands, Savannas and Shrublands |
| 4 | Altai Alpine Meadow and Tundra | Palearctic | Montane Grasslands and Shrublands |
| 5 | Altai Montane Forest and Forest Steppe | Palearctic | Temperate Conifer Forests |
| 6 | Daurian Forest Steppe | Palearctic | Temperate Grasslands, Savannas and Shrublands |
| 7 | Eastern Gobi Desert Steppe | Palearctic | Deserts and Xeric Shrublands |
| 8 | Gobi Lakes Valley Desert Steppe | Palearctic | Deserts and Xeric Shrublands |
| 9 | Great Lakes Basin Desert Steppe | Palearctic | Deserts and Xeric Shrublands |
| 10 | Juungar Basin Semi-Desert | Palearctic | Deserts and Xeric Shrublands |
| 11 | Manchurian Mixed Forests | Palearctic | Temperate Broadleaf and Mixed Forests |
| 12 | Sayan Montane Conifer Forests | Palearctic | Temperate Conifer Forests |
| 13 | Selenge-Orkhon Forest Steppe | Palearctic | Temperate Grasslands, Savannas and Shrublands |
| 14 | Alashan Plateau Semi-Desert | Palearctic | Deserts and Xeric Shrublands |
| 14 | East Siberian Taiga | Palearctic | Boreal Forests/Taiga |

Socond (Russia-Mongolia), Khuvsgul Lake National Park-Tunkinsky National Park (Russia-Mongolia), Cross-border protected areas such as Siilkhem Nature Reserves (Russia-Mongolia) have been negotiated and some have reached agreement on documents.

### 11.1.13 Ecoregions

An ecoregion is a contiguous area characterized by well-defined similarity in flora and fauna as well as geomorphology, climate and soils. Ecoregions are generally relatively large geographic units on the order of 50,000 square kilometers or more.

There are several alternative formal naming schemes for the Earth's ecoregions; one of the most widely used, developed by the World Wildlife Foundation, recognizes 867 separate ecoregions. Because of the very large scale of an ecoregion, the landscape is not monolithic, but may have pockets of ecological diversity. However, the ecoregion is defined by its preponderant vegetative, geological and meteorological composition (Axel and Andreas 2013).

Here is a list of most ecoregions that cover parts of Mongolia (Table 11.12). Altai Sayan and Mongol Daurian were selected and protected by WWF. The Altai-Sayan ecoregion is the mid-elevation portion of the highest mountain system in Siberia. It lies between the Siberian taiga and the Mongolian steppe. Relief and elevation are the major determinants of the native flora and fauna. This ecoregion is floristically diverse with as many as 800 species of plants (Bolor 2018).

## References


Axel B, Andreas B (2013) Mongolia's Ecoregions. http://birdsmongolia.blogspot.com/2013/02/mongolias-ecoregions-from-encyclopedia.html

Bat-Erdene T (2014) The methodology to update a land use layer of land information database by high and very high resolution satellite images. Ulaanbaatar

Birdlife International (2018) Data of Birdlife International. Birdlife.org's web site: http://datazone.birdlife.org/userfiles/file/IBAs/Ramsar/IBAs_Ramsar_Asia/19_Mongolia.pdf

Bolor R (2018) Central Asia: Central Mongolia. World Wildlife Fund's Web site data: https://www.worldwildlife.org/ecoregions/pa1007

Maja V, Kevan Z, Matthew M, Boris E, Micheal S, Tatjana MR (2015) Transboundary conservation: a systematic and integrated approach. Pannónia Nyomda, Gland, Switzerland

MNET (2017) Report on the State of the environment of Mongolia. Ministry of Nature, Environment and Tourism, Ulaanbaatar

MOFA (2018) Data of food, agriculture and light industry. Ministry of Food, Agriculture and Light Industry's web site: http://mofa.gov.mn/exp/blog/8/69

NSO (2017) Mongolian statistical yearbook. National Statistics Office of Mongolia, Ulaanbaatar

Oyungerel B (2004) The protected areas in Mongolia. Admon printing, Ulaanbaatar

Oyungerel B (2011) Specially Protected Natural Areas in Mongolia. Bull Buryat State Univ:188–189

Oyungerel B (2016) World Park Special Status on Some Representative Protected Areas of Mongolia. Korea Institute for International Economic Policy (KIEP) Research Paper Series, pp 317–341

Ramsar Convention (2019) The convention on wetlands. Ramsar's Web Site: https://www.ramsar.org/search?search_api_views_fulltext=Mongolia

Samiya P, Mühlenberg M (2006) Mongolian nature conservation. Munkhiin Useg Publishing, Ulaanbaatar

The Great Khural of Mongolia (2002) Land law of Mongolia

Tim H, Jennifer D (2001) Land reform in Mongolia: observations and recommendations. RDI, Seattle, Washington

Tsendeekhuu T (2001) Some traditions and customs of the nature protection of Mongolia. Ulaanbaatar

UNESCO (2017) Properties inscribed on the World Heritage List. UNESCO's Web site: https://whc.unesco.org/en/statesparties/mn

# Index


**A**

Accumulation, 12, 26, 27, 33, 44, 66, 109, 113, 140, 141, 144–146, 152, 155, 162
Achit lake, 37, 116, 165, 208
Active layer, 122, 124–126, 130
Active layer thickness, 124–126
Adaatsag, 46
Agricultural land, 136, 195–199
Airag lake, 91, 208
Air temperature variation, 111
Aj Bogd, 35, 190
Alag khairhan, 35
Alasha Gobi, 163, 165
Algae, 161, 166
Alluvial fans and sediments, 45, 46
Alluvial-proluvial plains, 27, 29
Alluvial soils, 145, 157
Alpine belts, 66, 171, 185
Alpine-type high mountains, 32
Alpine type relief, 44
Altai region, 5, 28, 35, 42, 65, 144
Altai-Sayan ecoregion, 210
Altai Tavan Bogd, 24, 35
Altankhukhii, 35
Altan Ulgii, 39
Altitudinal belts, 6, 163, 177, 182–185, 187, 190, 192
Angarkhai, 38
Animal, 4, 6, 7, 11, 12, 16, 33, 72, 145, 169, 171, 172, 197, 205
Annual precipitation, 53, 60, 61, 71, 86, 92, 121, 186, 188, 189, 192
Annual runoff, 83–85
Arable land, 136, 157, 196–199
Arctic Ocean basin, 5, 72, 77, 80, 83, 88, 89, 92, 109
Arsian pingo, 122
Artiodactyls, 173
Arts Bogd, 3, 43
Arvaikheer, 62, 71
Asain maximum, 62
Asgat, 38, 43, 47, 48, 104, 106
Asgat Gurvanbulag, 38
Asgatiin Saridag, 39
Asian summer monsoon, 55
Asralthairkhan, 3
Atas Bogd, 45
Average temperature, 4, 53, 55, 57, 87, 111, 162, 185
Average wind speed, 64

**B**

Baatarkhaihan, 35
Baga Bogd, 3, 38, 43, 188
Baga Buural, 47
Baga Gazriin Chuluu, 46, 47
Baga Khavtag, 45
Baga Khentii, 39, 80, 110
Baga Uul, 47
Baishin Tsav, 46
Baitag Bogd, 45
Baruun Khuurai depression, 28, 158, 181
Baruun Saikhan, 33, 43
Baruunturuun, 68, 136
Bayan, 3, 7, 34–36, 40, 69, 79, 88, 89, 91, 106, 109, 113
Bayanbor, 43
Bayan Bumbun Ranges, 35
Bayankhairhan, 39
Bayantsagaan, 42, 43, 47, 49
Bayan-Ulgii, 7, 69, 113
Biological diversity, 182
Birds, 161, 162, 169–175, 207, 208
Bogd, 3, 11, 34, 36, 38, 40, 42, 49, 101–103, 106, 181, 188, 204, 208
Bogd Ulaan, 49
Boreal, 6, 163, 164, 187, 210
Bor Khairhan, 39
Borzon Gobi, 158, 181
Botanical-geographical researches, 18, 166
Buduun Tolgoi, 47, 49
Buir lake, 48, 92, 208
Bumbat Khairhan, 35
Burged, 35
Busiin River, 30
Buuntsagaan, 44
Buun Tsagaan lake, 31

**C**

Carnivores, 173
Central Asia, 2, 10–13, 16, 18, 26, 34, 44, 72, 78, 103, 135, 161, 164, 167, 173, 180–182, 204


© Springer Nature Switzerland AG 2021

B. Yembuu (ed.), *The Physical Geography of Mongolia*, Geography of the Physical Environment,
https://doi.org/10.1007/978-3-030-61434-8

Central Asian Internal Drainage Basin, 5, 77, 80, 83, 87–89
Central Mongolia, 10, 13, 16, 24, 38, 62, 78, 166, 206
Chagatai Khanate, 10
Chernozem, 136, 138, 139, 143, 149–153, 166
Chinggis Khan, 10
Chiropteras, 173
Choiriin Bogd, 46
Chuluut, 28, 32, 38, 87, 95, 116, 124, 125, 132
Chuluut river, 5, 30, 39, 41, 109
Classification system, 5, 32, 68, 164
Climate, 1, 2, 4–6, 9, 12, 20, 21, 30, 31, 34, 51–54, 62, 64–69, 71, 72, 86, 87, 92, 98, 101, 104, 110–112, 119, 120, 124–128, 135, 137, 140, 154–157, 161, 162, 168, 171, 174, 183, 187, 188, 190, 191, 204, 210
Coniferous forest, 6, 187
Continental climate, 1, 53, 57, 68, 92, 110, 126, 161, 164, 167
Continuous permafrost, 97, 120–122, 124, 129
Cumulative ablation rate, 112

**D**
Daguur-Manchurian steppe, 167
Daguur-Mongol steppe, 180, 181
Daguur Mountain Forest Steppe, 164, 165
Dalanzadgad, 27, 58, 60, 69, 70, 192
Dariganga, 19, 27, 28, 33, 47, 49, 165, 180, 181
Dariganga Volcanic Plateau, 28
Darkhad depression, 57, 114, 116, 121, 122, 124, 125, 187, 189
Darkhan Khan, 47
Dartsag Khuren, 35
Darvi Range, 35
Dashbalbar, 68, 71
Davaanii lake, 116
Davst, 79
Dayan, 36, 89–91, 114, 115
Deflation lakes, 90
Degree of endemism, 169
Delgerkhangai, 46
Delger river valley, 188
Denudation plains, 29, 48
Denudation reliefs, 29
Desert steppe, 6, 57, 67, 136, 142, 144, 147, 148, 153, 162, 164, 165, 167, 170, 172, 177, 182, 184, 189, 190, 198, 210
Desert zone, 64, 66, 92, 162, 177, 182, 184–186, 189–192
Discontinuous permafrost, 119, 121, 125, 127
Dolon, 79
Dood tsagaan lake, 116
Dornod Mongolia, 4
Dornod steppe, 98
Drought frequency, 73, 190
Dry steppe, 66, 137–141, 144, 149, 153, 157, 167, 172, 182, 184, 188, 189
Dund Saikhan, 43
Dush, 49, 104

**E**
Eastern Arts, 38
Eastern Khangai, 24, 39, 113, 188
Eastern Mongolian steppe, 164, 165, 188
Ecoregion, 210
Ecosystems, 2, 6, 20, 21, 68, 71, 119, 131, 137, 157, 161, 168, 175, 179, 180, 183, 205, 209
Eej khairkhan, 207
Ekhiin Gol, 68, 192
Endemic bird areas, 174
Ereentsav low mountains, 48
Erhet Khairkhan, 38
Extreme continental, 110, 161, 164
Extremely cold days, 68
Extremely low temperatures, 54, 74
Eznii-Gol, 15

**F**
Forest steppe, 64, 66, 73, 92, 140, 161, 162, 164, 165, 169, 171, 177, 181–183, 185–188, 191, 210

**G**
Galba Gobi, 46, 181, 189
Ganga Tsagaan ovoo, 47
Gel Uul, 39
Geographical distribution, 51, 52, 77, 96, 101, 120, 137, 161, 170, 174, 177
Geographical regions, 23, 90, 135, 163–165, 168–170, 177
Gilbent, 45
Ginghis Uul, 45
Glacial complex, 18, 31, 108
Glacial geomorphology, 101, 110
Glacial valleys, 90
Glaciers, 3, 5, 9, 11, 14, 15, 17, 18, 24, 31, 34, 36, 38, 72, 77, 78, 81, 101–114, 121, 152, 185, 186
Global radiation, 55
Gobi-Altai, 2, 26–28, 32, 35, 42–45, 98, 155, 158, 163–165, 167, 170, 178, 181, 185, 186, 189
Gobi desert, 2, 4–6, 10, 56–58, 60, 64, 67, 68, 74, 80, 92, 103, 144, 158, 161, 167, 174, 182, 190, 191, 210
Govi Gurvan Saikhan, 43
Govi-Tien-Shan mountain system, 45
Great Khyangan Ridge, 3
Great Lakes depressions, 4, 6, 35, 44, 54, 64, 71, 136, 140–142, 155, 157, 158, 189, 190
Great Mongol Empire, 10
Great National Khural, 10
Groundwater, 5, 20, 71, 77, 79, 83–85, 96–98, 126, 131, 144–146, 154, 157

**H**
Hay-pasture, 136
High mountain (alpine) belt, 182, 184–186, 190
High mountain meadow, 182, 185

High mountains, 2–4, 6, 27, 29–33, 42, 43, 53, 57, 66, 72, 73, 101, 109, 111, 121, 122, 128, 138, 148–150, 152, 153, 161, 162, 166, 171, 173, 180–187, 198
Hilly plain relief, 48
Holarctic, 164
Hot springs, 93–95
Hummocks, 32, 33, 119, 122, 123

I
Ikh Buural, 40, 47
Ikhes lake, 27
Ikh Gazriin Chuluu, 46
Ikh Khairkhan, 46
Ikh Khaltar, 47
Ikh Khavtag, 45
Ikh Khyangan, 26
Ikh Myangan, 38
Ikh-Tukhum, 79, 136
Ikh Uul, 34, 47
Ilkhanate, 10
Indian monsoons, 55
Ingen khuuvur, 158
Inner Mongolia, 10, 15
Insectivores, 173
Institute of geography, 20, 78, 79, 119, 137, 182
Internal basin, 72
Invertebrates, 169

J
Jargalant Khairhan, 35

K
Kastanozem, 138–143, 146, 149–155, 157
Khagiin khar, 42, 114
Khalkh, 10, 42, 48, 60, 80, 87, 164, 165, 178, 180, 188
Khalkh and Numrug rivers, 42, 60
Khalzan, 42, 48
Khanbogd, 68, 71
Khangai, 2–7, 10, 14, 16, 24–28, 30–33, 35, 37–40, 44, 46, 53, 54, 60, 61, 64–67, 69, 71, 78, 79, 83, 84, 87, 89, 91, 93, 95, 97, 98, 101, 105, 109, 110, 113–116, 121, 122, 124, 126–131, 140, 141, 144, 154–156, 163–167, 170–172, 178–182, 184–189, 199, 203, 210
Khangai mountain, 3, 5, 7, 14, 16, 27, 28, 31–33, 38, 40, 44, 60, 67, 79, 84, 89, 93, 109, 114, 124, 126–129, 131, 140, 163–165, 167, 178, 180, 181, 185, 186, 188, 210
Khangai plateau, 14, 25, 28, 31, 39, 78, 166
Khan Khukhii mountains, 38, 187, 189
Kharaat Uul, 47
Kharhiraa, 18
Khar Khot, 15, 16
Khar lake, 6, 42, 136, 158, 174, 181, 208
Khar-Us lake, 80, 90, 91, 174, 208
Kharuult, 46
Khasagt Khairhan, 35
Khatan Suudliin Khyar, 45
Khentii, 2–7, 11, 15, 17, 24–28, 30, 32, 35, 39, 40, 42, 46, 48, 60, 64–66, 80, 84, 85, 87, 90, 91, 93, 95, 97, 98, 109–111, 113, 114, 120–122, 124–131, 139–141, 143, 144, 155, 163–165, 170, 172, 178–181, 185–188, 204, 205
Kherlen river basin, 31
Khermen tsav, 20, 33, 191
Khiidiin saridag, 39
Khongor sand dunes, 190
Khoton, 89–91, 113–116
Khovd, 6, 14, 31, 37, 51, 55, 68, 71, 80, 84, 85, 89, 102, 103, 111, 113, 116, 136, 157, 163–165, 209
Khuiten, 2, 24, 35, 103, 104, 114, 208
Khukh lake, 16, 24, 87–90
Khukh Serkh Ranges, 35
Khungui, 31, 34, 38
Khurgan, 89, 113–116
Khurkh, 43, 208
Khurmen Ovoo, 39
Khuvsgul, 3, 6, 7, 15, 24, 27, 28, 30, 32, 35–37, 60, 64, 66, 67, 77, 78, 80, 83, 85, 87–89, 91–95, 97, 101, 105, 109, 110, 113, 115, 121, 122, 124, 126–130, 132, 137, 139, 144, 155, 163–165, 167, 170, 171, 178, 180, 181, 185–188, 199, 210
Khyangan mountains, 42, 48, 54, 60, 80, 140, 164, 178
Khyargas lake, 6, 29, 38, 39, 91, 181

L
Lake Baikal, 5, 10, 80, 92, 136
Lake depressions, 3, 24, 26–29, 31, 34, 35, 38, 39, 44, 45, 67, 90, 113, 155, 158, 167, 180
Lamiin Dush, 47
Landscape regionalization, 183
Land use, 1, 21, 137, 195–199, 202, 203, 205
Large-scale erosion processes, 25
Late Pleistocene, 110, 111, 113, 116
Latitudinal directions, 3, 39
Latitudinal variations, 2, 55, 57, 183, 185
Latitudinal zones, 177, 182, 183, 185, 187, 188
Lichen, 161, 166, 186, 187
Low mountains of the middle Khalkh, 48

M
Malchin, 35
Mammals, 18, 161, 162, 169–174
Man and Biosphere, 207–209
Maximum temperature, 57, 68, 86, 87
Mean temperature trend, 70
Middle Khalkh steppe, 164, 165, 188
Mid-elevation mountains, 30, 32, 33, 38, 40, 43, 45, 46
Mineral water deposit, 93
Mongol Altai, 2, 3, 6, 14, 16–18, 24–28, 30–32, 35–38, 45, 60, 66, 78, 83–85, 87, 91, 93, 101–103, 105, 109, 111, 113–115, 136, 140, 155, 163–165, 171, 180, 181, 185–189, 209

Mongol Daguur, 163, 164, 170, 174, 208, 209
Mongolian gazelle, 172
Mongolian-Russian joint geological expeditions, 7, 24
Mongolian Science Commission, 18
Morit, 47
Mountain steppe, 121, 140, 145, 150, 153, 164, 165, 170, 171, 173, 180–182, 184, 187, 188
Mountain taiga, 80, 121, 137–139, 150, 152, 153, 162, 164, 165, 177, 180–182, 184–187
Mountain tundra, 137, 138, 147, 150, 171, 186, 198
Munkhkhaan, 47
Munkhkhairkhan, 35, 104–108, 110, 111, 180
Munkhsaridag, 15, 36, 92, 103, 104, 106, 109, 110, 162
Mushroom, 161, 166
Must, 35, 60, 82, 113, 122

**N**

Namir river basin, 37
National Atlas of Mongolia, 20, 35, 56, 58, 59, 62, 63, 66, 104, 158, 159
National Park Gorkhi-Terelj, 206
National parks, 205–207, 209, 210
Nature conservation, 195, 203, 204
Nemegt, 20, 33, 45, 46
Nomgon, 43
Nomin, 43, 158
Nomyn Khaan, 47
North Unjuul, 46
Noyon-Uul, 16

**O**

Oigon lake, 37
Ongon Khairhan, 46
Onon, 5, 6, 30, 31, 40, 77, 80, 87, 95, 157, 165, 174, 209
Orkhon river, 10, 31, 38, 80
Orkhon-Selenge basin, 97, 137, 140, 157
Orog, 27, 31, 44, 89, 91
Otgontenger mountain, 38

**P**

Pacific monsoon, 54, 55, 60
Pacific Ocean basin, 77, 80, 83, 88
Paleogeomorphological scheme, 27, 29
Pasture, 8, 51, 52, 71, 72, 97, 136, 157, 164, 168, 191, 195–199, 203, 205
Perissodactyls, 173
Permafrost monitoring sites, 120, 127
Permafrost temperature, 119, 127–130
Physical geography, 9, 11, 18–21, 24, 62, 135–137, 177, 179, 180, 182, 189
Physiographic classifications, 178
Physiographic regionalization, 177–180
Physiographic regions, 177–181
Phytogeographical regions, 163
Pingos, 121–124
Plain, 2–6, 13, 15, 24, 25, 27, 28, 30, 32–34, 44, 46–49, 60, 62, 65, 67, 90, 91, 110, 137, 138, 150, 155, 161, 163, 166, 169, 171, 172, 178–181, 183, 188, 189
Plant communities, 164, 167
Podzolic, 149, 150, 153
Potanin glacier, 14, 78, 102, 103, 111, 112
Potential evapotranspiration, 60, 71
Precipitation changes, 71, 191
Prevailing wind direction, 53
Protected areas, 4, 173, 195, 203–209

**Q**

Quaternary glaciations, 36, 107
Quaternary sediments, 29, 33

**R**

Ramsar sites, 207, 208
Rare birds, 173
Red list, 170
Remote sensing, 21, 24, 192, 200
Resident, 173, 203
Rinchinlhumbe, 55, 56, 71
Rivers runoff, 83, 84
Rock fall, 171
Rocky massif, 171
Rodents, 171–173
Russian geography, 21

**S**

Sainshand, 46, 55–57, 60, 136, 158, 192
Sair Uul, 35
Salkhit, 42
Sand dune, 4, 6, 24, 34, 45, 49, 64, 208
Sangiin dalai lake, 78, 89, 136
Sayan basin, 92, 163, 209
Sayan mountains, 103, 109, 162, 180, 181
Sayan variant, 163
Secret History of the Mongols, 77
Segs Tsagaan Bogd, 45
Selenge river, 7, 24, 31, 35, 72, 80, 87, 88, 92, 97, 180
Semidesert brown soil, 139
Semidesert zone, 162
Sevrei, 45
Shaazant, 47
Shar bulagiin saridag, 39
Sharga depression, 30
Sharga-Ganga river, 124
Shargaljuut, 77, 93–96
Sharga valley, 44
Shargiin Gobi, 4, 36
Shar Khulsnii Nuruu, 45
Shar us, 38
Shiliin Bogd, 34, 49
Shishkhed river basin, 189

Siberia-Mongolian High, 54, 61
Siberian continuous permafrost, 120
Siberian steppe, 167
Snowy mountain, 113
Soil classification, 146, 148–150, 152
Soil criteria, 146, 149, 152
Soil forming bedrocks, 136, 144, 155
Soil salinization, 91, 146, 157
Soil vertical zones, 138
Soil water erosion, 158
Soil zone, 137, 138, 142, 144, 154
Solar irradiation, 55
Solar zoning, 167
Solonchak, 142, 143, 145, 151, 153–155, 149, 152
Solonets, 144, 146, 151, 152, 155
South and North Tamir river, 38
Southeast Asia, 64
South-Khangai plateau, 28, 31, 39
South Unjuul, 46
Soviet-Mongolian palaeontological expedition, 20
Spatial differences, 55, 61, 64
Species, 6, 7, 161–174, 189, 205, 208, 210
Sporadic permafrost, 120, 121
Squirrels, 169, 172
Steppe region, 6, 57, 60, 80, 91, 144, 166, 169, 180
Steppe zone, 58, 60, 92, 110, 140, 141, 144, 157, 158, 162, 170, 182, 184–186, 188–192
Strictly protected areas, 173, 204, 205, 207–209
Summer precipitation, 5, 59
Sutai, 35, 104, 106
Suvarga Hairhan, 38

**T**

Taatsiin tsagaan lake, 27, 31, 38, 44
Taiga, 1, 6, 36, 37, 92, 109, 121, 138, 139, 148, 150, 153, 161, 171, 172, 185, 187, 191, 210
Taishir, 3, 27, 35
Taishir Range, 35
Takhiin Shar mountain range, 45
Takhillaga, 46
Tarvagatai, 3, 7, 38, 39, 116, 180, 187
Tavan Bogd, 2, 14, 24, 35, 36, 101–104, 106, 111, 113, 114
Tectonic lakes, 88–90
Telmen lake, 78
Terkhiin tsagaan lake, 38, 89, 92, 208
Tes river, 67
Thermokarst lakes, 119, 120, 122–124, 130
Togtokhiin Shil, 39
Toson, 78, 79, 174
Tosontsengel, 54, 62, 68
Trans-Altai Gobi, 28, 29, 35, 45, 61, 64, 67, 68, 158, 163–165, 178, 181, 190
Tsagaan Khairhan, 38, 39
Tsagaan lake valley, 27, 31, 37, 38, 44, 208
Tsagaannuur, 128
Tsagaan Shiveet, 37
Tsagaan shuvuut, 35
Tsahir Khaalgyn Nuruu, 42
Tsambagarav, 35, 103, 106, 111, 112, 174
Tsengelhairhan, 35
Tumen, 10, 42
Tumentsogt, 47
Tundra species, 171
Turgen, 21, 35, 37, 52, 103, 104, 106–108, 111–114, 180
Tuul, 2, 5, 10, 16, 21, 40, 67, 77, 80, 83–85, 87, 137, 174, 180, 181, 189, 200
Typology, 163, 167, 179

**U**

Ugii lake, 90, 208
Ulaan Am valley, 111
Ulaanbaatar, 1–4, 7, 10, 12, 15, 52, 57, 68, 69, 80, 83–85, 120, 123, 136, 174, 199–204, 209
Ulaangom, 54, 55, 57, 58, 62, 68–70, 192
Uliastai, 35, 38, 43, 51, 52
Ulz river, 48, 77, 80–83, 85–88, 180
Undershil, 46
Undur Jargalant, 38
Undurkhaan, 68, 84, 180, 189
Ungulates, 170–172
Untaa Jamaat, 38
Uriankhai region, 162
Uvs lake basin, 61, 209
Uyat, 47

**V**

Valley of Lakes, 29, 31, 35, 42, 44, 157, 163–165, 181, 189, 190
Vascular, 161
Vascular plant, 161, 163, 166
Vegetation, 2, 4, 6, 21, 34, 51, 52, 64, 65, 72, 120–122, 124–126, 131, 132, 135, 136, 154, 157, 158, 161–168, 177, 181, 182, 185–191, 197
Volcanic lakes, 89

**W**

Westerlies, 53, 54, 64
Western arts, 38
Western Europe, 53
Western Khangai, 38, 39, 113, 185
Western Mongolia, 13, 14, 18, 26, 71, 78, 91, 98, 109, 110, 114, 116, 122, 136, 162, 167, 174, 189, 198, 207

**Z**

Zaan shiree, 47
Zamiin-Uud, 52, 55, 56, 158
Zaraa Khairkhan, 45
Zavhan, 31, 71
Zonality, 155, 163, 166, 167, 177, 188

Zones, 1, 6, 21, 24–28, 30, 31, 39, 52, 56, 64, 66, 67, 70, 73, 92, 98, 110, 116, 119–129, 131, 135–141, 145, 148, 154, 157, 162, 167, 169, 171, 172, 177, 181–192, 196, 203, 207
Zoogeographical, 20, 161, 168, 169, 171
Zoogeographical regions, 168–170
Zoogeographical zoning, 169
Zoology, 16, 162
Zorgol Khairkhan, 46
Zotol Khan, 49
Zud, 52, 72, 74
Zuulun, 45
Zuunbayan, 46
Zuungar depression, 45
Zuungariin Gobi, 165
Zuun Saikhan, 33, 43
Zuun Yargait, 49